Population Dynamics
New Approaches and Synthesis

EDITED BY

Naomi Cappuccino
Department of Zoology
University of Texas
Austin, Texas

Peter W. Price
Department of Biological Sciences
Northern Arizona University
Flagstaff, Arizona

ACADEMIC PRESS
San Diego   New York   Boston   London
Sydney   Tokyo   Toronto

Front Cover Photographs: (Top) conifer sawflies (Hymenoptera: *Diprion idae*) on jack pine. Group feeding sawflies such as these regularly reach densities at which they defoliate their host plants. Photo by Naomi Cappuccino. (Center and bottom) tiger swallowtail butterflies (*Papilio glaucus*) puddling. Although tiger swallowtail adults are frequently seen nectering and puddling, their larvae are rare relative to the amount of host–plant resources in the environment. Photo by Hans Damman.

Academic Press, Inc.
A Division of Harcourt Brace & Company
525 B Street, Suite 1900, San Diego, California 92101-4495

United Kingdom Edition published by
Academic Press Limited
24-28 Oval Road, London NW1 7DX

Library of Congress Cataloging-in-Publication Data

Population dynamics : new approaches and synthesis / edited by Naomi
  Cappuccino, Peter W. Price.
        p.   cm.
  Includes bibliographical references and index.
  ISBN 0-12-159270-7
  1. Population biology.   2. Insect populations.   3. Insect pests.
  I. Cappuccino, Naomi,   II. Price, Peter W.
  QH352.P626   1995
  574.5'248--dc20   95-14616
                                                       CIP

PRINTED IN THE UNITED STATES OF AMERICA
95 96 97 98 99 00 EB 9 8 7 6 5 4 3 2 1

Contents

Contributors   xix
Preface   xxi

PART I

INTRODUCTION   1

Chapter 1

Novel Approaches to the Study of Population Dynamics   3
NAOMI CAPPUCCINO

  I. Population Dynamics: A Brief Historical Review of the Major Concepts   3
  II. Traditional and Novel Approaches to the Study of Population Dynamics   5
    A. Observations: The Search for Patterns in Time Series   5
    B. Mechanisms: The Factors Driving Population Change   8
    C. The Varied Processes of Population Regulation: Testing Old Theories and Building New Ones   10
    D. Case Studies: Building and Testing Theory in Well-Studied Systems   11
  III. The Role of Herbivorous Insects in the Study of Population Dynamics   11

IV. Population Dynamics: The New (Pluralist) Synthesis   12
        References   12

PART II

OBSERVATION AND
COMPARATIVE APPROACHES   17

Chapter 2

Population Regulation: Old Arguments and a New Synthesis   19
PETER TURCHIN
    I.   Introduction: The Never-Ending Debate   19
    II. Definition of Population Regulation   22
         A. Relationship to Previous Definitions   25
         B. Relationship to Stochastic Boundedness of Chesson
             (1978, 1982)   26
         C. Finite Population Sizes and the Question of Scale   26
    III. Density-Dependence Tests   27
    IV. So What Do the Data Say?   30
    V.   Beyond Density-Dependence Tests: Investigating the
         Structure of Population Regulation   31
         A. Population Dynamics in the Noisy World   33
    VI. Conclusion: Where Do We Go from Here?   35
         References   37

Chapter 3

*Ecology, Life History, and Phylogeny of Outbreak and
Nonoutbreak Species*   41
ALISON F. HUNTER
    I.   Introduction   41
         A. Phylogenetic Analysis   41
    II. Methods   42
         A. Scope and Data Sources   42
         B. Trait Definitions   42

C. Phylogenies   45
D. Data Analysis   45
III. Results   47
A. Proportions of Outbreaking and Nonoutbreaking Species   47
B. Number of Generations per Year   47
C. Overwintering Stage and Feeding Phenology   50
D. Larval Defenses   50
E. Larval Length at Maturity   51
F. Host Breadth   51
G. Maximum Fecundity   52
H. Female Flying Ability   52
I. Egg Dispersion   53
J. Larval Lifestyle (Solitary/Gregarious)   54
K. Clustering of Offspring   54
L. Taxonomic Distribution of Variation   54
IV. Discussion   56
A. Differences between Outbreaking and Nonoutbreaking Forest Macrolepidoptera   56
B. Phylogenetic Patterns and Constraints   56
C. Clustering of Offspring and Limited Dispersal: A Unifying Theme?   57
D. Life-History Traits and Outbreaks   58
E. Clustering of Offspring, Dispersal, and Spatially Complex Models   58
F. Generality?   59
G. Caveat   60
References   60

Chapter 4

Spatial Behavior and Temporal Dynamics of Outbreak and Nonoutbreak Species   65

NAOMI CAPPUCCINO
HANS DAMMAN
JEAN-FRANÇOIS DUBUC

I. Introduction   65
II. Comparative Studies   67
A. Goldenrod Aphids   67

B. Goldenrod Beetles   68

C. Birch-Feeding Caterpillars   73

III. Toward an Explanation for Greater Temporal Variability in
Species with Higher Spatial Variability   76

A. The Consequences of Feeding in Aggregations or Alone   76

B. Connecting Spatial Behavior and Population Dynamics   78

References   79

Chapter 5

Minor Miners and Major Miners:
Population Dynamics of Leaf-Mining Insects   83

MICHAEL J. AUERBACH
EDWARD F. CONNOR
SUSAN MOPPER

I. Introduction   83

II. Abundance Patterns   84

III. Sources of Mortality   84

A. Vertical Sources of Mortality   86

B. Horizontal Sources of Mortality   87

C. Abiotic Sources of Mortality   88

IV. Regulatory Effects of Variation in Mortality   88

V. Variation in Natality   91

VI. Case Studies   92

A. *Phyllonorycter tremuloidiella*, an Eruptive Species on
Populus tremuloides   92

B. *Cameraria hamadryadella*, an Eruptive Species on
Quercus alba   95

C. *Stilbosis quadricustatella*, a Latent Species on
Quercus geminata   99

VII. Conclusions   102

A. Comparison of Latent and Eruptive Leaf Miners   102

B. Are Leaf-Miner Populations Regulated?   102

C. Host-Plant Phenology and Leaf-Miner
Population Dynamics   104

References   105

PART III

MECHANISMS AND PROCESSES OF POPULATION DYNAMICS 111

Chapter 6

Density-Dependent Dispersal and Its Consequences for Population Dynamics 113

ROBERT F. DENNO
MERRILL A. PETERSON

  I. Introduction 113
 II. Evidence for Density-Dependent Dispersal 114
III. Consequences of Density-Dependent Dispersal for Local Population Dynamics 119
IV. Conclusions and Prospectus 124
   References 125

Chapter 7

Using Density-Manipulation Experiments to Study Population Regulation 131

SUSAN HARRISON
NAOMI CAPPUCCINO

  I. Introduction 131
 II. Review of Experimental Studies of the Past Twenty-Four Years 132
III. Guidelines for Experimental Studies of Regulation 135
   A. How Long Do Experiments Need to Run? 136
   B. How Large Should Experimental Units Be and Should They Be Caged? 136
   C. What Treatment Levels Should Be Used? 137
   D. Statistical Analyses 137
   E. Bottom-Up Regulation 138
   F. Top-Down Regulation and the Problem of Predator Mobility 138
   G. Lateral Regulation 139

IV. Beyond Local Regulation   140
    V. How Far Can Experiments Get Us?   141
        References   141

Chapter 8

Butterfly Metapopulation Dynamics   149

ILKKA HANSKI
MIKKO KUUSSAARI

I. Introduction   149
    II. Four Necessary Conditions for Metapopulation-Level Regulation
        and Persistence of Species   150
    III. The Incidence Function Model   151
    IV. The Glanville Fritillary *Melitaea cinxia*   154
        A. The Necessary Conditions for Metapopulation-
            Level Persistence   154
        B. Metapopulation Patterns and Processes   156
    V. Other Butterfly Metapopulations   159
        A. Three Necessary Conditions for Metapopulation-Level Persistence
            in Finnish Butterflies   159
        B. Butterfly Metapopulation Studies   161
        C. Migration Rate in Butterflies   164
    VI. Discussion   166
        References   168

Chapter 9

Internal Dynamics and Metapopulations:
Experimental Tests with Predator–Prey Systems   173

SANDRA J. WALDE

I. Introduction   173
    II. Experimental Manipulations of *Panonychus ulmi* on Apple   174
        A. Biology of *Panonychus ulmi* and *Typhlodromus pyri*   175
        B. Experimental Manipulations in a Young Orchard   175
        C. Experimental Manipulations in a Mature Orchard   177
        D. Conclusions   183

III. Metapopulation, Structure, Theoritical Predictions, and Spider
Mite Populations   183
IV. Conclusions   188
References   190

Chapter 10

Herbivore–Natural Enemy Interactions in Fragmented and Continuous Forests   195

JENS ROLAND
PHILIP D. TAYLOR

I. Introduction   195
II. Methods   197
   A. General Approach   197
   B. Detailed Methods   200
III. Results   201
   A. Isolates   201
   B. Megagrid   202
IV. Discussion   203
   A. Medium-Scale Dynamics of Tent Caterpiller   203
   B. Large-Scale Dynamics of Tent Caterpillar   204
References   207

Chapter 11

Simple Models and Complex Interactions   209

GREG DWYER

I. Introduction   209
II. Simple Mathematical Models with Nonintuitive Results   210
   A. Host–Parasitoid Interactions   210
   B. Host–Pathogen Interactions   212
   C. Inducible Plant Defenses   215
   D. Maternal Effects   217
   E. Overview   218

III. Using a Simple Model to Make Quantitative Predictions:
Gypsy Moth and Its Virus   219
References   224

Chapter 12

Field Experiments to Study Regulation of
Fluctuating Populations   229

JUDITH H. MYERS
LORNE D. ROTHMAN

I. Population Regulation: A Critique of the
Descriptive Approach   229

II. Population Regulation: The Experimental Approach   232
A. Suppressing the Population Peak: Spray Programs   232
B. Suppressing the Population Peak: Cropping Experiments   234
C. Attempting to Create an Outbreak   235
D. Biological Control of Introduced Forest Lepidoptera:
Experimental Evaluation of Parasitoids   237
E. Evaluating Geographic Variation in
Population Dynamics   240
F. Large-Scale Environmental Variables as
Experimental Perturbations   241
G. Experiments to Identify Mechanisms Affecting
Population Density   242

III. Conclusions   246
References   247

Chapter 13

Impact of Life-History Evolution on Population Dynamics:
Predicting the Presence of Maternal Effects   251

MARYCAROL ROSSITER

I. Importance of Population Quality Parameters   251
II. Maternal Effects Hypothesis of Herbivore Outbreak   253

III. Predicting the Presence of Maternal Effects 256
IV. Criteria for Choice of Life-History Traits for ME Score 256
V. Criteria for Choice of Phytophagous Insect Species 262
VI. Protocol for Scoring Herbivore Characteristics 262
VII. Predictions 263
VIII. Comparison of Inferences with Other Studies 266
IX. Gregarious Larval Behavior 267
X. Other Criteria for Predicting Maternal Effects 268
XI. Closing Remarks 270
References 271

PART IV

CASE STUDIES 277

Chapter 14

Long-Term Population Dynamics of a Seed-Feeding Insect in a Landscape Perspective 279

CHRISTER SOLBRECK

I. Introduction 279
II. Biology and Habitats 280
III. Methods 282
IV. Dynamics of Habitats and Seed Resources 283
V. Patterns of Abundance in *Lygaeus equestris* Populations 283
VI. What Factors Determine Fluctuations in
Lygaeus equestris Populations? 286
VII. General Discussion 292
A. Habitat Patterns and Insect Distribution 292
B. Factors Affecting Population Dynamics 294
C. Long-Term Population Change 296
D. Geographical Variation 296
VIII. Conclusions 297
References 298

Chapter 15

Adaptive Behavior Produces Population Stability in Herbivorous Lady Beetle Populations   303

TAKAYUKI OHGUSHI

I. Introduction   303
II. A Search for Causal Mechanisms   304
III. Preference–Performance Linkage: A Key to Understanding Population Dynamics in Insect Herbivores   304
IV. Case Study: The Thistle Lady Beetle   305
   A. Population Stability at Low-Density Level   306
   B. Density-Dependent Processes   308
   C. How Population Stabilization Proceeds throughout the Reproductive Season   308
   D. Oviposition Traits as Mechanistic Bases for Population Stability   309
   E. Top-Down Influences of Natural Enemies   312
   F. Oviposition Tactics Improve a Female's Reproductive Success   312
V. Future Directions   315
   References   316

Chapter 16

Working toward Theory on Galling Sawfly Population Dynamics   321

PETER W. PRICE
TIMOTHY P. CRAIG
HEIKKI ROININEN

I. Introduction   321
II. Studies on a Stem-Galling Sawfly   322
   A. Life History   322
   B. Preference–Performance Linkage   324
   C. Chemical Ecology   324
   D. Behavior   325
   E. Facilitation and Resource Regulation   325
   F. Third Trophic Level   326
   G. Stability in Natural Populations   326

III. The Search for Generality   329

    A. Patterns of Preference and Performance   329

    B. Patterns in Shrub and Tree Attack, Common and Rare Species   331

    C. The Role of Carnivores   332

    D. Resource Regulation   332

    E. Chemical Ecology   332

IV. Theory on *Euura* Population Dynamics   333

    A. Phylogenetic Constraints Hypothesis   333

    B. Plant Vigor Hypothesis   334

V. Toward More General Theory   334

    References   335

Chapter 17

Host Suitability, Predation, and Bark Beetle Population Dynamics   339

JOHN D. REEVE
MATTHEW P. AYRES
PETER L. LORIO, JR.

I. Introduction   339

II. Moisture Stress and Host Suitability   341

    A. Models of Host Suitability Applied to Bark Beetle Populations   341

    B. An Alternative Model for the Effects of Water Availability on Pines   342

    C. Experimental Results from 1990   343

    D. Experimental Results from 1991   344

    E. Future Studies   346

III. Predation and SPB Dynamics   347

    A. Background on Interactions between *Thanasimus dubius* and SPB   347

    B. Functional Response of *Thanasimus dubius*   348

    C. Numerical Response of *Thanasimus dubius*   348

    D. Life History of *Thanasimus dubius*   350

IV. Conclusions   352

    References   353

Chapter 18

*The Dominance of Different Regulating Factors for
Rangeland Grasshoppers*   359

GARY E. BELOVSKY
ANTHONY JOERN

I. Introduction   359
II. Grasshoppers, a First Visit   360
   A. Predators Eliminated and the Initial Grasshopper Density and Plant Availability Varied   362
   B. Elimination of Avian Predators   364
   C. Our Observations and Support from Elsewhere in the Western United States   364
III. Graphical Depictions of Population Mechanisms Identified in Our Studies   366
   A. Probability of Surviving from Hatching to the Adult Stage in the Absence of Predators   366
   B. Production of Hatchlings per Adult Female   368
   C. Probability That an Individual Grasshopper Is Killed by Predators   368
IV. An Integrated Model of Population Limitation   370
   A. Basic Modeling Strategy   371
   B. Scenario 1: Adult Density without Predators Declines as Initial Hatchling Density Increases   371
   C. Scenario 2: Adult Density without Predators Is Constant as Initial Hatchling Density Increases   373
   D. Operation of the Ricker Curve Model   373
V. Grasshoppers Revisited: Ricker Curves   376
   A. Observed Ricker Curves   376
   B. Implications for Our Grasshopper Populations   378
VI. Insights for Population Ecology   379
   A. Study Design   379
   B. Fallacy of Searching for Single Explanations   379
   C. Misconception of Invariant Density Dependence and Equilibrium   380
   D. Multiple Stable Equilibria   381
VII. Conclusions   381
   References   382

PART V

CONCLUSION 387

Chapter 19

Novelty and Synthesis in the Development of Population Dynamics 389

PETER W. PRICE
MARK D. HUNTER

I. Introduction   389
II. Elements of a New Synthesis   391
    A. Plant–Herbivore Interactions   391
    B. Chemical Ecology   392
    C. Microbial Ecology   392
    D. Life-History Evolution   393
    E. Behavioral Ecology   393
    F. Broad Comparative Ecology   394
    G. Experimental Biology   395
    H. Integration of Empirical and Theoretical Ecology   396
    I. Genetics, Heritability, and the Evolution of Populations   397
    J. International Integration   398
III. Novel Approaches and the Beginnings of Synthesis   398
    A. Historical Perspective   398
    B. What's New?   400
IV. The New Synthesis   402
    A. Scale   402
    B. Getting Empirical Facts into Theory   403
    C. "A New Reductionism?"   404
    References   404

Index   413

Contributors

Numbers in parentheses indicate the pages on which the authors' contributions begin.

Michael J. Auerbach (83)   Department of Biology, University of North Dakota, Grand Forks, North Dakota 58202

Matthew P. Ayres (339)   Department of Biological Sciences, Dartmouth College, Hanover, New Hampshire 03755

Gary E. Belovsky (359)   Department of Fisheries and Wildlife, and The Ecology Center, Utah State University, Logan, Utah 84322

Naomi Cappuccino (3, 65, 131)   Department of Zoology, University of Texas, Austin, Texas 78712

Edward F. Connor (83)   Department of Environmental Sciences, University of Virginia, Charlottesville, Virginia 22903

Timothy P. Craig (321)   Life Sciences Program, Arizona State University West, Phoenix, Arizona 85069

Hans Damman (65)   Department of Biology, Carleton University, Ottawa, Ontario, Canada K1S 5B6

Robert F. Denno (113)   Department of Entomology, University of Maryland, College Park, Maryland 20742

Jean-François Dubuc (65)   Groupe de Recherche en Écologie Forestière, Université du Québec à Montréal, Montréal, Québec, Canada H3C 3P8

Greg Dwyer (209)   Department of Entomology, University of Massachusetts, Amherst, Massachusetts 01003

Ilkka Hanski (149)   Department of Ecology and Systematics, Division of Population Biology, University of Helsinki, FIN-(000)14 Helsinki, Finland

Susan Harrison (131)   Division of Environmental Studies, University of California at Davis, Davis, California 95616

Alison F. Hunter (41)   Department of Biology, Dalhousie University, Halifax, Nova Scotia, Canada B3H 4J1

Mark D. Hunter (389)   Institute of Ecology, University of Georgia, Athens, Georgia 30602

Anthony Joern (359)   School of Biological Sciences, University of Nebraska, Lincoln, Nebraska 68538

Mikko Kuussaari (149)   Department of Ecology and Systematics, Division of Population Biology, University of Helsinki, FIN-(000)14 Helsinki, Finland

Peter L. Lorio, Jr. (339)   Southern Research Station, United States Department of Agriculture Forest Service, Pineville, Louisiana 71360

Susan Mopper (83)   Department of Biology, University of Southwestern Louisiana, Lafayette, Louisiana 70504

Judith H. Myers (229)   Departments of Plant Science and Zoology, Center for Biodiversity Research, University of British Columbia, Vancouver, British Columbia, Canada V6T 1Z4

Takayuki Ohgushi (303)   Institute of Low Temperature Science, Hokkaido University, Sapporo 060, Japan

Merrill A. Peterson (113)   Department of Entomology, University of Maryland, College Park, Maryland 20742

Peter W. Price (321, 389)   Department of Biological Sciences, Northern Arizona University, Flagstaff, Arizona 86011

John D. Reeve (339)   Southern Research Station, United States Department of Agriculture Forest Service, Pineville, Louisiana 71360

Heikki Roininen (321)   Department of Biology, University of Joensuu, FIN-80101 Joensuu, Finland

Jens Roland (195)   Department of Biological Sciences, University of Alberta, Edmondon, Alberta, Canada T6G 2E9

MaryCarol Rossiter (251)   Institute of Ecology, University of Georgia, Athens, Georgia 30602

Lorne D. Rothman (229)   Department of Zoology, Center for Biodiversity Research, University of British Columbia, Vancouver, British Columbia, Canada V6T 1Z4

Christer Solbreck (279)   Department of Entomology, Swedish University of Agricultural Sciences, S-750 07 Uppsala, Sweden

Philip D. Taylor (195)   Department of Biology, Acadia University, Wolfville, Nova Scotia, Canada B0P 1X0

Peter Turchin (19)   Department of Ecology and Evolutionary Biology, University of Connecticut, Storrs, Connecticut 06269

Sandra J. Walde (173)   Department of Biology, Dalhousie University, Halifax, Nova Scotia, Canada B3H 4J1

Preface

   Population dynamics has been one of the central areas of study in ecology since the early days of the discipline. Although it was a vital area of research during the 1950s and 1960s, it was unfortunately neglected during the 1970s, eclipsed by exciting developments in chemical, behavioral, and community ecology. The 1970s were also a period when the gulf between theory and experimentation grew, further hindering the development of population ecology.

   The past few years, however, have seen a resurgence of interest in population dynamics, spurred on by theoretical alternatives to the traditional equilibrium viewpoint of population regulation. Creative new techniques for analyzing long-term data are providing a more powerful way to view dynamics. The study of mechanisms of population change that have traditionally been ignored, such as pathogens, dispersal, and maternal effects, has increased our understanding of how populations fluctuate. Bold experiments, done on a larger scale than those attempted in the past, are enabling us to address directly the processes of population change. Population fluctuations are becoming the subject of comparative phylogenetic analyses, as authors seek to understand the life-history correlates of eruptive population dynamics and stability. Finally, the gap between theory and empiricism appears to be closing as many more authors address population dynamics with a mix of modeling and experimentation.

   We felt that the time was right to document these new advances. Our aim was to provide a representative cross section of the field, including examples of observational, experimental, and theoretical approaches, as well as studies that combine two or more of these approaches. In choosing chapter authors, we included several younger workers, who bring a fresh perspective to the field.

   Our goal in organizing this book goes beyond merely cataloging the current approaches to population dynamics. We asked authors to provide a critical assessment of their approach, by pointing out how past obstacles to progress have been overcome and by indicating questions that remain to be explored. In this way, our aim was to provide both a useful guide to the state of the art in population dynamics and a catalyst to future research. We especially hope that our book will be useful to graduate students and other young researchers, with whom we wish to share our excitement for the study of population dynamics.

Naomi Cappuccino
Peter W. Price

PART I

INTRODUCTION

Chapter 1

Novel Approaches to the Study of Population Dynamics

Naomi Cappuccino

I. Population Dynamics: A Brief Historical Review of the Major Concepts

The study of population dynamics is an old discipline with roots that ante-date the modern science of ecology. The relative stability of most natural populations, as well as the occasional wild fluctuations of a few, has interested philosophers and natural historians since the beginning of historical time (Edgerton, 1973). In 1798, the Rev. T. R. Malthus launched the modern study of the control of population growth, noting that animal populations increase geometrically up to the limit set by their resources. Although the study of population dynamics has come a long way since Malthus, explaining the stability and persistence of populations remains one of the most difficult challenges confronting twentieth-century ecologists.

Population dynamics is a subject with a history of provoking heated debate among ecologists. Throughout the twentieth century, interest in the debate has ebbed and flowed, the exact focus has shifted, and the terminology has evolved. Early authors were interested in whether biotic factors (Howard and Fiske, 1911) or climatic factors (Uvarov, 1931) controlled populations. Nicholson (1933) was the first to insist on the important point that in order for populations to be regulated ("controlled" or "balanced" in his terms), they must be subject to factors that act in a density-dependent manner (he preferred the term "density-governing"). Density-dependent factors impinge more severely upon the population when its density is high, and provide a mechanism by which the population returns to its equilibrium density following a perturbation. For the past four decades, the debate has mainly centered on the issue of density-dependent regulation, how frequent or strong it is in nature (Milne, 1958; Dempster, 1958; Strong, 1984), and whether it is indeed necessary to explain the persistence of populations (Andrewartha and Birch, 1954; Den Boer, 1968).

Early approaches to the question of regulation were mostly phenomenological: analyses of long-term abundance data (e.g., Morris, 1963) and short-term, more detailed measures of the mortality factors (k-factors) impinging upon populations (Varley and Gradwell, 1960). Although this descriptive approach was quantitative and quite rigorous, the tests designed to detect density dependence from time-series data and from k-factor analysis were often unreliable (Murdoch and Walde, 1989). Although recent advances in time-series analysis have greatly improved this technique as a tool for understanding regulation (Turchin, Chapter 2, this volume), the earlier problems with this approach probably contributed to the feeling among many ecologists that population regulation was an intractable topic for study.

At the same time that ecologists were wearying of the population regulation debate, emphasis in ecology was shifting to areas of inquiry with more appeal and apparent explanatory power. The 1970s saw the rise of community ecology, spurred on by the exciting ideas of Robert MacArthur (Kareiva, 1989). Evolutionary ecology, especially the study of life-history theory, and topics in behavioral ecology such as optimal foraging theory also flourished during this decade. The field of plant–herbivore interactions got off the ground in the early 1970s (Feeny, 1970; Root, 1973), initiating two decades of research that would lay the groundwork for an improved understanding of insect population dynamics. Yet it was not until the mid-1980s that population ecology would regain its place at the center of ecological interest (Stiling, 1994).

Throughout the late 1980s and into the current decade, the study of population dynamics has been steadily gaining momentum. General interest in a field is often stimulated by the consolidation of ideas that have been in the air, and the consequent formation of a new paradigm. In population ecology, a great deal of excitement has been provided by ideas that fall under the rubric "nonequilibrium theory." Although over the decades authors have questioned both the equilibrial notion of local density-dependent regulation as well as the usefulness of the very concept of equilibrium (e.g., Andrewartha and Birch, 1954; Den Boer, 1968), the assumption of a balance of nature has nevertheless dominated ecological thinking (Edgerton, 1973). Within the past decade, however, nonequilibrial ideas have come into vogue, especially in North America (Koetsier et al., 1990; Murdoch, 1994).

Ecological concepts often have hazy definitions (Peters, 1991) and the buzzword "nonequilibrium dynamics" is no exception. Purely nonequilibrial populations are not subject to density-dependent factors that return their densities toward an equilibrium; they fluctuate randomly and eventually reach extreme densities (Nisbet and Gurney, 1982). Of course, populations cannot continue to increase infinitely; all populations have an upper limit imposed by essential resources such as food or space. Eventually, a purely nonequilibrial local population will randomly walk its way to extinction.

While randomly walking populations are unambiguously nonequilibrial, the term "nonequilibrium dynamics" is most often used to refer to other types of dynamics in which local populations do not trend toward a point equilibrium. This includes notions such as density-vagueness (Strong, 1984) and stochastic boundedness (Chesson, 1978). In these cases, populations experience density dependence only at density extremes ("ceilings" and "floors"). However, because the fluctuations of such populations are bounded, these forms of dynamics are consistent with the notion of regulation. In this sense they are misplaced in the category "nonequilibrium" (Murdoch, 1994).

The other form of "nonequilibrium" dynamics that has attracted a great deal of attention is the concept of metapopulations dynamics (Gilpin and Hanski, 1991). Building on ideas present since Nicholson (1954) and Birch (1958) and formalized by Den Boer (1968) and Levins (1969), metapopulation theory attempts to explain the persistence of metapopulations (ensembles of subpopulations) as a consequence of spatially heterogeneous local dynamics coupled with dispersal between subpopulations. Individual subpopulations are not assumed to be regulated, and indeed they may go extinct. Because subpopulations fluctuate asynchronously, however, extinctions and outbreaks do not occur at the level of the entire metapopulation. Dispersal allows for the recolonization of vacant habitat islands following local extinction. There is no explicit assumption of local density dependence in metapopulation models, hence the inclusion under the rubric "nonequilibrial." However, a metapopulation composed of purely nonequilibrial (i.e., randomly walking) subpopulations will also randomly walk to extinction (Chesson, 1981).

II. Traditional and Novel Approaches to the Study of Population Dynamics

A. Observations: The Search for Patterns in Time Series

1. The Search for Evidence of Regulation

The study of population dynamics has been approached from a variety of angles. As already mentioned, the traditional approach has been observational, based on the analysis of time-series data. Although earlier analyses were often unable to distinguish reliably between random-walk dynamics and regulation in simulated time series, as Turchin points out (Chapter 2, this volume), newer analytical techniques and longer runs of data have improved this situation. Evidence for regulation is now commonly found in time series that are sufficiently long. Abundant evidence for regulation is also revealed by calculating the cumulative variance of natural populations (Murdoch and Walde, 1989). A regulated population, whose fluctuations are bounded, has a cumulative temporal variance that levels off after a certain length of time.

Using the criterion for regulation that a population must be bounded (Murdoch, 1994) or have a stationary probability distribution (Turchin, Chapter 2, this volume), the definition of regulation becomes quite broad and includes complex dynamical behaviors such as cycles and chaos (Murdoch, 1994). The only dynamical patterns that do not fit these definitions are those of populations undergoing random walks or increasing oscillations, or those with a trending equilibrium. Thus the question is no longer "is it regulated?" but "*how* is it regulated?"

A good example of two organisms that may both be considered regulated but that exhibit radically different dynamics is provided by the spruce budworm (*Choristoneura fumiferana*) and the tiger swallowtail butterfly (*Papilio glaucus*). The spruce budworm is a particularly destructive pest in the boreal forest of North America. Its population fluctuations are bounded, however, since their cumulative variance has been shown to level off with time (Murdoch and Walde, 1989). The upper limit to spruce budworm density is very close to that imposed by the abundance of its food plant, balsam fir. On the lower end, it is barely detectable in the habitat. If we consider the spruce budworm to be regulated, we need to specify how its regulation is different from that of the tiger swallowtail, whose larvae are quite rare relative to their food resources. We must distinguish between the many forms that regulation can take: simple local regulation, metapopulation regulation, complex dynamics. To help field ecologists in this endeavor, several chapters of this book may serve as guides. Harrison and Cappuccino (Chapter 7) address the problem of testing for local regulation via density-perturbation experiments. Walde (Chapter 9) and Roland and Taylor (Chapter 10) provide examples of how one might assess the effect of metapopulation structure on herbivore–natural enemy systems. Dwyer (Chapter 11) discusses the mechanisms resulting in complex dynamics in model systems; the sort of dynamics that Myers and Rothman (Chapter 12) and Rossiter (Chapter 13) seek to understand in populations of forest Lepidoptera.

2. The Comparative Approach

Another use of time-series data involves the comparative approach. This approach often sidesteps the question of regulation, concentrating instead on the magnitude of fluctuations, typically measured as the standard deviation of the log-transformed abundance. Once the temporal variability of a large number of species has been calculated, one may look for characteristics that are associated with variability. Theory often suggests species characteristics that might be correlated with a tendency to fluctuate greatly. For instance, Southwood (1981) has suggested that r-selected species—species that are smaller and have short generation times and high reproductive rates—should be more likely to erupt. The characteristics associated with r-selection have proven to be only weak or occasional predictors of temporal variability (Nothnagle and Schultz, 1987; Gaston and Lawton, 1988a; Morse *et al.*, 1988; Root and Cappuccino, 1992). Contradictory predictions have been made regarding the relative variability of specialists versus generalists. Watt (1964) suggested that generalists, which have less trouble finding food resources in the habitat, would more efficiently take advantage of occasional good growth conditions and outbreak. On the other hand, MacArthur (1955) predicted that specialists would be more variable because they are more susceptible to population crashes brought on by the vagaries of weather acting upon their single food resource. Analyses by Redfearn and Pimm (1988) and Gaston and Lawton (1988b) revealed weak support for Watt's hypothesis; however, Root and Cappuccino (1992) found no relationship between specialization and variability.

One ecological characteristic, the tendency to aggregate, has been shown to be consistently associated with high temporal variability. Outbreak insects often feed in groups or lay their eggs in masses or otherwise aggregate during some stage of their life cycle (Watt, 1964; Cappuccino, 1987; Hanski, 1987; Nothnagle and Schultz, 1987; Rhoades, 1985; Hunter, 1991; Root and Cappuccino, 1992). Capitalizing on this well-documented pattern, as well as other life-history characteristics consistently associated with high temporal variability, Hunter (Chapter 3, this volume) investigates population dynamics from an evolutionary perspective. In one of the rare studies to address population dynamics with phylogenetic analyses, she assesses the number of independent origins of characters associated with eruptive dynamics.

Although the link between spatial patterns and temporal dynamics has now been well documented, the causal link between clumping and outbreaking remains theoretical (Hanski, 1987). In Chapter 4, Cappuccino, Damman, and Dubuc discuss experimental manipulations of spatial pattern that may allow us to understand the mechanisms driving the increased temporal variability of clumped species.

Although the comparative method is typically used to reveal ecological and life-history characteristics associated with outbreak dynamics, it can also be used to determine whether sources of mortality or the prevalence of density dependence differ among outbreak and nonoutbreak species. This is the approach taken in Chapter 5 by Auerbach, Connor, and Mopper. Though they find neither differences in life-history characteristics nor differences in dominant mortality sources between leaf miners with outbreak or nonoutbreak dynamics, they document, as do several other authors of this volume, the importance of bottom-up effects and "horizontal" effects, such as interference competition and cannibalism, on leaf-miner dynamics. Competition and premature leaf abscission often impose density-dependent mortality on latent and eruptive miner species alike, and are thus likely candidates in the regulation of miner density.

The comparative method is a powerful approach to understanding population dynamics, especially when it is combined with rigorous phylogenetic analysis or experimentation on mechanisms. To take advantage of this approach, however, we need census data on nonoutbreak species. Though many insect pests are well studied, much of the information on the dynamics, ecological characteristics, and mortality sources of nonoutbreak species is anecdotal at best. We would make more rapid progress toward understanding the outbreaks of pests if every laboratory working on a pest species also had a team working on the comparative ecology of related nonpest species or other species using the same host plant.

B. Mechanisms: The Factors Driving Population Change

1. Three Less Commonly Studied Factors Influencing Population Change

Since the 1970s when field experimentation in ecology became popular, studies of mechanisms, the factors that influence population change, became increasingly prevalent. Perhaps as a result of the seminal paper of Hairston *et al.* (1960), in which they argued that herbivorous insects are controlled by their enemies, the emphasis in insect population ecology has mostly been on predation and parasitism as mechanisms driving population change. Models of herbivore–natural enemy dynamics focused on the behavioral details of predation and parasitism (Hassell, 1976); thus experimental studies mostly concentrated on mechanistic details such as the functional response of enemies when presented with different densities of prey or hosts (e.g., Waage, 1979).

Other mechanisms have received much less attention from experimentalists, even though their potential importance has been long recognized by theoreticians. Three less commonly studied mechanisms—dispersal, disease, and maternal effects—are given special consideration in this volume.

Dispersal has been seen as a central mechanism in population models (mathematical and conceptual), at least since Nicholson (1954) recognized that local host–parasitoid systems may go extinct, and that vacant habitat can be recolonized by dispersal from other patches. Despite the central role of dispersal in modern population models, this mechanism is one of the most poorly studied (Taylor, 1990). In Chapter 6, Denno and Peterson review studies of dispersal in a variety of insect taxa and argue that, for many species, dispersal is the single most important mechanism influencing population size and stability. Dispersal is also key to understanding population dynamics in fragmented landscapes, of the kind discussed by Hanski and Kuussaari (Chapter 8), Walde (Chapter 9), and Roland and Taylor (Chapter 10).

Despite the appealing theoretical work of Anderson and May (1980), suggesting that pathogens are responsible for the cyclic dynamics of forest lepidopterans, pathogens have traditionally taken a back seat to predators and parasites

Population Dynamics

in both theory and empirical work. As much as insect ecologists tend to shun the study of pathogens, insect pathologists typically ignore population ecology and concentrate instead on the physiology of the insect–pathogen interaction (Ignoffo, 1978). Slowly, however, the number of experimental ecologists addressing pathogen–host dynamics is growing. Roland and Taylor (Chapter 10), Dwyer (Chapter 11), and Myers and Rothman (Chapter 12) describe experiments on pathogen spread and host susceptibility in forest pests with complex dynamics.

With the exception of the earlier studies of Chitty (1960) and Wellington (1960), population ecologists have been concerned primarily with numbers of individuals and have typically glossed over variation in the quality of individuals in a population. However, newer theory has underscored the importance of variability in the quality of individuals within a population (Nisbet *et al.*, 1989). One mechanism responsible for potentially important differences in the quality of individuals within a population is the maternal effect. Although maternal effects have been well known to ecological geneticists, who strive to eliminate such nongenetical effects from their experimental designs, we are only beginning to appreciate their role in population dynamics. In Chapter 13, MaryCarol Rossiter describes how maternal effects can generate time lags and contribute to the complex dynamics of gypsy moths. She then surveys the literature for the importance of maternal effects in other insect taxa and discusses the circumstances under which maternal effects are likely to be important.

2. Linking Mechanisms to Dynamics through Models

The link between mechanisms and the dynamical patterns we see in nature is often made through mathematical models. However, the gulf between theory and empiricism (Kareiva, 1989) has often resulted in only a weak coordination between models and field studies. A prime example of both the importance of models and the confusion they often engender among empiricists is the vast literature on herbivore–parasitoid interactions in which spatial patterns in parasitism are central to the stability of the models (e.g., Hassell, 1985; Chesson and Murdoch, 1986; Pacala *et al.*, 1990). Countless empirical studies, many of which are inconclusive, have been performed to determine the spatial density dependence in parasitism rates (reviewed by Stiling, 1987; Walde and Murdoch, 1988). Most of these studies involve pest species (Kareiva, 1990), and thus are looking for stabilizing mechanisms in the populations least likely to show them. Furthermore, only recently have the inner workings of host–parasitoid models been laid bare in terms that are accessible to empiricists. In a clear review of the theory, Taylor (1993) explains how spatial patterns such as inversely density-dependent parasitism can lead, counterintuitively, to stability in host–parasitoid dynamics.

A superb example of how theory and fieldwork can, and should, be combined to understand population dynamics is provided by the work of Bill Murdoch and his colleagues (Reeve and Murdoch, 1985; Murdoch and Stewart-Oaten, 1989; Murdoch *et al.*, 1987, 1989; Murdoch, 1994). In a series of studies on a particularly stable system, California redscale *Aonidiella aurantii* and its parasitoid *Aphytis melinus,* they tested and rejected eight mechanisms that had the potential to stabilize the system, including spatial heterogeneity in attack rates. Most ecologists are not quite so effective at coupling ground-breaking theory with elegant experimentation. Nevertheless, to avoid performing experiments that only weakly pertain to theory, we need to bridge the gap between theoreticians and empiricists. Several chapters in this book, including those by Dwyer (Chapter 11), Hanski and Kuussaari (Chapter 8), and Belovsky and Joern (Chapter 18), represent major steps in bringing mathematical theory and empiricism closer together.

C. The Varied Processes of Population Regulation: Testing Old Theories and Building New Ones

The question of regulation has always been central to the study of population dynamics. Regulation is the process that translates mechanisms such as predation or dispersal into the long-term dynamical patterns of temporal constancy or persistence that are so often observed in nature. Although most studies testing for regulation have involved analysis of observational data, the most effective way to test for local regulation is to perform "convergence experiments" (Murdoch, 1970) in which density is manipulated to see if populations return to their previous levels. Perturbed populations of herbivorous ladybirds (Ohgushi and Sawada, 1985; Ohgushi, Chapter 15, this volume), tephritid flies (Cappuccino, 1992), and tussock moths (Harrison, 1994) have all shown tendencies to return to preperturbation densities. In Chapter 7, Harrison and Cappuccino review the literature on perturbation experiments and discuss ways in which the experimental approach to regulation can be improved.

Populations may also be stabilized as a result of metapopulation dynamics (Gilpin and Hanski, 1991). In Chapter 8, Hanski and Kuussaari explain metapopulation theory and provide evidence that many butterfly populations, while locally unstable, are indeed regulated at the level of the metapopulation. Most evidence for metapopulation regulation comes from the observations that are consistent with this theory: local populations fluctuate asynchronously, sometimes go extinct, and are later recolonized by dispersers from other subpopulations. An ideal way to test metapopulation theory would be either to manipulate dispersal, perhaps by putting up large barriers to movement (Reeve, 1990), or to alter the degree of fragmentation of the habitat. The large scale at which such experiments would have to be done makes them all but impossible in most systems, but may be possible for very small organisms such as phytophagous mites. In studying the predator–prey metapopulation dynamics of mites in apple orchards, Walde (Chapter 9) has performed one of the first field manipulations of metapopulation structure. Roland and Taylor (Chapter 10) take advantage of "experiments" performed by lumber companies in northern Alberta. They show how the longer duration of forest tent caterpillar outbreaks in more highly fragmented forest can be explained by both lower disease incidence and lack of parasitoid dispersal to habitat isolates.

D. Case Studies: Building and Testing Theory in Well-Studied Systems

The best ecology is often accomplished through a mix of observation, mechanistic studies, and process-oriented studies (Feinsinger and Tiebout, 1991). With the development of long-term, spatially explicit data sets, several researchers are in the position to address population stability using a combination of observation and experimentation. This book also features work on a variety of well-studied systems, including seed-feeders (Solbreck, Chapter 14), herbivorous coccinellids (Ohgushi, Chapter 15), tenthredinid sawflies (Price, Craig, and Roininen, Chapter 16), bark beetles (Reeve, Ayres, and Lorio, Chapter 17), and rangeland grasshoppers (Belovsky and Joern, Chapter 18).

These contributions tie together several of the themes addressed in earlier chapters. In Chapter 15, Ohgushi emphasizes the importance of oviposition physiology and behavior in the remarkably stable populations of the herbivorous lady beetle. His chapter, as well as those by Solbreck and Price *et al.*, underscores the prevalence of bottom-up and "lateral" or "horizontal" factors, also detected in the reviews by Auerbach *et al.* (Chapter 5) and Harrison and Cappuccino (Chapter 7). The herbivores described in these three chapters all track closely the carrying capacity set by the host plant, which may, however, be quite variable in space and time, as shown by Solbreck in Chapter 14.

The southern pine beetle populations examined by Reeve *et al.* in Chapter 17 exhibit the sort of complex dynamics described by both Turchin (Chapter 2) and Dwyer (Chapter 11). The delayed numerical response of a predatory clerid beetle with a long generation time is probably responsible for the cyclic dynamics. Rangeland grasshoppers can also reach outbreak densities, but do not do so in all locations, as described by Belovsky and Joern (Chapter 18). Their models, tested with a series of large-scale density manipulations, suggest that grasshopper densities may be attracted to two domains: one at high densities set by the interaction with the host plant, the other at lower densities imposed by enemies.

III. The Role of Herbivorous Insects in the Study of Population Dynamics

All of the chapter authors primarily study herbivorous insects. Herbivorous insects have been central to the question of population regulation and stability from the early debates on density dependence (Nicholson, 1958; Andrewartha and Birch, 1954; Birch, 1958) up through the modern discussions of the prevalence of different sorts of regulation in nature. Despite the central place that insects and insect ecologists have occupied in the development and testing of population dynamical theory, the approaches we take apply quite well to other types of organisms. Many of us have indeed addressed the question of regulation in other taxa, both in this book (Chapter 10) and in other publications (Myers and Krebs, 1971; Hanski *et al.*, 1991; Turchin, 1993).

IV. Population Dynamics: The New (Pluralist) Synthesis

One of the most refreshing aspects of the renaissance of population dynamics is the absence of polarizing debate. Early participants in the debates about density dependence and density independence were often arguing at cross-purposes (Sinclair, 1989), since density-dependent and density-independent factors are not mutually exclusive. Neither does the presence of agents capable of regulating local populations preclude the existence of metapopulation processes in the same or other species. The new synthesis in population dynamics is thus pluralist in nature, reflecting the diversity of the natural world and the range of ways in which populations can be regulated. A wide variety of subdisciplines in ecology have greatly contributed to our current understanding of population dynamics; in the concluding chapter of this volume, Price and Hunter discuss how this, too, contributes to the new pluralist synthesis.

Theory in population ecology has never been scarce. The time has come for empiricists to devise clever tests of these theories, to help refine theory by opening channels of communication with theorists, and to participate in the building of empirically based theory. This is the exciting challenge of population ecology as we enter the next century. We hope that the chapters presented in this volume will provide a guide to ways in which population dynamical questions may be addressed, as well as a stimulus to the creativity of the next generation of population ecologists.

Acknowledgments

I thank Peter Price, Mike Singer, and L. Ramakrishnan for helpful comments on this manuscript.

References

Anderson, R. M., and May, R. M. (1980). Population biology of infectious diseases. *Nature (London)* **280**, 361–367.

Andrewartha, H. G., and Birch, L. C. (1954). "The Distribution and Abundance of Animals." Univ. of Chicago Press, Chicago.

Birch, L. C. (1958). The role of weather in determining the distribution and abundance of animals. *Cold Spring Harbor Symp. Quant. Biol.* **22**, 203–218.

Cappuccino, N. (1987). Comparative population dynamics of two goldenrod aphids: Spatial patterns and temporal constancy. *Ecology* **68**, 1634–1646.

Cappuccino, N. (1992). The nature of population stability in *Eurosta solidaginis,* a nonoutbreaking herbivore of goldenrod. *Ecology* **73**, 1792–1801.

Chesson, P. L. (1978). Predator–prey theory and variability. *Annu. Rev. Ecol. Syst.* **9**, 288–325.

Chesson, P. L. (1981). Models for spatially distributed populations: The effect of within-patch variability. *Theor. Popul. Biol.* **19**, 288–323.

Chesson, P. L., and Murdoch, W. W. (1986). Aggregation of risk: Relationships among host–parasitoid models. *Am. Nat.* **127**, 696–715.

Chitty, D. (1960). Population processes in the vole and their relevance to general theory. *Can. J. Zool.* **38**, 99–113.

Dempster, J. P. (1958). The natural control of populations of butterflies and moths. *Biol. Rev. Cambridge Philos. Soc.* **58**, 461–481.

Den Boer, P. J. (1968). Spreading of risk and stabilization of animal numbers. *Acta Biotheor.* **18**, 165–194.

Edgerton, F. N. (1973). Changing concepts of the balance of nature. *Q. Rev. Biol.* **48**, 322–350.

Feeny, P. P. (1970). Seasonal changes in oak leaf tannins and nutrients as a cause of spring feeding by winter moth caterpillars. *Ecology* **51**, 565–581.

Feinsinger, P., and Tiebout, H. M., III (1991). Competition among plants sharing hummingbird pollinators: Laboratory experiments on a mechanism. *Ecology* **72**, 1946–1952.

Gaston, K. J., and Lawton, J. H. (1988a). Patterns in the distribution and abundance of insect populations. *Nature (London)* **331**, 709–712.

Gaston, K. J., and Lawton, J. H. (1988b). Patterns in body size, population dynamics, and regional distribution of bracken herbivores. *Amer. Nat.* **132**, 662–680.

Gilpin, M., and Hanski, I. (1991). "Metapopulation Dynamics: Empirical and Theoretical Investigations." Academic Press, San Diego, CA.

Hairston, N. G., Smith, F. E., and Slobodkin, L. B. (1960). Community structure, population control, and competition. *Am. Nat.* **44**, 421–425.

Hanski, I. (1987). Pine sawfly population dynamics: Patterns, processes, problems. *Oikos* **50**, 327–335.

Hanski, I., Hansson, L., and Henttonen, H. (1991). Specialist predators, generalist predators, and the microtine rodent cycle. *J. Anim. Ecol.* **60**, 353–367.

Harrison, S. (1994). Resources and dispersal as factors limiting a population of the tussock moth (*Orgyia vetusta*), a flightless defoliator. *Oecologia* **99**, 27–34.

Hassell, M. P. (1976). Arthropod predator–prey systems. *In* "Theoretical Ecology" (R. M. May, ed.), pp. 105–131. Blackwell, Oxford.

Hassell, M. P. (1985). Insect natural enemies as regulating factors. *J. Anim. Ecol.* **54**, 323–334.

Howard, L. O., and Fiske, W. F. (1911). The importation into the United States of the Parasites of the gipsy moth and the brown-tail moth. *Bull. USDA Bur. Entomol.,* **91**.

Hunter, A. F. (1991). Traits that distinguish outbreaking and nonoutbreaking Macrolepidoptera feeding on northern hardwood trees. *Oikos* **60**, 275–282.

Ignoffo, C. M. (1978). Strategies to increase the use of entomopathogens. *J. Invertebr. Pathol.* **31**, 1–3.

Kareiva, P. (1989). Renewing the dialog between theory and experiments in population ecology. *In* "Perspectives in Ecological Theory" (J. Roughgarden, R. M. May, and S. A. Levin, eds.), pp. 68–88. Princeton Univ. Press, Princeton, NJ.

Kareiva, P. (1990). Stability from variability. *Nature (London)* **344**, 111–112.

Koetsier, P., Dey, P., Mladenka, G., and Check, J. (1990). Rejecting equilibrium theory—A cautionary note. *Bull. Ecol. Soc. Am.* **71**, 229–230.

Levins, R. (1969). Some demographic and genetic consequences of environmental heterogeneity for biological control. *Bull. Entomol. Soc. Am.* **15**, 237–240.

MacArthur, R. H. (1955). Fluctuations of animal populations and a measure of community stability. *Ecology* **36**, 533–536.

Milne, A. (1958). Theories of natural control of insect populations. *Cold Spring Harbor Symp. Quant. Biol.* **22**, 253–271.

Morris, R. F., ed. (1963). The dynamics of epidemic spruce budworm populations. *Mem. Entomol. Soc. Can.* **31**, 1–332.

Morse, D. R., Stork, N. E., and Lawton, J. H. (1988). Species number, species abundance and body–length relationships of arboreal beetles in Bornean lowland rain forest trees. *Ecol. Entomol.* **13**, 25–37.

Murdoch, W. W. (1970). Population regulation and population inertia. *Ecology* **51**, 497–502.

Murdoch, W. W. (1994). Population regulation in theory and practice. *Ecology* **75**, 271–287.

Murdoch, W. W., and Stewart-Oaten, A. (1989). Aggregation by parasitoids and predators: Effects on equilibrium and stability. *Am. Nat.* **134**, 288–310.

Murdoch, W. W., and Walde, S. J. (1989). Analysis of insect population dynamics. *In* "Towards a More Exact Ecology" (P. J. Grubb and J. B. Whittaker, eds.), pp. 113–140. Blackwell, Oxford.

Murdoch, W. W., Nisbet, R. M., Blythe, S. P., Gurney, W. S. C., and Reeve, J. D. (1987). An invulnerable age class and stability in delay-differential parasitoid–host models. *Am. Nat.* **129**, 263–282.

Murdoch, W. W., Luck, R. F., Walde, S. J., Reeve, J. D., and Yu, D. S. (1989). A refuge for red scale under control by *Aphytis:* Structural aspects. *Ecology* **70**, 1707–1714.

Myers, J. H., and Krebs, C. J. (1971). Genetic, behavioral, and reproductive attributes of dispersing field voles *Microtus pennsylvanicus* and *Microtus ochrogaster*. *Ecol. Monogr.* **41**, 53–78.

Nicholson, A. J. (1933). The balance of animal populations. *J. Anim. Ecol.* **2**, 132–178.

Nicholson, A. J. (1954). An outline of the dynamics of animal populations. *Aust. J. Zool.* **2**, 9–65.

Nicholson, A. J. (1958). The self-adjustment of populations to change. *Cold Spring Harbor Symp. Quant. Biol.* **22**, 153–173.

Nisbet, R. M., and Gurney, W. S. C. (1982). "Modeling Fluctuating Populations." Wiley, Chichester.

Nisbet, R. M., Gurney, W. S. C., Murdoch, W. W., and McCauley, E. (1989). Structured population models: A tool for linking effects at individual and population levels. *Biol. J. Linn. Soc.* **37**, 79–99.

Nothnagle, P. J., and Schultz, J. C. (1987). What is a forest pest? *In* "Insect Outbreaks" (P. Barbosa and J. C. Schultz, eds.), pp. 59–80. Academic Press, San Diego, CA.

Ohgushi, T., and Sawada, H. (1985). Population equilibrium with respect to available food resource and its behavioural basis in an herbivorous lady beetle, *Henosepilachna niponica. J. Anim. Ecol.* **54**, 781–796.

Pacala, S. W., Hassell, M. P., and May, R. M. (1990). Host–parasitoid associations in patchy environments. *Nature (London)* **344**, 150–153.

Peters, R. H. (1991). "A Critique for Ecology." Cambridge Univ. Press, New York.

Redfearn, A., and Pimm, S. L. (1988). Population variability and polyphagy in herbivorous communities. *Ecol. Monogr.* **58**, 39–55.

Reeve, J. D. (1990). Stability, variability and persistence in host–parasitoid systems. *Ecology* **71**, 422–426.

Reeve, J. D., and Murdoch, W. W. (1985). Biological control by the parasitoid *Aphytis melinus,* and population stability of the California red scale. *J. Anim. Ecol.* **55**, 1069–1082.

Rhoades, D. F. (1985). Offensive–defensive interactions between herbivores and plants: Their relevance in herbivore population dynamics and ecological theory. *Am. Nat.* **125**, 205–238.

Root, R. B. (1973). Organization of a plant–arthropod association in simple and diverse habitats: The fauna of collards (*Brassica oleracea*). *Ecol. Monogr.* **43**, 95–124.

Root, R. B., and Cappuccino, N. (1992). Patterns in population change and the organization of the insect community associated with goldenrod. *Ecol. Monogr.* **62**, 393–420.

Sinclair, A. R. E. (1989). Population regulation in animals. *In* "Ecological Concepts" (J. M. Cherrett, ed.), pp. 197–241. Blackwell, Oxford.

Southwood, T. R. E. (1981). Bionomic strategies and population parameters. *In* "Theoretical Ecology" (R. M. May, ed.), pp. 30–52. Blackwell, Oxford.

Stiling, P. D. (1987). The frequency of density dependence in insect host–parasitoid systems. *Ecology* **68**, 844–856.

Stiling, P. D. (1994). What do ecologists do? *Bull. Ecol. Soc. Am.* **75**, 116–121.

Strong, D. R. (1984). Density-vague ecology and liberal population regulation in insects. *In* "A New Ecology: Novel Approaches to Interactive Systems" (P. W. Price, C. N. Slobodchikoff, and W. S. Gaud, eds.). Wiley, New York.

Taylor, A. D. (1990). Metapopulations, dispersal and predator–prey dynamics: An overview. *Ecology* **71**, 429–433.

Taylor, A. D. (1993). Heterogeneity in host–parasitoid interactions: "Aggregation of risk" and the "$CV^2 > 1$ rule." *Trends Ecol. Evol.* **8**, 400–405.

Turchin, P. (1993). Chaos and stability in rodent population dynamics: Evidence from nonlinear time-series analysis. *Oikos* **68**, 167–172.

Uvarov, B. P. (1931). Insects and climate. *Trans. Entomol. Soc. London* **79**, 1–249.

Varley, G. C., and Gradwell, G. R. (1960). Key factors in population studies. *J. Anim. Ecol.* **29**, 399–401.

Waage, J. K. (1979). Foraging for patchily distributed hosts by the parasitoid *Nemeritis canescens. J. Anim. Ecol.* **48**, 353–371.

Walde, S. J., and Murdoch, W. W. (1988). Spatial density dependence in parasitoids. *Annu. Rev. Entomol.* **33,** 441–466.

Watt, K. E. F. (1964). Comments on fluctuations of animal populations and measures of community stability. *Can. Entomol.* **96,** 1434–1442.

Wellington, W. G. (1960). Qualitative changes in natural populations during changes in abundance. *Can. J. Zool.* **38,** 289–314.

PART II

OBSERVATION AND COMPARATIVE APPROACHES

Chapter 2

Population Regulation: Old Arguments and a New Synthesis

Peter Turchin

I. Introduction: The Never-Ending Debate

Population regulation is one of the central organizing themes in ecology (Dennis and Taper, 1994; Murdoch, 1994). Yet, ever since the concept of population regulation by density-dependent mechanisms was formulated by Nicholson (1933), regulation has been the subject of an acrimonious debate (or actually a number of debates about its various aspects and implications), which continues to this very day.

In 1949, Charles Elton wrote: "It is becoming increasingly understood by population ecologists that the control of populations, i.e., ultimate upper and lower limits set to increase, is brought about by density-dependent factors" (p. 19). Subsequent history showed that this statement was somewhat premature. In a very influential book, Andrewartha and Birch argued that density-dependent factors "are not a general theory because, as we have seen . . . they do not describe any substantial body of empirical facts" (Andrewartha and Birch, 1954, p. 649). Andrewartha and Birch then proposed an alternative theory of population limitation by density-independent factors. The arguments of Andrewartha and Birch set the stage for the ensuing controversies, which could be grouped around two general themes, one logical and the other empirical: (1) What does population regulation mean, and are density-dependent factors a necessary condition for regulation? (2) Can we detect density-dependent regulation in real populations, and if yes, with what frequency does it operate?

Here are some highlights from the debate, with quotes showing how vitriolic the exchange occasionally became. Milne (1957) criticized Nicholson's theory as mistaken, because "(1) it is based on a false assumption, namely, that enemy action is perfectly density-dependent . . . and (2) it asserts that this density-dependent action is responsible for natural control of decrease as well as increase, which is ridiculous." Milne advanced his own theory of "imperfect" density dependence, which is largely forgotten now (Berryman, 1992). Similar concepts are "density vagueness" of Strong (1986) and "regulation by ceilings and floors" of Dempster (1983), according to which a population can fluctuate in a largely density-independent manner for most of the time, until it approaches either a lower or an upper extreme. Milne, Dempster, and Strong felt that they were departing greatly from the density-dependence school of thought, but they were basically proposing that density dependence involved nonlinearities, and that there could be a great deal of noise—hardly controversial, in retrospect. In a more significant departure from Nicholson's theory, Andrewartha and Birch felt that density dependence was not at all necessary to prevent outbreaks of organisms. This idea was later developed by Den Boer (1968), who suggested that density-independent fluctuations in natural populations can become stabilized by stochastic processes (i.e., by chance) via a mechanism that he called "spreading of risk." These proposals have now been rejected on logical grounds, and it is generally accepted that population regulation cannot occur in the absence of density dependence (Murdoch and Walde, 1989; Hanski, 1990; Godfray and Hassell, 1992). But what is regulation? Wolda (1989) wrote a paper with a rather plaintive title, "The equilibrium concept and density dependence tests: What does it all mean?", where he concluded that the concept of an equilibrium was "fundamentally impractical and unusable in the analysis of field data." Berryman (1991) countered with a paper entitled "Stabilization or regulation: What it all means!", arguing that, on the contrary, equilibrium has a well-defined meaning and can be estimated by an appropriate analysis of data.

While the debates about the logical underpinnings of regulation raged, there was a less visible, but intense activity to develop statistical methods for detecting density dependence in data (Morris, 1959; Varley and Gradwell, 1960; Reddingius, 1971; Bulmer, 1975; Slade, 1977; Vickery and Nudds, 1984; Pollard *et al.*, 1987; Reddingius and Den Boer, 1989; Turchin, 1990; Dennis and Taper, 1994; Holyoak, 1994). Which methods are best has also been much debated, but in the process several conclusions have emerged. One was disillusionment with the conventional life table analyses (Hassell, 1986). Finding a density-regulating mechanism operating at one life stage did not guarantee overall population regulation, since a directly density-dependent mechanism in one stage could be counteracted by an inversely density-dependent mechanism acting at a different stage. For this and other reasons, current analyses of density dependence focus on the overall population change from one generation to the next. Another conclusion was a realization that making tests for regulation more general and assumption-free carries the price of greatly reduced statistical power. For an example, an alternative to density-dependence tests are direct tests for regulation, also known as tests for boundedness, limitation, or attraction (Bulmer, 1975; Murdoch and Walde, 1989; Reddingius and Den Boer, 1989; Crowley, 1992). Such direct tests appear to constitute a more robust and generic approach, but it turns out that a large proportion of time series generated by a density-independent model of random walk cannot be distinguished by these methods from regulated time series (Murdoch, 1994). Thus, direct tests for regulation have very little power to distinguish regulated from random-walk populations.

Tests for density dependence have their share of problems. Standard F-values are inappropriate, observation errors cause biases in the estimation of the slope, and nonlinearities and lags can obscure density-dependent relationships. Most importantly, the question of statistical power has to be foremost in comparing various tests to each other. Recent developments, however, send a largely upbeat message: tests can be devised that address all of these concerns. In addition, the fodder for density-dependence tests is getting better all the time, both in quantity (the number of time-series data) and in quality (the length of time series). Two tests (Pollard *et al.*, 1987; Dennis and Taper, 1994) are gaining popularity for detecting direct (undelayed) density dependence, and Woiwod and Hanski (1992) and Wolda and Dennis (1993) applied them to a large number of data sets, with very interesting results that will be summarized presently.

Population regulation, however, remains an extremely contentious field. Two critical responses to Wolda and Dennis (1993) appeared even before the publication of the test on which their paper was based (Dennis and Taper, 1994)! In one, Holyoak and Lawton (1993) focused primarily on the statistical approach used by Wolda and Dennis (1993). In another, entitled "Density dependence, population persistence, and largely futile arguments," Hanski *et al.* (1993a) took Wolda and Dennis to task for suggesting that no valid conclusions about population regulation can be drawn on the basis of statistical tests for density dependence. Not to be outdone, Wolda *et al.* (1994) responded with a paper entitled "Density dependence tests, and largely futile comments."

Is there no end to the density-dependence debate? Have population ecologists "somehow managed to muddle the basic issues of the very existence of their study objects" (Hanski *et al.*, 1993a)? After five decades and several generations of participants in this debate, many ecologists are tired of it (actually, expressions of fatigue can be found going back at least two decades, yet the debate continues to excite passions). The study of regulation has apparently gone out of fashion among U.S. ecologists (Krebs, 1992; Murdoch, 1994). Is regulation a "bankrupt paradigm" (Krebs, 1991)?

The answer to all of these questions is emphatically in the negative. There is actually a growing consensus about key issues in population regulation, temporarily obscured by still on-going arguments, which, however, are increasingly focusing on subsidiary issues. It is becoming increasingly apparent that *the fundamental issues of the population regulation debate have been resolved.* This resolution has come as a result of two factors. First, the injection into population theory of healthy doses of mathematics, especially from fields such as stochastic processes and nonlinear dynamics. Second, a continuous increase in the quantity of time-series data. In fact, the data base has reached a critical mass (at least in insect data) that, together with theoretical developments, is precipitating a synthesis.

In the following sections, I will describe what I see as the growing consensus on basic issues of population regulation, and will suggest where we should go from there. My purpose is to capture the major themes of agreement emerging in recent publications, rather than emphasize the remaining disagreements (which are, in my view, becoming more and more minor). A caveat is in order. Most rapid progress in science is achieved by combining approaches using observational data, mathematical models, and experiments. Experiments in particular play the central role in elucidating the biological mechanisms that are responsible for regulation. However, my focus in this chapter is primarily on time-series (observational) data and somewhat on mathematical models; the experimental approach is covered by Harrison and Cappuccino (Chapter 7, this volume).

II. Definition of Population Regulation

Ecological populations are dynamical systems—systems in which state variables (such as population density, age and size distribution, and densities of interacting species) change with time, or fluctuate. I will frame the following discussion in terms generic to all kinds of dynamical systems. This allows us to benefit from recent developments in the nonlinear dynamics theory, and will also help to clear some minor controversies.

The best way to define population regulation (and regulation in any dynamical system) is to equate it with the presence of *a long-term stationary probability distribution of population densities* [Dennis and Taper, 1994; this is the same as May's (1973) stochastic equilibrium probability distribution and Chesson's (1982) convergence in distribution to a positive random variable]. The key word in this definition is *stationarity*. It implies that there is some mean level of density around which a regulated population fluctuates. Additionally, as time goes on, population density does *not* wander increasingly far away from this level. Using more precise terms, the variance of population density is bounded (Royama, 1977).

An unregulated system is not characterized by any particular mean level around which it fluctuates. One example of an unregulated system, the stock exchange market, is shown in Fig. 1a (I chose a nonecological example on purpose to illustrate the generality of the concept). This is an interesting example, because it is known that fluctuations in the Dow Jones Industrial Average (DJIA) index are somewhat affected by endogenous factors. For example, the market can be overvalued, which usually leads to a decline in the DJIA, although there may be a lag time before this occurs. Similarly, periods when the market is

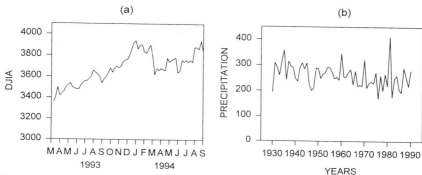

Figure 1. (a) Fluctuations in the Dow Jones Industrial Average (DJIA) index over 18 months. (b) Fluctuation pattern of annual rainfall (in cm) over 67 years on Barro Colorado Island, Panama. (Data from Wolda and Dennis, 1993.)

undervalued are eventually followed by a rally. Nevertheless, DJIA is not a stationary system. In 1994, DJIA mostly fluctuated between 3600 and 4000, but in the next year, or a year after that, DJIA will increase beyond 4000, and then eventually beyond 5000. (Although the "bears" are currently predicting a long-term decline.) Whatever happens in the future, there is nothing special about the current range of fluctuations, and in general there is no mean level around which DJIA fluctuates.

Examples of unregulated ecological systems include populations undergoing exponential growth or decline (with or without a stochastic component). Random walk is a special case in which birth and death processes just happen to cancel each other out (an unlikely event to happen purely by chance), so that there is no long-term tendency to either increase or decrease. Despite this, random walk is not a regulated process. It does not have a stationary probability distribution of population density, but rather a distribution that becomes wider with time (i.e., its variance is unbounded). A population fluctuating around a trend is not stationary (since the mean level of fluctuations is changing with time) and therefore is not, strictly speaking, regulated.

An example of a *regulated* system is shown in Fig. 1b. It is the observed pattern of rainfall on Barro Colorado Island, Panama, taken from Fig. 8 of Wolda and Dennis (1993). These data were described by the density-dependent model far better than by the density-independent model ($P<0.0001$). Wolda and Dennis, however, used this result as an example of how density-dependence tests may lead one astray, saying that "an ecological interpretation in terms of density-dependent regulation is hardly appropriate." It is true that the data come not from an ecological system, but from another kind of dynamical system. The result of the density-dependence test, however, makes perfect sense, because the earth's climate (of which rainfall is a manifestation) is a well-regulated system. The mechanistic equations governing its behavior are well known (weather forecasts are poor because it is a chaotic system, not because our understanding of weather is poor), and it is ultimately regulated by the balance between the flow of energy from the sun and the energy dissipated as heat into space. For this reason, annual precipitation levels in any locality are characterized by long-term averages, in other words, they comply with the stationarity definition of regulation. Indeed, it would be troubling if a density-dependence test would fail to detect regulation in this long data set.

The interpretation of this result became somewhat muddled because Wolda and Dennis compared it to a sequence of random numbers out of some probability distribution, whereas Holyoak and Lawton (1993) appeared to misunderstand the difference between a random sequence and a random walk. Actually, the matter is very simple.

First, drawing from a random distribution is not a random walk (Royama, 1977). A series of independent, identically distributed random variables is a null hypothesis for certain kinds of nonecological time-series models, and is written as $N_t = \epsilon_t$. This is very different from the appropriate null hypothesis for ecological applications (random walk), $r_t = \epsilon_t$ [r_t will be defined in Eq. (1)]. Written in terms of density, this becomes $\ln N_t = \ln N_{t-1} + \epsilon_t$.

Second, a dynamical system obtained by drawing random numbers from a probability distribution is regulated by definition because these numbers come from the same (therefore stationary) probability distribution time after time.

Third, real-life ecological populations may behave just like such a system. It is well known that density-dependent factors may cause population density to overshoot equilibrium ("overdamping"). Alternatively, it may take several steps to reach equilibrium after a perturbation ("underdamping"). An intermediate case is when the equilibrium is reached in one time step ("perfect damping"). A time series coming from a perfectly damped population is indistinguishable from a sequence of random numbers. Consider the following example: a perfectly damped system is affected first by endogenous and then by exogenous factors. Wherever population density starts at time t, the endogenous processes will immediately take it to the long-term mean, and then density will be perturbed in a random direction by exogenous processes, at which point we measure it. As we repeat this process, we see that each subsequent measurement has the same mean and probability distribution. Moreover, where population is at time t gives us no information about where it will be at time $t + 1$. Thus, in the case of perfect damping, there are no autocorrelations between subsequent population densities, and the system behaves as a sequence of random numbers. I emphasize that perfect damping is not an exotic condition, but is, in fact, typically observed when endogenous population dynamics are fast compared to the time interval at which we "read" the density.

A. Relationship to Previous Definitions

The "stationarity" definition of regulation, advocated here, is more general and precise than previous generally accepted definitions. Den Boer (1990) cites two such definitions given by Varley et al. (1973): "a regulated population [is] . . . one which tends to return to an equilibrium density following any departure from this level" (p. 19), and "density-dependent mortality serves to regulate the population and keeps it within limits" (p. 112). Both of these standard definitions have problems. The second one, taken literally, implies presence of some rigid "floors" and "ceilings" that are never exceeded by populations. This scenario is a special case of the stationarity definition, in which the stationary probability distribution of population density has a special form: rather flat in the middle, and abruptly decreasing to zero at the values of the floor and the ceiling. "Soft" floors and ceilings, on the other hand, are implied by the stationarity distribution. As we move away from the mean level of fluctuations, the probability distribution of density has to decrease to zero; otherwise its integral would be infinite and it would not be a true probability distribution. In other words, the stationarity definition makes more precise the notion of population limitation.

The other standard definition of regulation, relying on return to the equilibrium, historically had even more trouble. Empirical ecologists, from Andrewartha and Birch to Wolda, have always had difficulties accepting the reality of the equilibrium. Their position has much merit, if by "equilibrium" we understand a stable point equilibrium that arises in deterministic models, such as the Lotka–Volterra or the discrete logistic. All populations are influenced by exogenous variables and fluctuate with various degrees of irregularity. Even if the endogenous structure of the model with which we represent population fluctuations is characterized by a stable equilibrium, this point has no special status, and will be visited by the population trajectory as frequently (or possibly less frequently) as other points nearby. Furthermore, in the presence of noise, the return tendency of perturbed density is not exactly to the equilibrium, since the average population density can be quite different from the deterministic equilibrium (Chesson, 1981; these two quantities are the same only if endogenous dynamics are linear, and noise is additive). This fact was understood at least as far back as 1957, as evidenced by an exchange of letters between Andrewartha and Nicholson (Andrewartha, 1957).

It is also important to keep in mind that even in purely deterministic models, stable point equilibria are just one kind of qualitative dynamics. As a result of infusion of ideas from nonlinear dynamics, the notion of equilibrium in ecological theory has been replaced with a more general concept of *attractor* (e.g., Schaffer and Kot, 1985). In addition to point equilibria, there are periodic (limit cycles), quasiperiodic, and chaotic attractors (e.g., Turchin and Taylor, 1992).

The notion of attractor can be generalized even further, so that it is applicable to mixed deterministic/stochastic systems (in fact, it is even possible to define chaos for systems with random components, see Ellner and Turchin, 1995). In the presence of noise, a deterministic attractor becomes a stationary probability distribution of population density (May, 1973), which is the defining characteristic of a regulated dynamical system. Put in intuitive terms, the equilibrium is not a point, but a cloud of points (Wolda, 1989; Dennis and Taper, 1994). Thus, if we define equilibrium broadly as a stationary probability distribution, then *being regulated* and *having an equilibrium* are one and the same thing. The whole issue of equilibrium versus nonequilibrium dynamics becomes a semantic argument (see also Berryman, 1987).

B. Relationship to Stochastic Boundedness of Chesson (1978, 1982)

Chesson (1978, 1982) discusses several notions of species persistence: positive boundary growth rates, zero probability of converging to 0, stochastic boundedness, and convergence in distribution to a positive random variable. The first two do not actually result in persistence (see Chesson, 1982), and thus need not concern us here. The last one is identical to the stationarity definition of regulation. The idea of stochastic boundedness, however, is more general than stationarity, because it does not imply convergence in distribution (Chesson, 1982). This occurs because Chesson (1982) is concerned not with regulation, but only with persistence, and thus with lower bounds on population density (an example of a stochastically bound but not regulated process is exponential stochastic growth). Even if by stochastic boundedness we understand that population is bounded from both below and above (e.g., see Fig. 1 in Murdoch, 1970), then this is still not mathematically equivalent to having a stationary probability distribution (P. Chesson, personal communication).

C. Finite Population Sizes and the Question of Scale

An implicit assumption of the previous discussion is that we are dealing with effectively infinite population sizes. The situation is much more complex in finite population models. In particular, such models generally have an unhelpful property that eventual extinction is certain (Chesson, 1982). This shifts our attention from persistence to expected times until extinction, and whether extinction is likely to occur on an ecological time scale. Furthermore, we need to consider what the relevant population scale is, since the waiting time for extinction is greatly affected by population scale (Chesson, 1982). Spatial scale, in turn, will affect population scale, as well as mechanisms of population regulation (e.g., importance of movement relative to birth–death processes).

Consideration of multifarious effects of scale on population fluctuations is beyond the scope of this review. However, I need to address one issue: how stationarity is affected by temporal scale, since stationarity plays such a key role in the definition of regulation. The problem is the environment is constantly changing, so how can we talk meaningfully about stationary population fluctuations? To answer this question, let us consider a "typical" data set that is the grist for density-dependence tests. Such a data set is usually collected at yearly intervals and typically has 10–30 data points (fewer is not very useful, more is helpful, but rare). There are three temporal scales of importance: *Scale 0* of 1 year or less, *Scale 1* of decade, and *Scale 2* of century or more. Short-term environmental fluctuations occurring at Scale 0 are explicitly modeled as the stochastic component ϵ_t [see Eq. (1)]. Long-term changes in the environment at Scale 2 occur so slowly that we can ignore them, effectively assuming *quasi-stationarity*. Most problematic are environmental changes at Scale 1. They are not slow enough so that we can ignore them, but neither are they fast enough so we can model them as noise. If such environmental changes are strong enough, they will induce nonstationarity.

III. Density-Dependence Tests

Again, I begin with a definition. *Density dependence is a dependence of per capita population growth rate on present and/or past population densities* (Murdoch and Walde, 1989). There are many other definitions, but this one is the best I know (see the thoughtful discussion in Murdoch and Walde, 1989), and is becoming generally accepted in the literature (Hanski, 1990; Holyoak, 1994). Density dependence is considered to be *direct* if population growth rate is negatively affected by density, and *inverse* if it is positively affected. Different factors working with different time lags may cause a mixture of direct and inverse density-dependent effects, so it is useful to summarize them by whether or not there is an overall *return tendency*. Thus, return tendency is a generalized notion of direct density dependence.

As I stated earlier, it is now generally accepted that population regulation cannot occur in the absence of density dependence, or to be more precise, return tendency. Simply put, if population density has no effect on the per capita growth rate, then there could be no special range of population densities to which the population would return again and again. Therefore, there could be no stationary probability distribution of population densities and, by definition, no regulation. It is also well known that the presence of density dependence is not a sufficient condition of regulation. For regulation to occur, three general conditions must hold: (1) density dependence must be of the right sign (i.e., there has to be a return tendency), (2) return tendency needs to be strong enough to counteract the disruptive effects of density-independent factors, and (3) the lag with which return tendency operates must not be too long—otherwise it may cause diverging oscillations. To summarize, *regulation is not equivalent to return tendency,* since return tendency is a necessary but not sufficient condition of regulation. Precise quantitative conditions for regulation can be worked out by specifying a model of population dynamics (including the form and the strength of exogenous factors). Nisbet and Gurney (1982) provide an overview of relevant mathematical methods.

In practice, the population regulation debate has focused primarily on whether density dependence in the per capita growth rate can be demonstrated statistically and is of the right sign. Typically, little thought is given to the two key issues determining whether density dependence results in regulation—what are the strength and lag structure of the density-dependent component (I will have more to say about this in a later section). Thus, most current approaches to detecting density-dependent regulation look for immediate (undelayed) density dependence in the per capita growth rate. The population model underlying the usual density-dependence tests is

$$r_t \equiv \ln N_t - \ln N_{t-1} = f(N_{t-1}) + \epsilon_t, \tag{1}$$

where the function f is typically a linear function of either N_{t-1} (e.g., Dennis and Taper, 1994) or log N_{t-1} (e.g., Pollard *et al.,* 1987), and ϵ_t is the term representing random density-independent factors. As a population dynamics model, Eq. (1) can be criticized on many grounds: (1) its linear functional form, (2) lack of delayed density-dependent terms, and (3) the assumption that errors are additive and independent. In addition, population data are often collected at time intervals that do not coincide with generation time. It is often stated that Eq. (1) can be used only with data on animals that are both semelparous and univoltine (e.g., Den Boer, 1990). All of these criticisms are correct, and if our goal is to understand the structure of density dependence, then we need to approach data analysis with more sophisticated models. However, if our goal is more limited—simply to test for regulation—then Eq. (1) is adequate for the job. To see this, consider a regulated dynamical system, which (by definition) is characterized by an attractor (in the general sense of a stationary probability distribution). Wherever the trajectory starts at time t, after a long enough period, T, it can be found anywhere within the attractor. This means that the expectation of N_{t+T} is equal to the long-term mean (in other words, if the density is sampled at intervals of T units, then it will behave as a perfectly damped system). If N_t is above the long-term mean, then the expected change will be downward, and if N_t is below, the expected change will be upward. Given a long enough run of data, we should be able to detect this return tendency with model (1). Nonlinearities and lags (i.e., when T is much longer than one unit at which observations are made) will, to some degree, obscure this relationship, but will not invalidate it. In other words, we can still use model (1) if its various assumptions are violated, but we pay the price of reduced power, or ability to detect regulation if it is actually present.

I now turn to specific tests for statistical detection of the return tendency. Several points of agreement about such tests have emerged during the last few years. One is that the dependent variable in the test should be r_t rather than N_t. Reasons are purely biological rather than statistical, since the definition of density dependence is framed in terms of the per capita rate of change. Many ecologists dislike tests based on r_t, because its definition involves the independent variable N_{t-1}. It is frequently but incorrectly argued that such tests are tainted by spurious correlations due to a lack of independence between the dependent and independent variables. For a good discussion of this common misconception, see Prairie and Bird (1989). There is, however, a legitimate concern that measurement errors may introduce a bias in the estimated slope. In fact, Walters and Ludwig (1987) have calculated the magnitude of this bias for model (1) assuming that $f(N_{t-1}) = a + bN_{t-1}$.

Another point of agreement is about the appropriate null model for density independence. Of the two possibilities, the unbiased random walk, $f(N_{t-1}) = 0$, is a special case of the stochastic growth/decline model $f(N_{t-1}) = a$. Both of these models are density independent, but the second one is more general and thus should be used as the explicit alternative to density dependence. In practice, there will be little difference between the two models, because the estimate of the bias term a will be close to zero when data come from a regulated population.

The final and most important point is that standard tests of statistical significance are grossly inappropriate when testing for population regulation. When r_t (or N_t) is regressed on N_{t-1} using short data sequences generated by a randomly walking population, the estimate of the slope is biased downward, yielding a spurious indication of density dependence (e.g., Maelzer, 1970; Pielou, 1974). A statistical test for density dependence must take this unpleasant fact into account. Fortunately, this can be done in a conceptually straightforward manner. The basic idea is to somehow obtain a probability distribution of the regression slope estimates assuming the null hypothesis, and then to determine if the slope estimate from the data is more negative than would be expected if the null hypothesis were true. Since it is not possible to derive this theoretical distribution analytically, it must be approximated using computer-intensive methods.

Currently there are two competing tests for detecting the return tendency in time-series data. Both assume the same null hypothesis of density-independent stochastic growth/decline, $f(N_{t-1}) = a$. The density-dependent alternatives are somewhat different: the test statistic proposed by Pollard *et al.* (1987) is the correlation coefficient between r_t and log-transformed population density, whereas the test of Dennis and Taper (1994) regresses r_t on untransformed density. The Pollard *et al.* test is based on a randomization procedure, whereas Dennis and

Taper's test uses the parametric bootstrap. Conceptually, the differences between the two tests appear to be rather minor. Both perform well when used on data generated using model (1) (with or without density dependence). It remains to be seen whether either test has an advantage in sensitivity or power when applied to data generated with a variety of ecological models with noise. For now, there is little reason to prefer one over the other, and either (or, even better, both) can be used in data analyses.

IV. So What Do the Data Say?

The primary engine driving the density-dependence controversy has always been a lack of empirical support for population regulation. This is apparent from writings of Andrewartha and Birch (1954), as well as the spate of negative findings during the 1980s (Dempster, 1983; Gaston and Lawton, 1987; Stiling, 1987, 1988; Den Boer and Reddingius, 1989). Until it is shown that regulation is a common state of field populations, and that it can be reliably detected, the controversy is doomed to drag on. The conceptual and statistical issues reviewed here provide the necessary groundwork, but ultimately the controversy can only be resolved empirically. Fortunately, ecologists have not been idle during the four decades since Andrewartha and others mounted an attack on density-dependent regulation. We now have enormous amounts of data, especially those document-ing fluctuations of insect populations. An emphasis on insects is especially appropriate because historically the doubters of regulation came from the ranks of insect ecologists.

The following general answer has been crystallizing during the last five years: the longer the time-series data, the higher the probability of detecting statistically significant density dependence (Hassell et al., 1989; Woiwod and Hanski, 1992; Godfray and Hassell, 1992; Holyoak, 1993; Wolda and Dennis, 1993). For example, Woiwod and Hanski (1992) analyzed a massive data set of fluctuations of nearly 6000 populations of aphids (94 species) and moths (263 species). Using the test of Pollard et al. (1987), they found significant density dependence in 69% of aphid and 29% of moth time series. However, when short time series (less than 20 years) were excluded from the analysis, the frequency of density dependence rose to 84% and 57%, respectively. The same pattern holds no matter what test is used. Holyoak (1993) applied seven different tests to a data base of 171 time series and found that the frequency of detection of density dependence rose significantly with time-series length in results from all seven tests. The latest paper showing the same pattern found that the incidence of statistically significant density dependence increased with time-series length, paralleling nicely the power curves (relations between test power and series length) (Wolda and Dennis, 1993; see their Fig. 5). Wolda and Dennis concluded that such a pattern is *entirely consistent with the hypothesis of universal applicability of the density-dependence model.*

We are thus led to an inescapable conclusion. Early results suggesting lack of regulation were not due to absence of density dependence in the systems studied, but were simply a result of inadequate data. The problem is that there is a high probability that for short periods of time a random-walk model will behave in a way indistinguishable from a regulated system. For example, the probability of successfully detecting regulation in a 10-year-long data set is only about 10%, as is suggested by both theoretical power curves and empirical detection rates (Fig. 5 in Wolda and Dennis, 1993). Thirty or more data points are needed to increase the detection rate to 60–70% (roughly speaking; in actuality, power depends a lot on the strength of the return tendency and the magnitude of the exogenous component).

It is necessary to caution here against a wholesale acceptance of the idea that all populations are regulated at all times. Logic and data suggest that lack of regulation can temporarily occur in real populations. Examples are numerous, of which species extinctions and invasions are the most dramatic. Less dramatic but still common are cases of population density trends caused by a gradual or an abrupt change in the species environment, often caused by human activity. Thus, any particular population, observed during a particular period of time, can be either regulated or not. To model fluctuations of this population, for explanatory or predictive purposes, we need to know whether the appropriate model should be one with or without a stochastic equilibrium. A density-dependence test allows us to make this choice.

V. Beyond Density-Dependence Tests: Investigating the Structure of Population Regulation

A test for density dependence (or, more precisely, for return tendency) is a statistical tool for determining whether the dynamics of the studied population are stationary or not. Getting some answer to this question, however, is not the end of the investigation, but should rather be its beginning. Depending on the test outcome, we know that we need to model fluctuations of the population as either a stationary or nonstationary dynamical process. If it is nonstationary, then we need to make another decision: Should we model this population as a density-independent stochastic growth/decline process, or as a density-dependent process in which some parameter is following a temporal trend? If stationarity is indicated, then what is the qualitative type of its dynamics? Time-series data contain a wealth of information about the pattern of population fluctuations, and they can be used to make inferences about possible mechanisms driving these fluctuations. It is true that manipulative experiments are the most powerful tool for distinguishing between mechanistic hypotheses, but experiments are time-consuming, expensive, and, for many systems, impracticable. Time-series data, however, are already available (there are hundreds if not thousands of data sets for insects, mammals, birds, fish, and other organisms). Limiting time-series analysis to a test for density dependence does not utilize these data to the full.

As was discussed earlier, most tests for density dependence assume linear functional forms and focus exclusively on direct (undelayed) density dependence (Turchin, 1990). However, population dynamics are inherently nonlinear (Chesson, 1981; Royama, 1992; Turchin and Millstein, 1993), and nonlinearities are necessary for complex dynamical behaviors such as limit cycles and chaos. Thus, if we would like to investigate complex population dynamics using time-series data, we need to use nonlinear methods of analysis. Population interactions, for example, the interaction between specialist predators and prey, can also introduce lags in regulatory feedback. Detecting (or failing to detect) such lags provides a valuable indication of what kinds of mechanisms are driving fluctuations (Berryman, 1991). For example, a time-series analysis of southern pine beetle data (Turchin *et al.,* 1991) suggested that population oscillations in this beetle are driven by some delayed density-dependent mechanism (a simple regression on lagged density explained 55% of the variation in the per capita rate of change). This result contradicts the previously popular hypothesis that pine beetle outbreaks were caused by weather fluctuations, and instead points to an interaction with some biological factor, such as natural enemies (Reeve *et al.,* Chapter 17, this volume).

A natural extension of Eq. (1) that accounts for nonlinearities, lags, and exogenous influences (including trends) is the following nonlinear time-series model:

$$r_t = f(N_{t-1}, N_{t-2}, \ldots, U_t^1, U_t^2, \ldots). \tag{2}$$

Here N_{t-1}, N_{t-2}, . . . represent the influences of endogenous (density-dependent) factors, and U_t^i are the various exogenous (density-independent) factors. Exogenous influences are typically modeled as an independently distributed random variable. As discussed in Section II, this procedure assumes that exogenous factors act on a fast time scale (Scale 0). In addition to noise, certain endogenous variables may be modeled mechanistically. Suppose we have a rainfall time series for the locality where population data were collected. The influence of rainfall on population fluctuations, then, may be statistically explored with a transfer-function model (see Poole, 1978). Exogenous variables that exhibit trends can also be explicitly modeled within the framework of Eq. (2). Thus, a model may include an exogenous term for rainfall (e.g., U_t^1), another one for increasing habitat fragmentation (U_t^2), and a third for the influence of all other exogenous factors, modeled phenomenologically as noise (U_t^3).

In most current analyses, however, model (2) is simplified by assuming that exogenous factors will act in an additive way and can be modeled as independently distributed normal random variables (this may be too simplistic, but it certainly is better than including no exogenous factors at all). The model then becomes

$$r_t = f(N_{t-1}, N_{t-2}, \ldots) + \epsilon_t. \tag{3}$$

The basic idea for analysis is, then, to choose a family of functional forms for f and to use nonlinear regression to fit time-series data to model (2). This approach has been followed by Berryman and Millstein (1990), who employed a lagged logistic model, by Turchin and Taylor (1992), who used response surface methodology, and by Ellner *et al.* (1991), who used neural nets for fitting data (see Ellner and Turchin, 1995, for more details).

A. Population Dynamics in the Noisy World

One primary motivation for developing nonlinear methods for time-series analysis is to be able to determine the qualitative type of stationary population dynamics that may be characterizing the studied system. As has been popularized by Robert May, in the absence of noise the general model (2) is capable of several dynamical behaviors, ranging from stability (exponential or oscillatory), through limit cycles and quasiperiodicity, to chaos (Schaffer and Kot, 1985; Turchin and Taylor, 1992). In the presence of noise, however, the nice progression from stability to periodic behavior and then to chaotic motion breaks down. As discussed earlier, deterministic attractors such as stable points or fractal chaotic structures become probability clouds. However, there is a certain degree of carryover from the purely deterministic to the stochastic world. For example, if noise-free dynamics are periodic (which can include limit cycles, quasiperiodicity, and oscillatory stability), then it is highly likely that dynamics with noise will also be characterized by some degree of statistical periodicity. Thus, one important way to characterize real-life population dynamics is by the presence and magnitude of statistical periodicity.

The second important aspect of noisy population dynamics is the issue of stability versus chaos. The definition of chaos can be extended to dynamical systems affected by noise (Ellner and Turchin, 1995). Chaos is defined as sensitive dependence on initial conditions (Eckmann and Ruelle, 1985). In deterministic models, we detect sensitive dependence by starting two trajectories very close to each other, solving the model for both sets of initial conditions, and observing whether the two trajectories diverge (implying sensitive dependence and chaos) or converge (implying lack of sensitive dependence and stability). For systems influenced by stochastic factors, we follow the same procedure, but ensure that both trajectories are influenced by the same sequence of random perturbations (Ellner and Turchin, 1995). The exponential rate of convergence/divergence is measured by the dominant Lyapunov exponent (for more details, see Turchin and Millstein, 1993).

This brief overview suggests the following classification of population dynamics patterns that we observe in nature. The simplest case is when endogenous dynamics do not have a tendency to overshoot the equilibrium (i.e., exponentially stable equilibrium point). Depending on the relative strength of the endogenous versus exogenous components, we will have a spectrum of "simple dynamics" ranging from a tight regulation around a stable equilibrium at one end to stochastic population growth/decline (this includes random walk as a special case) at the other end.

Complex dynamics can also be characterized by the relative strength of the endogenous versus exogenous components. For example, it would be interesting to estimate the degree to which endogenous factors contribute to population fluctuations. In addition, complex dynamics have two other major aspects. The first is the degree of periodicity, which can be measured, for example, by the autocorrelation coefficient at the dominant period. Values close to 1 indicate strong periodicities, and values close to 0 suggest absence of a periodic component. One of the best examples of periodic dynamics in ecology is the larch budmoth (Fig. 2). The autocorrelation function at lag 9, corresponding to the average period of larch budmoth oscillations, is about 0.7, indicating a very high degree of periodicity.

The second characteristic of complex dynamics is the degree of stability, as measured by the Lyapunov exponent (the long-term average divergence rate). A positive value of the Lyapunov exponent indicates chaotic dynamics, whereas a negative value indicates stability. The more negative the Lyapunov exponent is, the more stable the dynamics. In the presence of noise, the transition between stability and chaos is gradual. If the Lyapunov exponent is near zero, the system will typically fluctuate between positive and negative "local Lyapunov exponents" (see Ellner and Turchin, 1995), that is, it will behave in a manner intermediate between chaos and stability. I will call such dynamical behaviors *quasi-chaos* (i.e., "almost chaos"). This may be an important kind of qualitative dynamical behavior. Although we do not have any generally accepted examples of "strong chaos," there are at least two examples of quasi-chaotic or weakly chaotic dynamics. The first one is the measles epidemics (Schaffer and Kot, 1985; Ellner *et al.*, 1994), and the other one is boreal vole populations (Turchin, 1993; Hanski *et al.*, 1993b).

Periodicity and sensitive dependence are two separate attributes of complex dynamics. Both stable and chaotic fluctuations can have or not have a periodic component. For a discussion of the relationships between chaos and periodicity, see Kendall *et al.* (1993).

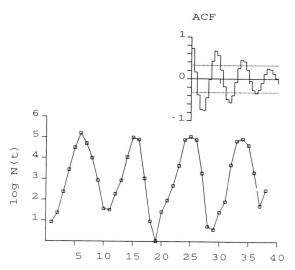

Figure 2. Population oscillations of the grey larch budmoth *Zeiraphera diniana* in the Engadine Valley. ACF is the autocorrelation function calculated for these data, with dotted lines indicating approximate 95% confidence limits. (Data from Baltensweiler and Fischlin, 1989.)

To summarize, novel approaches, influenced by recent developments in nonlinear dynamics, give us quantitative tools to probe the structure of population regulation. These approaches rely on fitting time-series data with models explicitly incorporating nonlinearities and delayed density dependence. Such analyses can yield insights into possible mechanisms that may impose regulation. They allow us to quantify relative contributions of endogenous versus exogenous factors and presence and strength of periodicity, and to classify dynamics into stable, quasi-chaotic, or strongly chaotic oscillations.

VI. Conclusion: Where Do We Go from Here?

My main goal in this chapter is to point out that there is now a broad consensus among ecologists, both theoretical and empirical, about two major issues in the population regulation debate. First, it has become clear that when we speak about "regulation around an equilibrium," we mean an equilibrium in a general sense, that is, a stationary probability distribution of population density. An equilibrium is not a particular point, but a cloud of points. Perhaps the strongest evidence of growing consensus is the close similarity (in fact logical identity) between various modern notions of regulation: May's (1973) stochastic equilibrium probability distribution, Chesson's (1982) convergence in distribution to a positive random variable, and Dennis and Taper's (1994) long-term stationary probability distribution.

Second, all recent empirical analyses agree that the frequency of detection of density dependence (return tendency) increases with the length of data time series. Thus, most field populations are regulated, and previous failures to show this were due to inadequate data sets (and, in some degree, to poor methods). This conclusion is in agreement with the results of experimental analyses, summarized by Harrison and Cappuccino (Chapter 7, this volume). At the same time we should not be carried away and claim universal applicability of regulation, since there are numerous counterexamples demonstrating that regulation does not always operate in all populations at all times.

The debate continues. Some of it is purely nit-picking, whereas the rest addresses valid areas of disagreement that are, however, of secondary importance. Thus, we still do not have one "best" method for detecting density dependence in time-series data. There are also important issues to be resolved about effects of observation errors, nonlinearities, and lags on the performance of density-dependence tests. Any new developments in these areas, however, are unlikely to change the general conclusion that regulation is widespread.

Demonstrating density dependence is no longer a key issue in population ecology. But plenty of unanswered questions remain. Most importantly, although we now know that most field populations are characterized by equilibria (in the general sense), we know little about the relative frequency of mechanisms that bring these equilibria about. How often is regulation a result of local processes, and how often is it due to metapopulation dynamics (Murdoch, 1994)? Of local factors, what is the relative importance of intrinsic factors? Bottom-up versus top-down forces? How often do we need to invoke interactions between multiple factors to explain regulation? These questions must be approached by detailed life table studies, and especially by field experiments. The review by Harrison and Cappuccino (Chapter 7, this volume) makes the first step in synthesizing such experiments. I hope that we will not need another 40 years to develop consensus about mechanisms bringing about regulation.

Of equal importance is to go beyond the simplistic "regulated versus unregulated" dichotomy. A regulated population might exhibit simple dynamics, or various combinations of periodic behavior and chaos. "One key role of population dynamics theory is to elucidate and define these classes of population behavior, and to specify the sorts of biological conditions, species interactions and environments likely to give rise to each kind" (Lawton, 1992). We are just beginning to address this task.

Finally, we should approach ecological questions with a synthetic approach that blends together statistical analyses of time-series data, experiments, and mathematical models. The density-dependence debate was to a large degree driven by mutual misunderstandings between theoretically minded and empirical ecologists. To bridge this gap, empirical ecologists should become more theoretical and make an effort to read and understand theoretical papers. Theoretical ecologists, on the other hand, should become more empirical and make an effort to understand and address in their models the concerns of empiricists. One of the hopeful signs is that graduate training in ecology today places a greater emphasis than before on quantitative skills. In a recent poll of readers of *Trends in Ecology and Evolution* (Sugden, 1994), readers under 36 showed greater enthusiasm for mathematical ecology than the general sample. The gap between theory and empiricism in population ecology may soon be bridged.

Acknowledgments

I thank Mike Auerbach, Alan Berryman, Joe Elkinton, Ilkka Hanski, Susan Harrison, and John Reeve for their comments and suggestions on the manuscript, Peter Chesson for a very helpful clarification of the difference between stationarity and stochastic boundedness, and Naomi Cappuccino for doing an excellent job as an editor.

References

Andrewartha, H. G. (1957). The use of conceptual models in population ecology. *Cold Spring Harbor Symp. Quant. Biol.* **22,** 219–232.

Andrewartha, H. G., and Birch, L. C. (1954). "The Distribution and Abundance of Animals." Univ. of Chicago Press, Chicago.

Baltensweiler, W., and Fischlin, A. (1988). The larch budmoth in the Alps. *In* "Dynamics of Forest Insect Populations" (A. A. Berryman, ed.), pp. 332–351. Plenum Press, New York.

Berryman, A. A. (1987). Equilibrium or nonequilibrium: Is that the question? *Bull. Ecol. Soc. Am.* **68,** 500–502.

Berryman, A. A. (1991). Stabilization or regulation: What it all means! *Oecologia* 86, 140–143.

Berryman, A. A. (1992). Vague notions of density-dependence. *Oikos* **62,** 252–254.

Berryman, A. A., and Millstein, J. A. (1990). "Population Analysis System. POPSYS Series 1. Single Species Analysis." Ecological Systems Analysis, Pullman, WA.

Berryman, A. A., Stenseth, N. C., and Isaev, A. S. (1987). Natural regulation of herbivorous forest insect populations. *Oecologia* **71,** 174–184.

Bulmer, M. G. (1975). The statistical analysis of density dependence. *Biometrics* **31,** 901–911.

Chesson, P. L. (1978). Predator–prey interactions and variability. *Annu. Rev. Ecol. Syst.* **9,** 323–347.

Chesson, P. L. (1981). Models for spatially distributed populations: The effect of within-patch variability. *Theor. Popul. Biol.* **19,** 288–325.

Chesson, P. L. (1982). The stabilizing effect of a random environment. *J. Math. Biol.* **15,** 1–36.

Crowley, P. H. (1992). Density-dependence, boundedness, and attraction: Detecting stability in stochastic systems. *Oecologia* **90**, 246–254.

Dempster, J. P. (1983). The natural control of populations of butterflies and moths. *Biol. Rev. Cambridge Philos. Soc.* **58**, 461–481.

Den Boer, P. J. (1968). Spreading of risk and stabilization of animal numbers. *Acta Biotheor.* **18**, 165–194.

Den Boer, P. J. (1990). On stabilization of animal numbers. Problems of testing. 3. What do we conclude from significant test results? *Oecologia* **83**, 38–46.

Den Boer, P. J., and Reddingius, J. (1989). On the stabilization of animal numbers. Problems of testing. 2. Confrontation with data from the field. *Oecologia* **79**, 143–149.

Dennis, B., and Taper, B. (1994). Density dependence in time series observations of natural populations: Estimation and testing. *Ecol. Monogr.* **64**, 205–224.

Eckmann, J.-P., and Ruelle, D. (1985). Ergodic theory of chaos and strange attractors. *Rev. Mod. Phys.* **57**, 617–656.

Ellner, S., and Turchin, P. (1995). Chaos in a "noisy" world: New methods and evidence from time series analysis. *Am. Nat.* **145**, 343–375.

Ellner, S., Gallant, A. R., McCaffrey, D., and Nychka, D. (1991). Convergence rate and data requirements for Jacobian-based estimates of Lyapunov exponents from data. *Phys. Lett. A* **153**, 357–363.

Ellner, S., Gallant, A. R., and Theiler, J. (1994). Detecting nonlinearity and chaos in epidemic data. *In* "Epidemic Models: Their Structure and Relation to Data" (D. Mollison, ed.). Cambridge Univ. Press, Cambridge, UK.

Elton, C. (1949). Population interspersion: An essay on animal community patterns. *J. Ecol.* **37**, 1–23.

Gaston, K. J., and Lawton, J. H. (1987). A test of statistical techniques for detecting density dependence in sequential censuses of animal populations. *Oecologia* **74**, 404–410.

Godfray, H. C. J., and Hassell, M. P. (1992). Long time series reveal density dependence. *Nature (London)* **359**, 673–674.

Hanski, I. (1990). Density dependence, regulation and variability in animal populations. *Philos. Trans. R. Soc. London, Ser. B* **330**, 141–150.

Hanski, I., Woiwod, I., and Perry, J. (1993a). Density dependence, population persistence, and largely futile arguments. *Oecologia* **95**, 595–598.

Hanski, I., Turchin, P., Korpimaki, E., and Henttonen, H. (1993b). Population oscillations of boreal rodents: Regulation by mustelid predators leads to chaos. *Nature (London)* **364**, 232–235.

Hassell, M. P. (1986). Detecting density dependence. *TREE* **1**, 90–93.

Hassell, M. P., Latto, J., and May, R. M. (1989). Seeing the wood for the trees: Detecting density dependence from existing life-table studies. *J. Anim. Ecol.* **58**, 883–892.

Holyoak, M. (1993). The frequency of detection of density dependence in insect orders. *Ecol. Entomol.* **18**, 339–347.

Holyoak, M. (1994). Identifying delayed density dependence in time-series data. *Oikos* **79**, 296–304.

Holyoak, M., and Lawton, J. H. (1993). Comments arising from a paper by Wolda and Dennis: Using and interpreting the results of test for density dependence. *Oecologia* **95**, 92–594.

Kendall, B. E., Schaffer, W. M., and Tidd, C. W. (1993). Transient periodicity in chaos. *Phys. Lett. A* **177**, 13–20.

Krebs, C. J. (1991). The experimental paradigm and long-term population studies. *Ibis* **133**, 3–8.

Krebs, C. J. (1992). Population regulation revisited. *Ecology* **73**, 714–715.

Lawton, J. H. (1992). There are not 10 million kinds of population dynamics. *Oikos* **63**, 337–338.

Maelzer, D. A. (1970). The regression of log N_{n+1} on log N_n as a test of density dependence: An exercise with computer-constructed, density-independent populations. *Ecology* **51**, 810–822.

May, R. M. (1973). Stability in randomly fluctuating versus deterministic environments. *Am. Nat.* **107**, 621–650.

Milne, A. (1957). The natural control of insect populations. *Can. Entomol.* **89**, 193–213.

Morris, R. F. (1959). Single-factor analysis in population dynamics. *Ecology* **40**, 580–588.

Murdoch, W. W. (1970). Population regulation and population inertia. *Ecology* **51**, 497–502.

Murdoch, W. W. (1994). Population regulation in theory and practice. *Ecology* **75**, 271–287.

Murdoch, W. W., and Walde, S. J. (1989). Analysis of insect population dynamics. *In* "Towards a More Exact Ecology" (P. J. Grubb and J. B. Whittaker, eds.), pp. 113–140. Blackwell, Oxford.

Nicholson, A. J. (1933). The balance of animal populations. *J. Anim. Ecol.* **2**, 132–178.

Nisbet, R. M., and Gurney, W. S. C. (1982). "Modeling Fluctuating Populations." Wiley, Chichester, UK.

Pielou, E. C. (1974). "Population and Community Ecology: Principles and Methods." Gordon & Breach, New York.

Pollard, E., Lakhani, K. H., and Rothery, P. (1987). The detection of density-dependence from a series of annual censuses. *Ecology* **68**, 2046–2055.

Poole, R. W. (1978). The statistical prediction of population fluctuations. *Annu. Rev. Ecol. Syst.* **9**, 427–448.

Prairie, Y. T., and Bird, D. T. (1989). Some misconceptions about the spurious correlation problem in the ecological literature. *Oecologia* **81**, 285–288.

Reddingius, J. (1971). Gambling for existence. A discussion of some theoretical problems in animal population ecology. *Acta Biotheor.* **20** Suppl., 1–208.

Reddingius, J., and Den Boer, P. J. (1989). On the stabilization of animal numbers. Problems of testing. 1. Power estimates and estimation errors. *Oecologia* **78**, 1–8.

Royama, T. (1977). Population persistence and density dependence. *Ecol. Monogr.* **47**, 1–35.

Royama, T. (1992). "Analytical Population Dynamics." Chapman & Hall, London.

Schaffer, W. M., and Kot, M. (1985). Do strange attractors govern ecological systems? *BioScience* **35**, 342–350.

Slade, N. A. (1977). Statistical detection of density dependence from a series of sequential censuses. *Ecology* **58**, 1094–1102.

Stiling, P. (1987). The frequency of density dependence in insect host–parasitoid systems. *Ecology* **68**, 844–856.

Stiling, P. (1988). Density-dependent processes and key factors in insect populations. *J. Anim. Ecol.* **57,** 581–594.

Strong, D. (1986). Density-vague population change. *TREE* **1,** 39–42.

Sugden, A. M. (1994). 100 issues of *TREE*. *TREE* **9,** 353–354.

Turchin, P. (1990). Rarity of density dependence or population regulation with lags? *Nature (London)* **344,** 660–663.

Turchin, P. (1993). Chaos and stability in rodent population dynamics: Evidence from nonlinear time-series analysis. *Oikos* **68,** 167–172.

Turchin, P., and Millstein, J. A. (1993). "EcoDyn/RSM: Response Surface Modeling of Nonlinear Ecological Dynamics. I. Theoretical Background." Applied Biomathematics, Setauket, NY. (Available from the senior author on request.)

Turchin, P., and Taylor, A. D. (1992). Complex dynamics in ecological time series. *Ecology* **73,** 289–305.

Turchin, P., Lorio, P. L., Taylor, A. D., and Billings, R. F. (1991). Why do populations of southern pine beetles (Coleoptera: Scolytidae) fluctuate? *Environ. Entomol.* **20,** 401–409.

Varley, G. C., and Gradwell, G. R. (1960). Key factors in population studies. *J. Anim. Ecol.* **29,** 399–401.

Varley, G. C., Gradwell, G. R., and Hassell, M. P. (1973). "Insect Population Ecology: An Analytical Approach." Univ. of California Press, Berkeley.

Vickery, W. L., and Nudds, T. D. (1984). Detection of density-dependent effects in annual duck censuses. *Ecology* **65,** 96–104.

Walters, C. J., and Ludwig, D. (1987). Effects of measurement errors on the assessment of stock-recruitment relationships. *Can. J. Fish. Aquat. Sci.* **38,** 704–710.

Woiwod, I. P., and Hanski, I. (1992). Patterns of density dependence in moths and aphids. *J. Anim. Ecol.* **61,** 619–629.

Wolda, H. (1989). The equilibrium concept and density dependence tests. What does it all mean? *Oecologia* **81,** 430–432.

Wolda, H., and Dennis, B. (1993). Density dependence tests, are they? *Oecologia* **95,** 581–591.

Wolda, H., Dennis, B., and Taper, M. (1994). Density dependence tests and largely futile comments: Answers to Holyoak and Lawton (1993) and Hanski, Woiwod and Perry (1993a). *Oecologia* **98,** 229–234.

Chapter 3

Ecology, Life History, and Phylogeny of Outbreak and Nonoutbreak Species

Alison F. Hunter

I. Introduction

Research on outbreaking species has been central to the study of population dynamics of insect herbivores. Yet species that cause noticeable defoliation over large areas are a minority. For example, in forests, I estimate that outbreaking species comprise at most 1–3% of the macrolepidopteran fauna (see Section III, A). Outbreaking species are unusual and must differ in some way from nonoutbreaking species that share the same habitat and type of hosts. Since detailed ecological studies of nonoutbreaking species are rare, we cannot directly compare the factors affecting their populations. Comparing life-history and ecological traits of species with contrasting dynamics may provide clues about the nature of the differences in population dynamics. Nonoutbreaking species may respond differently to environmental changes, or they may experience a different biotic environment (natural enemies and hosts). Although this kind of comparison will not directly reveal which extrinsic factors are important, it may suggest hypotheses and patterns that can be investigated in natural systems.

In earlier work, I compared the traits of outbreaking and nonoutbreaking hardwood-feeding Macrolepidoptera of northern North America (Hunter, 1991). Here, I extend that work to include conifer-feeding and European species, and incorporate phylogenetic comparisons, which has not previously been done. The expanded scope and analysis of potential phylogenetic effects reinforces the inferences. Also, a new, potentially unifying pattern among outbreaking species is identified.

A. Phylogenetic Analysis

Why worry about phylogeny in comparative analyses? Closely related species are not independent in their traits: organisms that share common ancestry are generally more similar to one another than those that are more distantly related. For example, we might find that most outbreaking species are hairy and aposematic, and most nonoutbreaking species have smooth integuments and cryptic coloration. But if all the hairy caterpillars are members of the Lymantriidae and all the cryptic caterpillars are geometrids, then there is really only one origin of the association, and it is as likely due to chance as a connection between larval defenses and outbreaking dynamics. To avoid overestimating the degrees of freedom, phylogenetic analysis should be used (Harvey and Pagel, 1991).

Different methods of analysis are used for discrete and continuous characters (Harvey and Pagel, 1991). Generally the methods for continuous traits are better developed; unfortunately these traits are in the minority in the current data set. Good phylogenetic information, especially for the moths, is limited. However, I have proceeded with the philosophy that attempting to use the available information is preferable to ignoring potential phylogenetic effects, and may stimulate further research to fill the phylogenetic and life-history gaps.

II. Methods

A. Scope and Data Sources

The data were gathered from published literature on North American, British, and North European tree- and shrub-feeding Macrolepidoptera. Major sources were the Canadian Forest Insect Survey (CFIS: McGugan, 1958; Prentice, 1962, 1963), books on forest insects, identification guides, and taxonomic works. In addition to the references cited in Hunter (1991), I used Furniss and Carolin (1977), Rose and Lindquist (1984, 1985, 1992), and Bolte (1990) for North American coniferophagous species and Buckler (1896), Carter (1984), Emmet and Heath (1983, 1990, 1991), Pittaway (1993), Skou (1986), and Stokoe and Stovin (1948) for European species. Nomenclature follows Hodges *et al.* (1983) for North America and Emmet and Heath (1991) for Europe.

B. Trait Definitions

Population dynamics: Species were assigned to large, moderate, and restricted outbreak categories, and common or rare nonoutbreaking categories as follows:
Large outbreaks: for North American species, at least two outbreaks in 100 years, greater than 50% defoliation of hosts occurs for at least 2 years, and outbreaks cover more than 1000 ha (Mattson *et al.*, 1991). The ranking of European species was more subjective and based predominantly on comments in Carter (1984).

Moderate outbreaks: Reasonably frequent defoliation generally covering less than 1000 contiguous hectares (i.e., not matching criteria of Mattson *et al.,* 1991). Restricted outbreaks: Few (i.e., one or two) recorded episodes of defoliation, or of extremely restricted area.

Common: Found for at least 10 years in the CFIS, the criterion for inclusion in the data set of Redfearn and Pimm (1988). British butterflies were rated as common or rare by Hodgson (1993) based on occurrence in 10-km^2 areas. European moths appearing in the lists of Taylor and Woiwod (1980), or Spitzer *et al.* (1984), from light trap collections, were categorized as common. Spitzer *et al.* (1984) listed only the frequently collected noctuids from south Bohemia. Taylor and Woiwod included only species with records for at least 6 years and 15 sites in Britain.

Rare: Remaining species, or classified as rare by Hodgson (1993).

Feeding phenology: The first and last month of the larval stage was recorded from verbal descriptions and charts. These are the earliest and latest dates for populations and do not represent individual development. "Spring-feeding" means that larvae are feeding around the time of budburst (start feeding before June in eastern or central North America, before May in Britain or western North America, or overwinter as larvae).

Number of generations: All species were assumed to be univoltine unless otherwise reported in the literature. Species that are univoltine in the north but multivoltine in the south were categorized as multivoltine.

Life-style: I classified species as gregarious if the larvae feed in groups for at least part of their development. Thus species that feed in groups initially, but individually in later instars, are gregarious. Species that cluster their eggs but disperse before initiating feeding are not gregarious. All species were assumed to be solitary unless otherwise reported (because authors seldom comment on this for solitary feeders).

Egg dispersion pattern: "Masses" means that all reproductive effort occurs in one mass, or large clusters of greater than 100 eggs; "clusters" are groups of 5 to 100 eggs; "singly" refers to less than 5 eggs.

Female flying ability: Females of a few species have reduced wings and cannot fly. I classified other species only if some comment was made about flying ability (i.e., strong fliers or poor fliers), or if it was reported that females are seldom caught relative to males (poor fliers).

Fecundity: I used the maximum recorded fecundity; the mean is subject to reductions due to food quality, population density, and so on, and in many cases only the range is reported. However, extremes are biased depending on sample size, that is, the larger the sample, the larger the extreme.

Host breadth: I collated data on host genera from all sources. A categorical ranking (Table 1) was also made because of the difference in host genera available to conifer and hardwood feeders (Fig. 1). There are only 11 genera of

TABLE 1
Division of Species into Host Breadth Categories

	Hardwood feeders	Conifer feeders	Mixed feeders
Monophagous	1 genus	1 genus	NA[a]
Few genera	2–7 genera	2–4 genera	—
Many genera	8–15 genera	5–6 genera	Less than 5 genera
Heaps of genera	Greater than 15	Greater than 6	Greater than 5

[a]NA = not applicable.

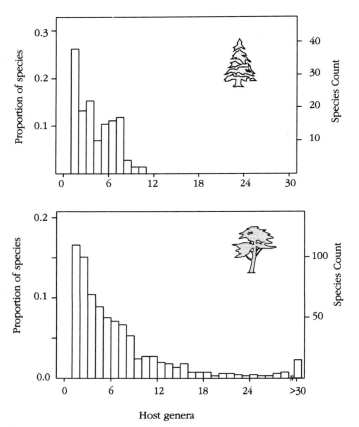

Figure 1. Frequency distribution of diet breadth (number of genera) of tree-feeding Macrolepidoptera. (A) Conifer-feeding species. (B) Hardwood-feeding species.

conifers but 79 genera of hardwoods. Additional recognition was given to mixed feeders on both hardwoods and conifers because of the chemical dissimilarity of these hosts (Holloway and Hebert, 1979).

Coloration: Larvae strikingly marked with combinations of black, yellow, red, and white were scored as warningly colored. Green, gray, and brown larvae are cryptic. This information was obtained from photographs, colored drawings, or descriptions of larvae.

Defense: Larvae with spines, sharp tubercles, or many setae were considered to have physical defenses. Leaf rolls or leaf ties and silk tents may also deter predators (Damman, 1987). Chemical defenses encompass urticating hairs, distastefulness to predators, and repellent scent glands with deterrent properties. Such defenses are probably underreported in the literature, especially for species of little economic importance (90% of species here).

Clustering of offspring: A common feature among outbreaking species seems to be spatial clustering of the larval stage. Gregarious species put all of their offspring on one or a few trees. Many spring-feeding species put their eggs in masses or cannot fly. I combined species with gregarious larvae, eggs in masses, or female flightlessness, versus species that fly and spread their eggs, to compare clustering of outbreak and nonoutbreak species.

C. Phylogenies

The phylogenies of these groups are mostly poorly known. I used published phylogenies where available and, failing that, taxonomic arrangement to define sister groups (Table 2). Comments like "is closely related to" and "is primitive" in taxonomies were interpreted to represent evolutionary relationships. The results reported here may be altered when new phylogenies become available.

D. Data Analysis

Nonphylogenetic: Tables were made for discrete characters, and the log-likelihood ratio (G) was used to test for differences. For continuous variables, I used analysis of variance. A nested analysis of variance was used to examine the distribution of variance among the taxonomic levels.

Phylogenetic. Character states were traced on the trees using MacClade (Maddison and Maddison, 1992). I used several phylogenetic methods to analyze these data because they contained a mixture of continuous and discrete traits, and because of the unknown robustness of these new procedures to violation of their assumptions. One character is considered independent, and the other (usually outbreaking) dependent.

1. The contingent states test (Sillén-Tullberg, 1993) determines whether the dependent trait changes equally often for two states of the independent trait. This

TABLE 2
Higher-Level Classification, Number of Species in the Families Worldwide, in North America, and in the Current Data Set, and Sources of Phylogenetic Information

Superfamily	Family	World	N.A.	Current data set	Phylogenetic reference[a]
Hesperioidea	Hesperiidae	3,650	263	3	p: Scott (1984)
Papilionoidea	Papilionidae	534	30	5	
	Nymphalidae	4,500	185	16	p: Scott (1984)
	Pieridae	1,100	58	3	
	Lycaenidae	4,700	142	15	
Bombycoidea	Lasiocampidae	1,000	35	23	t: Franclemont (1973)
	Saturniidae	1,300	68	24	p: Michener (1952)
					t: Ferguson (1972)
	Endromidae	1	1	1	
(Sphingoidea[b])	Sphingidae	1,000	124	30	t: Hodges (1971); Pittaway (1993)
(Drepanoidea[c])	Drepanidae	400	5	9	c: Hodges et al. (1983)
	Thyatiridae	180	16	10	c: Emmet and Heath (1991)
Geometroidea	Geometridae	20,000	1,414	363	p: McGuffin (1967, 1972, 1977, 1981, 1989)
Noctuoidea	Notodontidae[d]	3,500	136	64	p: Miller (1991)
	Lymantriidae	2,500	32	22	t: Ferguson (1978)
	Arctiidae	20,000	264	14	c: Hodges et al. (1983)
	(Nolidae[e])		16	4	c: Hodges et al. (1983)
	Noctuidae[f]	25,000	2,925	251	p: Kitching (1984)
					c: Hodges et al. (1983)
					c: Emmet and Heath (1991)

[a]Source type: p, phylogeny; c, classification; t, taxonomy.
[b]The Sphingidae have often been placed in their own superfamily, however, Nielsen (1989) and Scoble (1991) place them in the Bombycoidea.
[c]There is continuing debate over whether the Drepanidae, Thyatiridae, and related families should be included in the Geometroidea, or separated from the Geometridae as Drepanoidea.
[d]Some genera were not analyzed by Miller (1991) and could not be placed; these have been included only in the nonphylogenetic analyses.
[e]Nolidae are sometimes treated as a subfamily of the Noctuidae.
[f]The phylogeny of this large family is not well resolved (Kitching, 1984). I could place species in subfamilies and tribes, but little more. I excluded the Noctuidae and Nolidae from the analyses that required more precise phylogenetic information.

requires counting the number of internodes in which a change from nonoutbreak to outbreak occurs, and the number in which this change has the potential to occur (i.e., that are nonoutbreak), separately for the states of the independent trait. It assumes that changes are equally likely to occur in any of the branches. Sillén-Tullberg (1993) claims that this test is sensitive to the number of branches in the tree, but not the exact topology of the tree. However, it is sensitive to the locations of the changes in the traits on the tree, which change with different phylogenetic patterns.

2. The distribution test was developed by Proctor (1991) for use when there are no reversals in the trait of interest. It tests whether origins of the trait co-occur with an independent trait more often than would be expected given the distribution of the independent trait over the whole study group. She identified the taxonomic level at which most of the origins of the dependent trait occurred, then generated expected frequencies of the independent trait at the same taxonomic level. Here, the genus is the appropriate level. This method is less sensitive to tree topology than the contingent states test.

3. For continuous traits, I use paired sister-group comparisons (Møller and Birkhead, 1992). For each group of outbreaking species, the nearest relatives that are nonoutbreaking are found. Then a paired difference *t*-test is performed on the outbreaking versus nonoutbreaking sister pairs.

The noctuids pose a problem because of their poorly resolved phylogenies, and I excluded them when using the contingent states test, which requires a reasonably well-resolved tree. They are included in exploratory analyses, the distribution test, and sister-group comparisons (but only using intrageneric pairs).

III. Results

A. Proportions of Outbreaking and Nonoutbreaking Species

In this data set only 12% of species have outbreaks, but outbreaking species are still overrepresented. Many more species of forest Lepidoptera exist but were not included because too little is known about them. I estimate the true proportion of tree-feeding, outbreaking species in this region to be 2.6%, based on the lists of described species (Hodges *et al.*, 1983; Emmet and Heath, 1991), the probable number of undescribed species in these families in Canada (Munroe, 1979), and the proportion of species feeding on woody plants (Gaston and Reavey, 1989). Those with large or moderate outbreaks constitute only 1.6% of the tree-feeding macrolepidopteran fauna.

B. Number of Generations per Year

Multivoltine species comprise 22% of forest Lepidoptera. The proportion of univoltine and multivoltine species does not differ between outbreaking (24% multivoltine) and nonoutbreaking (22% multivoltine) groups, contrary to expectation based on rates of population increase. This may be partly because outbreaking species are more often spring feeders, and few spring-feeding species

TABLE 3
Analysis of the Associations between Outbreaking and Other Variables[a]

	Nonphylogenetic		Contingent changes		Distribution test	
	Nonoutbreak species	Outbreak species	Nonoutbreak to nonoutbreak branches	Nonoutbreak to outbreak branches	All genera	Outbreak genera
Phenology						
Summer	481	39	489	17	195	26
Spring	273	61	282	26	138	31
	$G_1 = 22.2$	$P < 0.001$	$G_1 = 9.51$	$P < 0.005$	$G_1 = 3.29$	$0.1 > P > 0.05$
Overwintering stage						
Egg	143	36	136	15	63	18
Larva	114	21			66	11
Pupa	436	40	673	28	188	26
Adult	52	3			14	2
	$G_3 = 20.2$	$P < 0.005$	$G_1 = 7.68$	$P < 0.01$	$G_3 = 4.49$	nsd
Larval defenses						
Cryptic	444	25	518	13	200	18
Other defenses	211	71	203	27	125	38
	$G_1 = 60.6$	$P < 0.001$	$G_1 = 25.0$	$P < 0.001$	$G_1 = 16.8$	$P < 0.001$
Host breadth class						
Monophagous	140	8			50	1

48

Few genera	373	28	605[b]	21	172	18
Many genera	160	26	252	25	75	17
Heaps of genera	78	37			43	21
	$G_3 = 51.4$	$P < 0.001$	$G_1 = 11.7$	$P < 0.005$	$G_3 = 28.7$	$P < 0.001$
Egg dispersion						
Singly	294	22	329	11	142	13
Clusters	151	32	198[c]	35	94	23
Masses	48	41			35	19
	$G_2 = 66.3$	$P < 0.001$	$G_1 = 25.97$	$P < 0.001$	$G_2 = 20.6$	$P < 0.001$
Larval life-style						
Solitary	724	66	653	62	317	40
Gregarious	32	34	32	14	24	17
	$G_1 = 71.8$	$P < 0.001$	$G_1 = 16.1$	$P < 0.001$	$G_1 = 20.9$	$P < 0.001$
Clustering of offspring						
Dispersed	375	21	669	18	166	23
Clustered	98	66	112	20	64	34
	$G_1 = 98.4$	$P < 0.001$	$G_1 = 28.6$	$P < 0.001$	$G_1 = 19.6$	$P < 0.001$

[a]The nonphylogenetic columns show the distribution of species. The contingent changes columns show the distribution of branches of the phylogenetic trees with no change in outbreaking state (nonoutbreaking ancestral), and of the changes to outbreaking. The distribution test columns show states of the genera (or most of the species in a genus) and of the genera with outbreaking species.

[b]Few genera = less than 10; many genera = 10 or more.

[c]Masses and clusters and together.

49

are multivoltine (9% of spring feeders versus 30% of summer feeders, $G_1 = 59.7$, $P < 0.001$; Gaston and Reavey, 1989). Looking only at summer feeders, 46% of outbreaking species are multivoltine, but just 29% of nonoutbreaking species ($G_1 = 4.72$, $P = 0.05$). No phylogenetic analysis was done on this trait.

C. Overwintering Stage and Feeding Phenology

Most Macrolepidoptera overwinter as pupae (56%) and are summer-feeding (61%). A disproportionate number of outbreaking species are spring-feeding, which is confirmed by phylogenetic analysis (Table 3). There are 43 origins of outbreaking dynamics, 26 of which occur in spring-feeding lineages, although there are fewer spring-feeding than summer-feeding lineages (Table 3). Similarly, using the distribution test, 195 genera are predominantly summer-feeding and 138 are spring-feeding, but among outbreaking genera, 31 feed in spring and only 26 are late-season feeders. A disproportionate number of outbreaking species overwinter as eggs or larvae (Table 3). This is expected from the fact that spring-feeding species generally overwinter as eggs or larvae (81% of spring-feeders) but summer-feeders overwinter as pupae (83%).

D. Larval Defenses

Over 67% of forest Macrolepidoptera, but only 26% of outbreaking species, are cryptic (Table 3). The rest are defended by setae, chemicals, tents, or leaf shelters or have warning coloration (probably indicating a chemical defense). Possession of setae and bright colors tends to be a family-level trait: most Sphingidae, Geometridae, and Noctuidae are hairless and cryptic, whereas Lymantriidae, Arctiidae, and Saturniidae are all hairy and generally brightly colored (Table 4). The Notodontidae, however, vary greatly in coloration of larvae and possession of defenses. As noted in the following, gregarious species in particular tend to be well defended and there is a loose correspondence between the occurrence of repellent defenses and gregariousness in families (Table 4). Within the cryptic families, outbreaking species tend to be well defended, with the exception of about half of the outbreaking geometrids. In the Noctuidae, all the outbreaking species except two have some repellent defense, and in the Notodontidae only one species is outbreaking and cryptic. Phylogenetic analysis shows a higher number of origins of outbreaking in defended lineages (Table 3), even within the Geometridae ($G_1 = 14.4$, $P < 0.001$). Defenses are also associated with egg clustering ($G_1 = 41.8$, $P < 0.001$), but this could be due to gregarious species. When gregarious species are removed, 52% of species with eggs in masses but that feed solitarily versus only 32% of species with eggs in clusters or singly have other defenses besides crypsis ($G_1 = 7.38$, $P < 0.01$).

TABLE 4
Percentages of Species in Defended and Cryptic Families That Have Outbreaks,
or Are Gregarious, in the Larger Macrolepidopteran Families

	% defended	% outbreaking	% gregarious	No. of species
Defended families				
Lasiocampidae	100	52	39	23
Saturniidae	100	42	63	24
Lymantriidae	100	70	13	23
Arctiidae	92	29	29	14
Nymphalidae	88	12	38	16
Notodontidae	56	16	21	64
Cryptic families				
Geometridae	14	9	<1	363
Noctuidae	40	4	4	251
Sphingidae	10	0	3	30
Lycaenidae	33	0	0	15

E. Larval Length at Maturity

Previously, I found no difference in size between outbreaking and non-outbreaking species (Hunter, 1991). With the added data, there is a difference: outbreaking species average 39 mm (across the three categories) and non-outbreaking species only 34 mm (Table 5). Phylogenetic analysis, however, gave no difference in size of larvae between outbreaking species and their sister species ($t_{39} = 0.61$, $P = 0.55$). This is not surprising given that there are substantial differences in size (and tendency to have outbreaks) among the families. Subtracting family mean size from each observation removes the differences in size among population dynamics classes (Table 5).

Across all Lepidoptera, species with larger larvae have higher host breadth ($R = 0.16$, $P < 0.05$, $N = 708$) and fecundity ($R = 0.32$, $P < 0.01$, $N = 155$). These relationships are less strong within taxonomic units: only 2 are significant for host breadth, and only 4 for fecundity (out of 8 units with more than 12 species). However, ANOVA using host breadth classes shows larger larvae in the high classes, even when family differences are removed ($F_{3,731} = 5.92$, $P < 0.001$).

F. Host Breadth

Forest Macrolepidoptera are relatively polyphagous (Fig. 1). Outbreaking species have greater host breadth than do nonoutbreaking species (Tables 3 and

TABLE 5
Mean Maximum Fecundity, Number of Host Genera, and Larval Length
for the Population Classes, with ANOVA Results from Log-Transformed
Raw Values or on Residuals from Family Means

	Rare	Common	Limited outbreak	Medium outbreak	Large outbreak	F	F on residuals
Larval length	34	34	41	40	37	2.49	0.65
(mm)	(377)[a]	(274)	(34)	(16)	(41)	$P = 0.04$	nsd
Host genera	4	8	10	12	18	48.8	31.2
	(455)	(267)	(38)	(19)	(42)	$P < 0.001$	$P < 0.001$
Fecundity	180	208	279	366	322	7.51	1.25
	(53)	(53)	(13)	(13)	(30)	$P < 0.001$	$P = 0.29$

[a]Number of observations.

6). There are differences in host breadth among families ($F_{16,809} = 4.46$, $P < 0.001$), but if these are removed there are still differences among outbreak classes (Table 5). Outbreaking arises more frequently in polyphagous lineages (Table 3). Sister group comparisons show that outbreaking species have on average seven more genera in their diets than do nonoutbreaking sister species ($t_{41} = 2.83$, $P = 0.007$).

Using the host breadth categories, significantly more spring-feeding species fall in the "many genera" category (30%), and fewer in the "few genera" category (38%), than do summer-feeding species (17 and 53%, respectively, $G_3 = 27.6$, $P < 0.001$). When host breadth is treated as a continuous variable, there is no difference between spring and summer feeders ($F_{1,807} = 1.10$, $P = 0.30$).

G. Maximum Fecundity

Data on fecundity are scarce ($n = 162$ out of 858), particularly for nonoutbreaking species. Fecundity of outbreaking species is higher (Table 5). Again, there are differences among families ($F_{11,157} = 8.16$, $P < 0.001$), but there are still differences among the outbreak classes when family effects are removed (Table 5). Given the large gaps in the data, further phylogenetic analyses were not possible.

H. Female Flying Ability

A thorough analysis of wing reduction has been done (Hunter, 1995) and will not be repeated here; the following is a summary of the results. Reduced wing size of females is related to outbreaking and to spring feeding. Reduced wings evolved only in spring-feeding lineages, but there are a few species that have shifted to summer feeding. Nonphylogenetic comparisons show greater host breadth and fecundity of species with wing reduction. Phylogenetic analysis shows that a disproportionate number of origins of reduced wings occurred in polyphagous lineages, but there have been many reversals to lower host breadth. There are too few data on fecundity to make solid conclusions. Many of the wing-reduced geometrids have winter-active adults (i.e., eggs are laid during the cold season).

I. Egg Dispersion

Whereas 76% of outbreaking species cluster their eggs, only 40% of non-outbreaking species do so, and this association also holds in phylogenetic analyses (Table 3). Egg dispersion pattern is naturally related to gregarious lifestyle; all gregarious species must have their eggs in clusters. However 40% of solitary species cluster their eggs. If gregarious species are excluded, there is still a larger proportion of egg-clustering species (63%) among outbreaking species than among nonoutbreaking species (38%; $G_1 = 19.6$, $P < 0.001$). Solitary spring-feeding species are more likely to have their eggs in a mass (15% of species) than summer-feeding species (6%), but the same proportion clusters them (33% and 30%,; $G_2 = 14.4$, $P < 0.001$).

There are strong patterns for the continuous variables: as the degree of aggregation increases from solitary, to clusters, to masses, the fecundity, host breadth, and larval length also increase (Table 6). When residuals from family means are used, the differences in fecundity are no longer significant (Table 6). Also, sister-group comparisons of the host breadth of species that have eggs in

TABLE 6
Mean Maximum Fecundity, Number of Host Genera, and Larval Length of Macrolepidoptera as Related to Egg Dispersion Pattern, with ANOVA Results from Log-Transformed Raw Values or on Residuals from Family Means

	Solitary	Clusters	Masses	F	F on residuals
Fecundity	159	255	369	17.9	2.33
	(67)[a]	(41)	(39)	$P < 0.001$	$P = 0.10$
Host genera	5.8	7.0	14.6	18.5	13.1
	(307)	(167)	(81)	$P < 0.001$	$P < 0.001$
Larval length	34.2	36.5	42.3	15.5	5.54
(mm)	(292)	(178)	(83)	$P < 0.001$	$P = 0.011$

[a]Number of observations.

clusters or masses and those that spread them singly showed no significant difference ($t_{49} = -1.3$, $P = 0.2$); however, comparison of those that lay eggs in masses versus in clusters or singly gave a slightly higher host breadth for those with eggs in masses ($t_{31} = 1.77$, $P = 0.09$).

J. Larval Lifestyle (Solitary/Gregarious)

Only 8% of all species, but 34% of outbreaking species, are gregarious (Table 3). Gregarious species may also be spring feeders, but not more often than expected ($G_1 = 0.24$, $P = 0.54$). They have a strong tendency to have warning colors or other defenses ($G_1 = 104.8$, $P < 0.001$). They are equally as polyphagous as solitary species ($G_3 = 4.84$, $P = 0.18$). Gregarious species are larger than solitary species (45 mm versus 33 mm; $F_{1,720} = 34.8$, $P < 0.001$).

There are 25 origins of gregariousness, resulting in 64 species. Phylogenetic analysis confirms an association with warning coloration ($G_1 = 28.3$, $P < 0.001$, and Sillén-Tullberg and Hunter, 1995) and with outbreaking (Table 3). Sister-group comparisons showed no difference in host breadth of gregarious species ($t_{24} = 0.64$, $P = 0.53$).

K. Clustering of Offspring

Whereas 76% of outbreaking species have clustered offspring because of gregariousness, poor female flying ability, or eggs deposited in masses, only 21% of nonoutbreaking species have spatially aggregated offspring (Table 3). Outbreaking arises much more frequently than expected in lineages with clustered offspring (Table 3). As would be expected, the previous differences for outbreaking and nonoutbreaking species hold for those with clustered offspring: more species than expected overwinter as eggs ($G_3 = 14.6$, $P < 0.005$), more are spring feeders ($G_1 = 34.2$, $P < 0.001$), and more have repellent defenses ($G_1 = 58.2$, $P < 0.001$).

Only 19 outbreaking species were classified as likely to have dispersed offspring (Table 3); of these 19, 11 are spring-feeding or have their eggs in clusters. Among the remaining eight, four have limited outbreaks: *Heterocampa guttivita, Lochmaeus manteo* (Heterocampini, Notodontidae), *Rheumaptera hastata,* and *Semiothisa sexmaculata* (Geometridae). Nothnagle and Schultz (1987) scored *H. guttivita* females as poor fliers but I could not confirm this; the species does, however, have a high host breadth (25 genera of deciduous trees). *Rheumaptera hastata* outbreaks occur in Alaska, although its range extends throughout Canada and the northern United States.

L. Taxonomic Distribution of Variation

The variance components from a fully nested analysis of variance reveal the distribution of variability among the levels of taxonomic classification (Fig. 2).

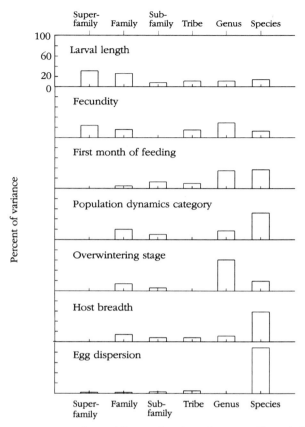

Figure 2. Variance components from a fully nested analysis of variance. The scale for larval length applies to all of the other variables.

Most of the variation in larval length occurs at the superfamily and family levels; the other variables have the bulk of the variation at the genus or species level (Fig. 2). Thus phylogeny plays the largest role in morphological features such as size, hairiness, and coloration (see earlier discussion). Behavioral traits, like egg dispersion and host breadth, vary greatly among closely related species and are not subject to constraints (Fig. 2). Host type (deciduous or coniferous) can also differ among closely related species. However, some groups appear to be constrained in host use: the Notodontidae use only deciduous trees, a few genera in other families are recorded only from conifers, and the large genera *Catocala* and *Acronicta* (Noctuidae) feed only on deciduous trees (while other genera of similar size usually contain members that feed on conifers).

IV. Discussion

A. Differences between Outbreaking and Nonoutbreaking Forest Macrolepidoptera

Outbreaking species of forest Macrolepidoptera differ significantly from nonoutbreaking species in all of the life-history and ecological traits tested here, except the number of generations per year. Though there is no single feature of all outbreaking Macrolepidoptera that is lacking in nonoutbreaking species, there are some common themes. Outbreaking species are more likely to be spring-feeding, polyphagous, overwinter as eggs, have gregarious larvae, have their eggs in one mass or a few clusters, have higher fecundity and host breadth and a larger size, and tend to be well defended against predators with bright coloration, hairs, and tents or leaf ties. Outbreaking species fall into two groups, spring-feeding and gregarious, with some overlap, indicated by correlations among the features. Spring-feeding species overwinter as eggs or larvae, may have females that do not fly, have greater host breadth, and are more likely to put their eggs in masses than summer-feeding species. Gregarious species are larger than solitary species and have eggs in masses or clusters, brightly colored larvae, and physical or chemical defenses against predators. These are the same patterns identified before (Hunter, 1991), and much of that discussion is relevant here (including comparisons to findings of other researchers) and will not be repeated. Instead, I wish to concentrate on the new-found links between the two major groups of outbreaking species, that is, clustering of offspring and limited dispersal. The current discussion concentrates on these findings and on the phylogenetic patterns.

B. Phylogenetic Patterns and Constraints

Some families have more outbreaking species than others (Table 4), suggesting that phylogenetic effects may be important. However, the phylogenetic analyses show that the associations between population dynamics and the traits arise repeatedly in independent lineages. Furthermore, the associations are repeated in the unrelated but ecologically similar sawflies (Haack and Mattson, 1993; Larsson et al., 1993). We can thus rule out shared ancestry or confounding variables related to phylogeny as a cause for most of the correlations. The exceptions are size and fecundity. It is true that outbreaking species are on average larger and more fecund than nonoutbreaking species, but this could just be due to the phylogenetic conservatism of size (Fig. 2) and the tendency of families with large species to have more outbreaking species. Within a family, outbreaking species are not larger or more fecund.

C. Clustering of Offspring and Limited Dispersal: A Unifying Theme?

The most consistent difference between outbreaking and nonoutbreaking species is greater clustering of offspring. Though direct measurements of dispersion of forest insects have not been made, I expect that most offspring of outbreaking species will be aggregated on one tree or a few adjacent trees (because of reduced wings, poor female flight, gregariousness, or eggs deposited in masses). Also, either because of flightlessness or heavy-bodied females (all eggs are produced at once), I expect outbreaking species to have short dispersal distances. Although many of the spring-feeding species have larval dispersal by ballooning, the vast majority of individuals travel only short distances by this method (Mason and McManus, 1981).

Assuming that outbreaking species do have limited dispersal, the correlation is surprising because low powers of dispersal have been associated with rarity (Hodgson, 1993; Kunin and Gaston, 1993), and spread from a focal area is one of the major types of outbreaks (Berryman, 1987). Strong dispersal ability was one of the traits identified by Hodgson (1993) as distinguishing common and rare butterflies in Britain. These traits facilitate exploitation of habitats created by modern land use, that is, disturbed, productive habitats with herbaceous food plants. Only one common butterfly species feeds on woody plants (Hodgson, 1993). Abundance was measured by the number of 10-km^2 areas occupied, so forest species could reach high densities but would not be considered common in this scheme owing to the small total area of forests. Differences in the taxonomic group, type of host plants, and definition of abundance account for the divergent conclusions here and in Hodgson (1993). More generally, many outbreaking species in agricultural systems have good dispersal abilities that may contribute to their pest status (e.g., locusts, the fall armyworm *Spodoptera frugiperda,* and the corn earworm *Heliothis zea*). Lepidoptera with larvae that feed on plants of disturbed habitats fly farther than species that feed on woody plants (Scott, 1975; Shreeve, 1981). Quite different ecological processes come into play in forest than in agricultural systems, where hosts are more ephemeral and difficult to locate.

Spread from focal areas has been used to describe many forest insect outbreaks. However, some outbreaks only appear to spread from focal areas. Liebhold and Elkinton (1990) showed that dispersal by ballooning could not explain the apparent spread of gypsy moth outbreaks. Similar spatial analyses have not been done for most other forest insects, but other flightless species have outbreaks that can cover enormous areas (e.g., *Alsophila pometaria:* 450,000 ha; *Operophtera bruceata:* 150,000 ha; *Orgyia pseudotsugata:* 324,000 ha; *Paleacrita vernata:* 10,900,000 ha; Mattson *et al.,* 1991). It seems unlikely that spread from loci accounts for these large areas; probably large-scale weather patterns synchronize the regional dynamics (Hanski and Woiwod, 1993).

D. Life-History Traits and Outbreaks

At least five explanations are consistent with the patterns discovered: (1) lack of female choice of host tissues leads to the evolution of a suite of traits that permit populations to build to high densities (Price *et al.*, 1990; Price *et al.*, Chapter 16, this volume); (2) maternal effects are most strongly expressed in these species and generate time lags that destabilize dynamics (Rossiter, Chapter 13, this volume); (3) species with low dispersal have individuals that concentrate their risks in one location, and high variance in success leads to high variance in population size (Hanski, 1987); (4) dispersal and spatial heterogeneity can dramatically affect dynamics in spatially complex models (Kareiva, 1990); and (5) clustered species are more affected by disease.

The conceptual model of Price *et al.* (1990) revolves around a proposed correlation between female choice and offspring performance on host plants. Nonoutbreaking species are thought to be more choosy and will emigrate if high-quality resources are not found, keeping population densities in line with resources. Outbreaking species are thought to oviposit off the host tissue, and usually in clusters because of phylogenetic constraints. Although many of the features postulated to differ between outbreaking and nonoutbreaking species by Price *et al.* (1990) are supported in the present analysis, there is limited evidence of phylogenetic constraints for these traits. Egg dispersion varies among closely related species, but egg clustering does not necessarily indicate lack of choosiness (Moore *et al.*, 1988). Another trait identified by Price *et al.* (1990) as a correlate of outbreaking dynamics was phenological dissociation between oviposition time and larval feeding time (overwintering as an egg or winter-active adults). This trait is ancestral to some families, but in others it varies among congenerics.

Hanski (1987) presented a model to account for the tendency of gregarious sawfly species to have outbreaks. He proposed that there is higher variance in the survival of gregarious species, because they put all eggs in one spatial "basket," which has either a very high payoff or a very low one. Solitary species are thought to have an intermediate, more constant survival rate. The arguments of Hanski (1987) may be extended beyond gregarious species to encompass flightless species and those that put eggs in a single mass. The offspring of these species are also clustered, which is the essential feature of Hanski's model. What is still unclear is the spatial scale over which the mechanism can operate, and its ability to explain the large spatial extent of many forest insect outbreaks. Measurements of the spatial patterning of forest insects and the responses of natural enemies to these patterns are urgently needed (see also Cappuccino *et al.*, Chapter 4, this volume).

E. Clustering of Offspring, Dispersal, and Spatially Complex Models

Mathematical models of population dynamics that incorporate a spatial dimension and dispersal have different behavior from models that lack these features (Kareiva, 1990). The differences between outbreaking and nonoutbreaking species in dispersal distances and dispersion of offspring suggest that spatial pattern, heterogeneity, or movement may be important determinants of population dynamics of forest species. There is a diversity of spatial models and many of them predict that sufficiently low dispersal rates will destabilize dynamics. Two examples are: in metapopulation dynamics models, population size variability depends on the amount of dispersal between local populations and the degree of synchronization among local populations (Vance, 1980; Hastings, 1982; Reeve, 1990); and in individual-based cellular automata models, local dispersal generates chaotic dynamics (Molofsky, 1994).

Metapopulation dynamics models are probably the best known of the spatially explicit models. However, they may not be applicable to forest systems, which are not subdivided into distinct patches. Forests are not naturally fragmented in this way, but isolated forest islands may have been created by logging. Another requirement of metapopulations, if they are to persist for significant durations, is asynchronous changes among local populations. Moth and aphid populations in Britain are synchronized over large distances, apparently by environmental correlation and not by migration (Hanski and Woiwod, 1993). However, small-scale spatial variability may be averaged out in the large-scale samples (Hanski and Woiwod, 1993), as is apparent for gypsy moth populations (Liebhold, 1992).

Fieldwork to test the applicability of spatially explicit models is sorely lacking (Kareiva, 1990). We lack knowledge of natural dispersal rates and spatial patterns of insect populations, and of local synchrony of density fluctuations. Contrasting responses of outbreaking and nonoutbreaking species to changes in host quality, or responses of natural enemies to changes in densities of these species, would also be fertile avenues for further research.

Disease transmission should be greater between contagiously distributed individuals, so the disease hypothesis is attractive. However, there is little direct evidence to support it. Hanski and Otronen (1985) found greater mortality of outbreaking sawfly species than of rare species, which they attributed to disease. Gregarious species, but not solitary species, show age-related resistance to disease, suggesting that colonial species have evolved a mechanism to reduce the impact of disease (Hochberg, 1991). More comparison of host–pathogen interactions of species with differing dynamics is needed.

F. Generality?

Similar patterns in the differences between outbreaking and nonoutbreaking species have been found for forest sawflies (Haack and Mattson, 1993; Larsson *et al.*, 1993), which are the nearest ecological relatives of Macrolepidoptera. Lepidoptera that feed on annuals or short-lived perennial plants require greater host location abilities than many forest species. Host plant abundance is liable to play a greater role in their dynamics. Other groups of herbivores are expected to show different patterns, depending on the temporal and spatial stability of their habitats.

The more general pattern of dispersal or clustering effects on dynamics may apply to a wide diversity of organisms. If dispersal from cyclic vole populations is prevented by enclosing populations in mouse-proof fences they will reach unusually high densities, so that most individuals die of starvation (Krebs, 1992). This fence effect is eliminated if exit gates are provided. Dispersal may also be important in red grouse populations (Watson, 1992). Generally, however, studies that test models incorporating spatial patterns and dispersal are lacking (Kareiva, 1990).

G. Caveat

One final caveat: although the number of species analyzed here may seem impressive, more than 3000 species occurring in this region were not included because too little is known about their biology. These species that have not attracted the attention of ecologists or natural historians are probably rare and may have quite different characteristics from those that are more common (Kunin and Gaston, 1993).

Acknowledgments

I thank Naomi Cappuccino, Hans Damman, Lorne Rothman, and Christer Solbreck for comments on earlier drafts. This research was supported by a Natural Sciences and Engineering Research Council of Canada postdoctoral fellowship, and by a Killam post-doctoral fellowship to the author.

References

Berryman, A. A. (1987). The theory and classification of outbreaks. *In* "Insect Out-breaks" (P. Barbosa and J. C. Schultz, eds.), pp. 3–30. Academic Press, San Diego, CA.

Bolte, K. B. (1990). Guide to the Geometridae of Canada (Lepidoptera). VI. Subfamily Laurentiinae 1. Revision of the genus *Eupithecia. Entomol. Soc. Can. Mem.* **151,** 1–253.

Buckler, W. (1896). "The Larvae of British Butterflies and Moths." Ray Society, London.

Carter, J. D. (1984). "Pest Lepidoptera of Europe." Dr. W. Junk Publishers, Dordrecht, The Netherlands.

Damman, H. (1987). Leaf quality and enemy avoidance by the larvae of a pyralid moth. *Ecology* **68,** 88–97.

Emmet, A. M., and Heath, J. (1983). "The Moths and Butterflies of Great Britain and Ireland," Vol. 10. Harley Books, Colchester, UK.

Emmet, A. M., and Heath, J. (1990). "The Moths and Butterflies of Great Britain and Ireland," Vol. 7(1). Harley Books, Colchester, UK.

Emmet, A. M., and Heath, J. (1991). "The Moths and Butterflies of Great Britain and Ireland," Vol. 7(2). Harley Books, Colchester, UK.

Ferguson, D. C. (1972). "The Moths of America North of Mexico Including Greenland," Fasc. 20.2. E. W. Classey and R. B. D. Publ., London.

Ferguson, D. C. (1978). "The Moths of America North of Mexico Including Greenland," Fasc. 22.2. E. W. Classey and Wedge Entomol. Res. Found., London.

Franclemont, J. G. (1973). "The Moths of America North of Mexico Including Greenland," Fasc. 20.1. E. W. Classey and R. B. D. Publ., London.

Furniss, M. M., and Carolin, V. M. (1977). Western forest insects. *USDA For. Serv. Misc. Publ.* **1339.**

Gaston, K. J., and Reavey, D. (1989). Patterns in the life histories and feeding strategies of British Macrolepidoptera. *Biol. J. Linn. Soc.* **37,** 367–381.

Haack, R. A., and Mattson, W. J. (1993). Life history patterns of North American tree-feeding sawflies. *In* "Sawfly Life History Adaptations to Woody Plants" (M. Wagner and K. F. Raffa, eds.), pp. 503–545. Academic Press, San Diego, CA.

Hanski, I. (1987). Pine sawfly population dynamics: Patterns, processes, problems. *Oikos* **50,** 327–335.

Hanski, I., and Otronen, M. (1985). Food quality induced variance in larval performance: Comparison between rare and common pine-feeding sawflies (Diprionidae). *Oikos* **44,** 165–174.

Hanski, I., and Woiwod, I. P. (1993). Spatial synchrony in the dynamics of moth and aphid populations. *J. Anim. Ecol.* **62,** 656–668.

Harvey, P. H., and Pagel, M. D. (1991). "The Comparative Method in Evolutionary Biology." Oxford Univ. Press, Oxford.

Hastings, A. (1982). Dynamics of a single species in a spatially varying environment: The stabilizing role of high dispersal rates. *J. Math. Biol.* **16,** 49–55.

Hochberg, M. E. (1991). Viruses as costs to gregarious feeding behaviour in the Lepidoptera. *Oikos* **61,** 291–296.

Hodges, R. W. (1971). "The Moths of America North of Mexico Including Greenland," Fasc. 21. E. W. Classey and R. B. D. Publ., London.

Hodges, R. W., Dominick, T., Davis, D. R., Ferguson, D. C., Franclemont, J. G., Munroe, E. E., and Powell, J. A. (1983). "Check List of the Lepidoptera of America North of Mexico." E. W. Classey and R. B. D. Publ., London.

Hodgson, J. G. (1993). Commonness and rarity in British butterflies. *J. Appl. Ecol.* **30,** 407–427.

Holloway, J. D., and Hebert, P. D. N. (1979). Ecological and taxonomic trends in macrolepidopteran host plant selection. *Biol. J. Linn. Soc.* **11,** 229–251.

Hunter, A. F. (1991). Traits that distinguish outbreaking and non-outbreaking Macrolepidoptera feeding on northern hardwood tress. *Oikos* **60,** 275–282.

Hunter, A. F. (1995). The ecology and evolution of reduced wings in forest Macrolepidoptera. *Evol. Ecol.* (in press).

Kareiva, P. (1990). Population dynamics in spatially complex environments: Theory and data. *Philos. Trans. R. Soc. London, Ser. B* **330,** 175–190.

Kitching, I. J. (1984). An historical review of the higher classification of the Noctuidae (Lepidoptera). *Bull. Br. Mus. (Nat. Hist.) Entomol.* **49,** 153–234.

Krebs, C. J. (1992). The role of dispersal in cyclic rodent populations. *In* "Animal Dispersal: Small Mammals as a Model" (N. C. Stenseth and W. Z. Lidicker, Jr., eds.), pp. 160–175. Chapman & Hall, London.

Kunin, W. E., and Gaston, K. J. (1993). The biology of rarity: Patterns, causes and consequences. *TREE* **8,** 298–301.

Larsson, S., Bjorkman, C., and Kidd, N. A. C. (1993). Outbreaks in diprionid sawflies: Why some species and not others? *In* "Sawfly Life History Adaptations to Woody Plants" (M. Wagner and K. F. Raffa, eds.), pp. 453–483. Academic Press, San Diego, CA.

Liebhold, A. M. (1992). Are North American populations of gypsy moth (Lepidoptera: Lymantriidae) bimodal? *Environ. Entomol.* **21,** 221–229.

Liebhold, A. M., and Elkinton, J. S. (1990). Models of the spatial dynamics of epidemic gypsy moth populations. *In* "Population Dynamics of Forest Insects" (A. D. Watt, S. R. Leather, M. D. Hunter, and N. A. C. Kidd, eds.), pp. 359–367. Intercept, Andover, UK.

Maddison, W. P., and Maddison, D. R. (1992). "MacClade 3.01: Analysis of Phylogeny and Character Evolution." Sinauer, Sunderland, MA.

Mason, C. J., and McManus, M. L. (1981). Larval dispersal of the gypsy moth. *USDA For. Serv. Tech. Bull.* **1584,** 161–202.

Mattson, W. J., Herms, D. A., Witter, J. A., and Allen, D. C. (1991). Woody plant grazing systems: North American outbreak folivores and their host plants. *USDA For. Serv. Gen. Tech. Rep.* **NE-153,** 53–84.

McGuffin, W. C. (1967). Guide to the Geometridae of Canada (Lepidoptera). I. Subfamily Sterrhinae. *Entomol. Soc. Can. Mem.* **50.**

McGuffin, W. C. (1972). Guide to the Geometridae of Canada (Lepidoptera). II. Subfamily Ennominae. 1. *Entomol. Soc. Can. Mem.* **86.**

McGuffin, W. C. (1977). Guide to the Geometridae of Canada (Lepidoptera). II. Subfamily Ennominae. 2. *Entomol. Soc. Can. Mem.* **101.**

McGuffin, W. C. (1981). Guide to the Geometridae of Canada (Lepidoptera). II. Subfamily Ennominae. 3. *Entomol. Soc. Can. Mem.* **117.**

McGuffin, W. C. (1988). Guide to the Geometridae of Canada (Lepidoptera). II. Subfamily Ennominae. 4. *Entomol. Soc. Can. Mem.* **138.**

McGuffin, W. C. (1989). Guide to the Geometridae of Canada (Lepidoptera). III, IV and V. Subfamilies Archierinae, Oenochrominae and Geometrinae. *Entomol. Soc. Can. Mem.* **145.**

McGugan, B. M. (1958). Forest Lepidoptera of Canada. Vol. 1. *Publ. Can. Dep. Agric.* **1034.**

Michener, C. D. (1952). The Saturniidae (Lepidoptera) of the Western Hemisphere: Morphology, phylogeny and classification. *Bull. Am. Mus. Nat. Hist.* **98,** 335–501.

Miller, J. A. (1991). Cladistics and classification of the Notodontidae (Lepidoptera: Noctuoidea) based on larval and adult morphology. *Bull. Am. Mus. Nat. Hist.* **204,** 1–226.

Møller, A. P., and Birkhead, T. R. (1992). A pairwise comparative method as illustrated by copulation frequency in birds. *Am. Nat.* **139,** 644–656.

Molofsky, J. (1994). Population dynamics and pattern formation in theoretical populations. *Ecology* **75**, 30–39.

Moore, L. V., Myers, J. H., and Eng, R. (1988). Western tent caterpillars prefer the sunny side of the tree, but why? *Oikos* **51**, 321–326.

Munroe, E. E. (1979). Lepidoptera. *Entomol. Soc. Can. Mem.* **108**, 427–481.

Nielsen, E. S. (1989). Phylogeny of major lepidopteran groups. *In* "The Hierarchy of Life" (B. Fernholm, K. Bremer, and H. Jörnvall, eds.), pp. 281–294. Elsevier, Amsterdam, The Netherlands.

Nothnagle, P. J., and Schultz, J. C. (1987). What is a forest pest? *In* "Insect Outbreaks" (P. Barbosa and J. C. Schultz, eds.), pp. 59–80. Academic Press, San Diego, CA.

Pittaway, A. R. (1993). "The Hawkmoths of the Western Palaearctic." Harley Books, Colchester, UK.

Prentice, R. M. (1962). Forest Lepidoptera of Canada. Vol. 2. *Can. Dep. For. Bull.* **128**.

Prentice, R. M. (1963). Forest Lepidoptera of Canada. Vol. 3. *Can. For. Branch, Dep. Publ.* **1013**.

Price, P. W., Cobb, N., Craig, T. P., Wilson Fernandes, G., Itami, J. K., Mopper, S., and Preszler, R. W. (1990). Insect herbivore population dynamics on trees and shrubs: New approaches relevant to latent and eruptive species and life table development. *In* "Insect–Plant Interactions" (E. A. Bernays, ed.), Vol. 2, pp. 1–38. CRC Press, Boca Raton, Fl.

Proctor, H. C. (1991). The evolution of copulation in water mites: A comparative test for nonreversing characters. *Evolution (Lawrence, Kans.)* **45**, 558–567.

Redfearn, A., and Pimm, S. L. (1988). Population variability and polyphagy in herbivorous insect communities. *Ecol. Monogr.* **58**, 39–55.

Reeve, J. D. (1990). Stability, variability, and persistence in host–parasitoid systems. *Ecology* **71**, 422–426.

Rose, A. H., and Lindquist, O. H. (1984). Insects of eastern pines. *Can. For. Serv., Publ.* **1313**.

Rose, A. H., and Lindquist, O. H. (1985). Insects of eastern spruces, fir and hemlock. *Can. For. Serv., Tech. Rep.*. **23**.

Rose, A. H., and Lindquist, O. H. (1992). Insects of eastern larch, cedar and juniper. *Can. For. Serv., Tech. Rep.* **28**.

Scoble, M. J. (1991). Classification of the Lepidoptera. *In* "The Moths and Butterflies of Great Britain and Ireland" (A. M. Emmet and J. Heath, eds.), Vol. 7, Part 2, pp. 11–45. Harley Books, Colchester, UK.

Scott, J. A. (1975). Flight patterns among eleven species of diurnal Lepidoptera. *Ecology* **56**, 1367–1377.

Scott, J. A. (1984). The phylogeny of butterflies (Papilionoidea and Hesperoidea). *J. Res. Lepid.* **23**, 241–281.

Shreeve, T. G. (1981). Flight patterns of butterfly species in woodlands. *Oecologia* **51**, 289–293.

Sillén-Tullberg, B. (1993). The effect of biased inclusion of taxa on the correlation between discrete characters in phylogenetic trees. *Evolution (Lawrence, Kans.)* **47**, 1182–1191.

Sillén-Tullberg, B., and Hunter, A. F., (1995). Evolution of larval gregariousness in

relation to repellent defenses and warning coloration in tree-feeding Macrolepidoptera: A phylogenetic analysis based on independent contrasts. *Biol. J. Linn. Soc.* In
press.

Skou, P. (1986). The geometroid moths of North Europe. *Entomonograph* **6,** 1–348.

Spitzer, K., Rejmanek, M., and Soldan, T. (1984). The fecundity and long-term variability in abundance of noctuid moths (Lepidoptera, Noctuidae). *Oecologia* **62,**
91–93.

Stokoe, W. J., and Stovin, G. H. T. (1948). "The Caterpillars of British Moths Including
the Eggs, Chrysalids, and Food Plants," Vols. 1 and 2. Frederick Warne and Co.,
London.

Taylor, L. R., and Woiwod, J. P. (1980). Temporal stability as a density-dependent species
characteristic. *J. Anim. Ecol.* **49,** 209–224.

Vance, R. R. (1980). The effect of dispersal on population size in a temporally varying
environment. *Theor. Popul. Biol.* **18,** 343–362.

Watson, A. (1992). A red grouse perspective on dispersal in small mammals. *In* "Animal
Dispersal: Small Mammals as a Model" (N. C. Stenseth and W. Z. Lidicker, Jr.,
eds.), pp. 260–273. Chapman and Hall, London.

Chapter 4

Spatial Behavior and Temporal Dynamics of Outbreak and Nonoutbreak Species

Naomi Cappuccino, Hans Damman, and Jean-François Dubuc

I. Introduction

The factors that cause populations of certain species to outbreak and, conversely, that keep other species from outbreaking have long interested population ecologists. Whereas some authors have stressed the importance of external forces, such as natural enemies, resource availability, and climatic conditions, in driving population dynamics, others have sought the answer to population dynamics in the characteristics of the species themselves. Comparative studies of outbreak and nonoutbreak herbivores have revealed a variety of life-history traits that are associated with a tendency to erupt (e.g., Spitzer et al., 1984; Gaston and Lawton, 1988a,b; Redfearn and Pimm, 1988; Pimm, 1991). Although the findings vary substantially from one review to the next, the one striking pattern that has emerged is the consistent correlation of feeding in aggregations and a greater tendency to outbreak (Watt, 1965; Knerer and Atwood, 1973; Hanski, 1987; Nothnagle and Schultz, 1987; Wallner, 1987; Hunter, 1991, this volume; Root and Cappuccino, 1992; Haack and Mattson, 1993; Larsson et al., 1993). The connection between feeding in aggregations and a greater tendency to outbreak appears predictably and conspicuously, but surprisingly little effort has been devoted to determining whether the association is merely coincidental or indicates that spatial behavior can drive population dynamics.

Levels of aggregation in herbivores may vary greatly. In species that lay their eggs in groups, the larvae may live truly gregariously and remain in a tight group as they feed, or they may disperse upon hatching (e.g., Tsubaki and Shiotsu, 1982; Damman, 1991). In the latter case, cooperation between larvae in feeding or defense becomes unlikely, but the generally poor dispersal abilities of larvae (Dethier, 1959; Cain et al., 1985) should keep local densities higher than those typically attained in species that lay their eggs singly.

Several consequences of aggregation by insect herbivores could explain the tendency of aggregating species to outbreak. Organisms that live in groups may enjoy greater protection from their natural enemies. Truly gregarious species may cooperate to increase the effectiveness of defensive behaviors (Vulinec, 1990; Codella and Raffa, 1993). Even species that loosely aggregate simply because they hatched from a cluster of eggs may dilute their risk of being killed by enemies that search in a density-independent or inversely density-dependent fashion and that therefore do not accumulate on patches of their victims (Hamilton, 1971; Cappuccino, 1988; Turchin and Kareiva, 1989). Stiling (1987) and Walde and Murdoch (1988) found that predation and parasitism often act in a density-independent or inversely density-dependent manner.

In contrast, density-dependent predation and parasitism may favor the evolution of egg-spreading and solitary feeding (Stamp, 1980). Disease, although less well studied than predation and parasitism, is likely to provide an even more important selective agent for egg-spreading, since pathogen transmission is necessarily density dependent (Dwyer, 1992). Brower (1958) has suggested that feeding as widely scattered individuals may also enhance the ability of herbivores to hide from specialist predators or parasites. However, these potential advantages of egg-spreading do not seem to translate into eruptive population dynamics.

Plant-mediated components of herbivore success create a second potential connection between aggregation and outbreaks. Rhoades (1985) hypothesized a relationship between aggregation, population dynamics, and the induction of host–plant defenses. The gist of Rhoades' (1985) view was that herbivores that were sensitive to chemical defenses induced by the plant in response to feeding damage were constrained to live at low densities: at sufficiently low densities they caused too little damage to elicit production of the defenses (Rhoades referred to these insects as "stealthy" herbivores). In contrast, "opportunistic" herbivores that either tolerated the induced defenses or avoided them by feeding on already weakened plants could reach much greater densities. For example, eruptive sawflies showed a much smaller variation in developmental rate as food-plant quality varied than did nonoutbreak species (Hanski and Otronen, 1985). Bark beetles, which rely on active aggregation to reach densities at which they can overwhelm the resin defenses of their coniferous hosts (e.g., Berryman, 1972; Reeve *et al.*, Chapter 17, this volume), may represent an extreme version of Rhoades' (1985) "opportunistic" species. Rhoades (1985) suggested that as a consequence of being freed from the constraint of feeding inconspicuously, "opportunistic" herbivores stood a much greater chance of outbreaking than "stealthy" ones.

Most investigations of the comparative advantages of aggregation and solitary living in insect herbivores have focused on the average performance of a female's offspring. However, the variability in the performance of broods deposited in different spatial patterns may prove equally important. When the suitability of sites for larval survival and development varies spatially and unpredictably, a female may benefit by "spreading the risk" and placing her eggs in several sites, thereby increasing the chances that at least some of her offspring will survive. The idea of "risk-spreading" or "bet-hedging," first formalized by den Boer (1968) as a way of explaining population persistence without recourse to density-dependent regulation, was applied by Strathmann (1974) to the dispersal strategies of marine organisms. Root and Kareiva (1984, 1986) invoked risk-spreading to explain the straight-line oviposition flight of *Pieris rapae,* which spreads the offspring of a single female widely in space.

In two of the rare attempts to consider the connection between variance in performance and population dynamics, Hanski (1987) and Nothnagle and Schultz (1987) applied the concept of risk-spreading to the cases of pine-feeding sawflies and to forest Lepidoptera, respectively. Both suggested that outbreak species tend to be "risk-concentrators." Species that clump their eggs have greater spatial variability in success, and this spatial variability may lead to increased temporal variability. Hanski (1987) illustrated this with a simple model of sawfly dynamics, in which the risk of escaping the control of predators increased as a function of between-group variance in survivorship.

In this chapter we attempt to evaluate the notion that spatial behavior can influence the potential for outbreaks. In our basic experimental design we compare the performance of related organisms that feed either individually or in aggregations on a shared host plant after manipulating their spatial patterns in the field. We will consider both the mean and the variance in response to either natural enemies or plant-based characteristics. Between us we have performed such experiments on three different groups of organisms: goldenrod-feeding aphids, goldenrod-feeding chrysomelid beetles, and birch-feeding moths. We specifically test three hypotheses linking spatial behavior to population dynamics: (1) Aggregating species should suffer lower mortality from natural enemies when placed onto plants at high densities as compared to low densities, whereas the reverse should hold for solitary feeders. (2) Aggregating species should show a weaker response to induced changes in plant quality than solitary feeders. (3) Both aggregating and nonaggregating species should exhibit more variable performance when placed in a clumped spatial pattern.

II. Comparative Studies

A. Goldenrod Aphids

Two aphid species, *Uroleucon nigrotuberculatum* (Olive) and *U. caligatum* (Richards), coexist on goldenrod (*Solidago* spp.) throughout much of eastern North America. The spatial patterns, survivorship, and population dynamics of these aphids have been studied extensively by one of us (Cappuccino, 1987,

1988) as well as by other authors (e.g., Kareiva, 1984; Edson, 1985; Pilson, 1989). Although *U. nigrotuberculatum* and *U. caligatum* have similar life histories and share many natural enemies, they differ markedly in their population dynamics and aggregative behavior. Both species feed in colonies, as is typical for aphids, but *U. nigrotuberculatum* forms much larger colonies and its distribution in goldenrod fields is much more clumped than that of *U. caligatum* (Edson, 1985; Cappuccino, 1988). The greater clumping of *U. nigrotuberculatum* results from both the active aggregation of winged forms and the more sedentary nature of the wingless forms (Cappuccino, 1987, 1988). In addition to being clumped in space, *U. nigrotuberculatum* was also more variable in time than *U. caligatum* (Cappuccino, 1988). Although local outbreaks of *U. nigrotuberculatum* occur commonly in upstate New York, as well as elsewhere in its range, we have yet to observe an outbreak of *U. caligatum*.

To understand the role of spatial behavior in determining aphid performance and, ultimately, population dynamics, Cappuccino (1988) imposed different spatial strategies on the two species in a series of field experiments. Colony size had little effect on the fecundity and growth rates of either species, but did affect survivorship. Nymphs clumped together on a single host plant suffered much less predation from generalist predators than the same number of nymphs scattered widely in small colonies. Also interesting, however, was the variability in survivorship of aphids subjected to these two treatments. The strategy of *U. nigrotuberculatum* adults, that of placing all offspring on a single goldenrod ramet, led to a highly variable success rate. The more mobile life-style of *U. caligatum*, leading to the scattering of offspring over many small colonies, although on average less rewarding, was nevertheless more predictable. Behaviorally, *U. caligatum* fit the pattern of a "risk spreader" well.

B. Goldenrod Beetles

Four leaf beetles (Coleoptera: Chrysomelidae) commonly attack goldenrod in eastern North America: *Trirhabda borealis* Blake, *Microrhopala vittata* Fabr., *Ophraella conferta* (LeConte), and *Exema canadensis* Pierce. *Trirhabda borealis*, common in Ontario, is replaced by *T. virgata* LeConte as the dominant leaf-chewing beetle on goldenrod farther south. During 10 years of sampling the goldenrod fauna at sites throughout upstate New York, Ontario, and northwestern Québec, three of these genera, *Trirhabda, Microrhopala,* and *Ophraella,* reached densities at which they severely damaged their host plant (Root and Cappuccino, 1992; Cappuccino, 1991; N. Cappuccino and H. Damman, unpublished data). In Ithaca, New York, a 6-year-long census of the goldenrod herbivores indicated that *T. virgata* and *Microrhopala* dominated the spring fauna and often attained herbivore loads at which their biomass ranged from 1 to 10% of the host plant's leaf biomass (Root and Cappuccino, 1992).

Exema, although present at most sites, had a much lower impact; on only two occasions did its biomass exceed even .01% of the leaf biomass. *Ophraella* was not included in this census because it fed at night, but a rough estimate of the impact of *Ophraella* based on its distinctive feeding damage suggested that it ranked just below *Microrhopala* in abundance.

The three outbreak species, *Trirhabda*, *Microrhopala*, and *Ophraella*, all aggregate at some stage of the life cycle. Female *Microrhopala* may cluster over 25 eggs on a single leaf in feces-covered stacks of 2 to 4. On emerging, the larvae mine the goldenrod leaves gregariously. *Trirhabda* and *Ophraella* lay their eggs in clumps of up to 25 and 35, respectively (H. Damman, unpublished data). Once the larvae emerge, however, they do not actively congregate, but rather feed independently of one another at densities that loosely reflect the distribution of eggs.

Exema, the nonoutbreak species, lives solitarily at all stages of development. During a 2-month-long oviposition period, females scatter up to 80 eggs over an entire field, suggesting that they spread the risk in both space and time. As larvae, *Exema* carry protective, portable cases resembling *Trirhabda* feces, and as adults they mimic caterpillar feces (Root and Messina, 1983). *Exema* larvae do small amounts of damage to many leaves, thus fitting Rhoades' (1985) "stealthy" feeding style. They feed infrequently, but efficiently, and move often (Damman, 1995).

1. Effect of Group Size on *Microrhopala* Larvae

The gregariousness of *Microrhopala* eggs and larvae only indirectly affects their vulnerability to natural enemies and does not influence how well they develop on their host plant. Increasing the number of egg masses laid on a plant did not alter the proportion of eggs destroyed by predators or parasitoids (Damman and Cappuccino, 1991). However, a greater proportion of larvae disappeared between hatching and pupation in egg masses in which all but one individual was killed than on plants having two unmanipulated egg masses (Damman, 1994). *Microrhopala* larvae mining leaves in groups of one, three, or nine larvae completed development equally quickly and attained indistinguishable adult weights (Damman, 1994). The main effect of group size on larval success is that within-species competition forces larvae feeding in larger groups to abandon the natal mine more quickly, increasing the chances of predation during the risky move over the plant surface to the next available, unmined leaf (Damman, 1994). These studies do not provide any information on the variability of survivorship on plants harboring *Microrhopala* at various group sizes.

2. Effect of Aggregation on *Ophraella* Egg Survivorship

We studied *Ophraella conferta* in an old field in northwestern Québec where the females laid eggs on *Solidago rugosa*. The mean number of eggs per cluster was 11.8 ± 0.58 ($\bar{X} \pm$ SE). Twelve replicates of nine egg clusters were found. One randomly chosen egg cluster in each replicate was assigned to the clumped (eight eggs together) treatment; eggs in excess of eight were culled from the cluster with a pin. We assigned the other eight clusters in a replicate to the solitary treatment; all eggs but one were removed from each of these clusters to result in eight solitary eggs. The eggs were checked every 2 days until hatching began.

Eggs in the clumped treatment showed a survivorship no better on average than that for solitary eggs ($\bar{X}_{\text{spread out}} = .438$, $\bar{X}_{\text{clumped}} = .344$; *t*-test: $t_{22} = .705$, $P = .48$). The variance in survivorship of the clumped eggs did not differ from that for the solitary eggs (Bartlett's test: $\chi^2_{\text{adj}} = 2.07$, $P = .14$).

3. Effect of Aggregation on *Trirhabda* and *Exema* Larvae

Trirhabda and *Exema* living in old fields in the vicinity of Ottawa, Ontario, Canada, use *Solidago canadensis* as their main host plant. We assigned *Solidago* ramets randomly to one of four beetle-density treatments in 12 replicate blocks: (1) clumped, uncaged treatment (1 ramet); (2) clumped, caged treatment (1 ramet); (3) solitary, uncaged (10 ramets); and (4) solitary, caged (5 ramets). The 10 larvae aggregated on ramets in the clumped treatments reflected the maximum densities reached by *T. borealis* in the Ottawa area. The single larvae placed on ramets in the solitary treatments corresponded to the density typical of *Exema*. The cages placed around the plants in the caged treatments excluded both aerial and terrestrial predators and parasites. *Trirhabda* larvae were collected and placed on ramets as second instars. *Exema* larvae were placed out at a fecal-case length of 3 mm. We followed the larvae for 10 days, after which we collected the survivors and reared them for parasites in the laboratory. In the field, *Trirhabda* and *Exema* larvae that disappeared from the original ramet either moved to adjacent ramets or were never seen again despite thorough searches of neighboring vegetation. Thus, larvae not found within 25 cm of the original ramet were recorded as dead.

Both *Trirhabda* and *Exema* larvae in the clumped treatment, whether caged or uncaged, disappeared at a higher rate than those in the solitary treatment (Fig. 1), suggesting that aggregation increased mortality. A greater proportion of the uncaged as compared to caged larvae disappeared (Fig. 1), indicating that natural enemies represented an important cause of death. A tachinid parasite of *Trirhabda, Opsomeigenia xylota* (Curran), killed an additional 50% of the larvae surviving to pupation (Fig. 2). Although we saw a chalcidoid wasp, *Eutetrastichus chlamytis* (Ashmead), ovipositing on *Exema* larvae in the field (twice on larvae feeding at high density and twice on larvae feeding solitarily), no adults of any parasite ever emerged from *Exema* larvae brought into the lab.

Although predation and parasitism represented important mortality sources, they could not account for the difference in mortality in the clumped and solitary

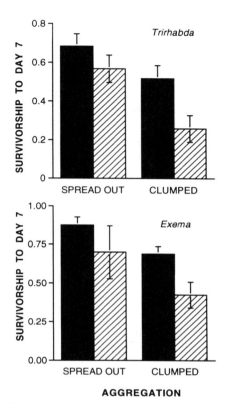

Figure 1. Survivorship of *Trirhabda borealis* and *Exema canadensis* larvae protected from (solid bars) and exposed to (hatched bars) predators when feeding spread out at a density of 1 per plant and clumped at a density of 10 per plant. For both species the Protection × Density interaction was not significant ($P > .5$), and the Protection and Density main effects were significant ($P < .05$). Bartlett's test for the homogeneity of variances for the exposed larvae only indicated no difference for *Trirhabda* ($\chi^2_{adj} = .68$, d.f. $= 1$, $P = .41$) and a significantly greater variance in the spread out treatment for *Exema* ($\chi^2_{adj} = 4.93$, d.f. $= 1$, $P = .03$).

treatments. The lack of an interaction between aggregation and caging indicated that predators took equal proportions of solitary and clumped larvae (Fig. 1). *Opsomeigenia* also attacked equal proportions of clumped and solitary larvae (Fig. 2). The greater disappearance of clumped larvae did not stem from cannibalism because larvae held in the lab at much higher densities never attacked each other. These results suggest that high densities of larvae either reduced plant quality or stimulated dispersal of both the outbreak and nonoutbreak species.

The variances in the likelihood of predation and parasitism for solitary and clumped larvae provide little support for the idea that either *Trirhabda* or *Exema*

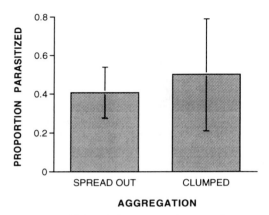

Figure 2. Parasitism rate of *Trirhabda borealis* larvae exposed to natural enemies when feeding spread out at a density of 1 per plant and clumped at a density of 10 per plant. Parasitism rates were equally high (ANOVA: $F_{1,10} = .11$, $P = .74$) and variable (Bartlett's test for homogeneity of variances: $\chi^2_{adj} = 1.63$, d.f. $= 1$, $P = .20$) at both densities.

could spread the risk by spreading their eggs (Fig. 1). In fact, solitary *Exema* larvae showed a greater variance in survivorship than the aggregated ones. In contrast to the aphid, *U. nigrotuberculatum*, aggregation proved more of a hindrance than a help to both the solitary and clumped individuals. For *Trirhabda* in particular, this is a curious result considering the egg-clumping behavior of the adults.

4. *Exema* as a Stealthy Feeder?

To examine the possibility that *Exema* larvae move around on the host plant to avoid damaged-induced changes, we compared the behavior and performance of larvae on ramets with simulated chewing damage to that of larvae on control ramets. Chewing damage was simulated by damaging the leaves with a hole punch over a 4-day period, ultimately removing 25% of the area of every second leaf. Adjacent undamaged stems served as controls. Once the damage was complete, we placed 4-day-old *Exema* larvae on the damaged and control ramets. Larvae on odd-numbered ramets were placed on a damaged leaf, whereas those on even-numbered ramets started out on an undamaged leaf. We checked the larvae daily, recording their position on the plant relative to the release point as well as the condition of the leaves on which we found them.

Exema larvae changed position on the plant every 3 or 4 days on average. They did not move more often on damaged plants (arcsin-transformed moves/day; ANOVA: $F_{1,26} = .787$; $P = .38$), nor was the mean net displacement greater for larvae on damaged plants (ANOVA: $F_{1,26} = .817$; $P = .37$). Roughly half of the larvae survived to Day 10, at which time survivorship did not differ significantly between the two treatments (G_{adj} = .0002, d.f. = 1, P = .98). On the damaged plants, 33% of the larvae starting out on a damaged leaf and 40% of the larvae placed on an undamaged leaf moved within the first 5 days (frequencies not significantly different; Monte Carlo R × C tests; P ≈ 1.00). Finally, of seven pairs of larvae remaining after Day 7, fecal cases of larvae on damaged plants were larger for three pairs, smaller for two pairs, and not distinguishably different from those on control plants for the remaining two pairs. Clearly, *Exema* larvae do not respond behaviorally or developmentally to damage. It remains possible that, although they cannot sense the difference between damaged and undamaged leaves, they are nevertheless programmed to move to avoid damage.

C. Birch-Feeding Caterpillars

Paper birch, *Betula papyrifera* Marsh, hosts many species of Lepidoptera, of which only a few are reported to cause damage that qualifies them as outbreak species (Rose and Lindquist, 1982). Little is known about the dynamics of the species not considered to be eruptive, although information is available on the oviposition habits of the adults and on larval feeding habits. Of nine outbreak species reported in Rose and Lindquist (1982) and Martineau (1985), 6 are either egg clumpers or group feeders, whereas only 3 of the 22 nonoutbreak species for which sufficient information was available aggregated in the egg or larval stage (Fisher's exact test: P = .03). In addition to feeding singly, many of the nonoutbreak species also disperse their damage: larvae generally move to a new feeding site well before depleting their food supply. In the following, we describe a series of experiments aimed at comparing the response to damage of three nonoutbreak species and one outbreak species on birch.

The forest tent caterpillar *Malacosoma disstria* Hübner (Lasiocampidae) is a widespread generalist defoliator that erupts somewhere in Canada each year (Rose and Lindquist, 1982). Females lay eggs in masses that ring the branchlets of the host plant. *Malacosoma* overwinters as first instars inside the egg, emerging at budburst in spring. The larvae initially feed in groups, completely defoliating one branch before moving on to another. Toward the end of their development the larvae become more solitary.

Enargia infumata Grote (Noctuidae), which feeds on birch and trembling aspen, is widespread but generally rare (Rose and Lindquist, 1982). The larvae, which even when common do not occur more than one to a branch (J.-F. Dubuc, unpublished data), tie pairs of expanding leaves together with silk to make sandwich-style shelters. Larvae change feeding sites often, making five or six shelters over the course of their development.

Two pyralid moths, *Ortholepis pasadamia* Dyar and *Acrobasis betulella*

Hulst, also commonly feed on paper birch. These two species do not outbreak (Rose and Lindquist, 1982), although *Acrobasis* has fluctuated to a greater extent than *Ortholepis* at our study sites (N. Cappuccino, unpublished data). These two species have similar life histories. Both overwinter as second-instar larvae in hibernacula on the branchlets. Both emerge at budburst and tie the leaves together to form shelters. *Ortholepis* larvae move to new leaf shelters several times during development. In contrast, *Acrobasis* larvae may feed from their original shelters throughout larval development, though they will move to a new feeding site if they deplete all the surrounding leaves. Both pyralids pupate in June in northwestern Québec and the adults emerge in July. As befits the more stealthy life-style of the larvae, *Ortholepis* females lay eggs in batches of two to four eggs. *Acrobasis,* on the other hand, lays batches of up to 25 eggs, often in the persistent leaf ties made by the postdiapause larvae or in shelters made by other species (Cappuccino, 1993).

Like *Exema* on goldenrod, *Enargia* and *Ortholepis* have behaviors that fit the "stealthy" feeding syndrome of Rhoades (1985), whereas the gregarious *Malacosoma* corresponds well to the "opportunistic" syndrome. Although not an eruptive species, the more gregarious and more variable *Acrobasis* also falls toward the opportunistic end of the spectrum, at least when compared to the confamilial *Ortholepis*. If the stealthy/opportunistic dichotomy accurately describes the differences between these species, then we would predict *Enargia* to be more sensitive to damage-induced reductions in plant quality than *Malacosoma* and for *Ortholepis* to be more sensitive than *Acrobasis*. To test these predictions, we reared larvae of all four species in the field on damaged and undamaged birch branches. As in the experiment on *Exema*, we used a hole punch to damage leaves, removing approximately 25% of the leaf area at every other leaf node over the course of 4 days. Second-instar larvae were introduced to the damaged branches and nearby controls and protected from predators with nylon sleeve-cages. *Malacosoma* and *Enargia* were placed three larvae to a branch, whereas the two pyralids were introduced one per branch. *Malacosoma* and *Enargia* were transferred to new damaged and undamaged branches when they had eaten about 40% of the foliage.

The three *Enargia* larvae on each branch avoided each other; only once were two larvae found feeding at the same leaf node. Moreover, on the damaged branches, they avoided damaged leaves, concentrating their feeding on undamaged leaf nodes (Table 1). The two pyralid species also showed a tendency to avoid damaged leaves. *Malacosoma*, on the other hand, showed no preference for either undamaged or damaged foliage (Table 1).

Although the birch-feeding caterpillars showed the predicted behavioral response to previous plant damage, we found no clear indication that the behaviors reflected a greater sensitivity of the stealthy feeders to plant quality. *Malacosoma* males and, surprisingly, females of the seemingly "stealthy" *Enargia* attained

TABLE 1

Response of the Birch-Feeding Lepidoptera *Malacosoma disstria*, *Enargia enfumata*, *Acrobasis betulella*, and *Ortholepis pasadamia* to Damaged and Undamaged *Betula papyrifera* Branches

| | No. branches where larvae had eaten more: | | | Pupal weight (mg) $\bar{X} \pm SE$ | | | | | |
| | | | | Males | | | Females | | |
	Damaged leaves	Undamaged leaves	P^a	Damaged leaves	Undamaged leaves	P^b	Damaged leaves	Undamaged leaves	P^b
Malacosoma	12	12	1.0	314 ± 10	284 ± 11	.044	509 ± 20	481 ± 26	.667
Enargia	6	16	.03	209 ± 6	193 ± 8	.179	242 ± 17	218 ± 11	.034
Acrobasis	7	15	.07	43 ± 1	40 ± 4	.722	37 ± 4	44 ± 2	.141
Ortholepis	6	13	.06				30 ± 1	30 ± 3	.569

[a]One-tailed binomial probability of a result equally or more extreme.
[b]Mann–Whitney U-test.

greater pupal weight on damaged branches (Table 1). There were no significant differences in developmental rates between species for any species (Mann–Whitney U-tests, $P > 0.1$).

III. Toward an Explanation for Greater Temporal Variability in Species with Higher Spatial Variability

A. The Consequences of Feeding in Aggregations or Alone

There exists a vast and varied literature on the effect of feeding alone or in aggregations on survivorship and fecundity (Stamp, 1980; Courtney, 1984; Vulinec, 1990; Codella and Raffa, 1993). As noted in the Introduction, a second literature clearly associates feeding in groups with a greater tendency to undergo dramatic population outbreaks. In this study, however, we found only inconsistent evidence of a causal connection between the aggregative behavior and the potential for herbivorous insects to outbreak.

Though aggregation can enhance the ability of normally aggregating species to avoid predators and parasites (Vulinec, 1990; Codella and Raffa, 1993), and should increase the vulnerability of normally solitary species to their natural enemies (Brower, 1958), we found little evidence to suggest that this represents a general pattern (Table 2). Only the outbreaking, group-feeding aphid, *Uroleucon nigrotuberculatum*, conformed to expectations in that it more successfully avoided predators when feeding in large aggregations. For the goldenrod beetles, we either found no effect of degree of aggregation on mortality from natural enemies or we found a pattern opposite to what we expected: *Exema*, normally a solitary species, suffered both a greater mean and variance in mortality when feeding alone than when feeding at atypically high densities.

As with the results of the enemy-avoidance experiments, no clear evidence emerged linking feeding in groups with an ability to circumvent or tolerate plant-based obstacles. All the solitary birch-feeding moths avoided damaged leaves, but the only solitary goldenrod beetle, *Exema*, did not. Although the outbreaking *Malacosoma* appeared to benefit from feeding on previously damaged birch foliage, so did the nonoutbreaking, solitary *Enargia*. Thus the notion that non-outbreak species may be more sensitive to damage-induced changes in plant quality was not supported in this study.

Avoiding damage may also reduce the risk of predation or parasitoids if enemies cue in on damaged leaves to find their prey or hosts (Heinrich, 1979; Heinrich and Collins, 1983). To date, however, studies of the effect of feeding damage on predation and parasitism rates have failed to reveal higher mortality on damaged plants (Bergelson and Lawton, 1988; Hawkins, 1988). We would

TABLE 2
Overview of the Consequences of Aggregation for Goldenrod and Birch Herbivores

		Enemy-mediated mortality[a]				Plant-mediated effects		
		Predators		Parasites				
	Population dynamics	$\bar{X}$	s^2	$\bar{X}$	s^2	Development	Fecundity	Other patterns
U. nigrotuberculatum	Outbreak	A > S	A > S	=		=	=	
U. caligatum	Nonoutbreak					=	S > A	
Trirhabda	Outbreak	=	=	=	=	=		Mine as a group
Microrhopala	Outbreak	=		0		=	=	
Ophraella	Outbreak	=	=			=		
Exema	Nonoutbreak		S > A	0		=		Ignores damage
Malacosoma	Outbreak					=	A > S (males)	Ignores damage
Enargia	Nonoutbreak					=	A > S (females)	Avoids damage
Acrobasis	Nonoutbreak					=	=	Avoids damage
Ortholepis	Nonoutbreak					=	=	Avoids damage

[a] A, aggregated; S, solitary; =, no difference; 0, factor had no effect on performance.

even predict the converse for the goldenrod beetle *Exema*. If *Exema* larvae benefit from resembling *Trirhabda* excrement and if adults are protected by mimicking caterpillar frass, this species may enjoy better survivorship on a plant bearing a great deal of frass, which would necessarily bear a great deal of damage as well. In fact, *Exema* and *Trirhabda* abundances correlated positively in a survey of the goldenrod fauna in central New York (R. B. Root and N. Cappuccino, unpublished data). Clearly, *Exema* does not feed stealthily to avoid damage-induced deterioration in host plant quality.

Hanski (1987) has suggested that the key to the link between temporal variability (outbreaks) and spatial variability (aggregation) may lie in the greater between-group variance in survival experienced by gregarious species. Gregarious feeders may, by chance, enjoy higher patch survivorship, allowing them to escape the control of their natural enemies more often. This may be the case for the aphid *U. nigrotuberculatum* (Cappuccino, 1987). However, *Trirhabda* and *Ophraella,* which clump their eggs, did not show a higher variance in survivorship when feeding in groups, and the normally solitary larvae of *Exema* experienced significantly more variable mortality when solitary than when aggregated.

B. Connecting Spatial Behavior and Population Dynamics

Although we can easily examine the costs and benefits of the gregarious and solitary life-styles, it is harder to forge the link between the consequences of group living for a female's offspring and the likelihood that populations of that species will erupt. Given the general lack of support for either enemy-based or plant-based causal links between aggregation and the tendency for outbreaks, the reasons why aggregation so consistently correlates with patterns of population dynamics remain elusive. One potential explanation is that the tendency to aggregate or to occur singly may correlate with other aspects of the life history that drive the population dynamics. For example, Gilbert (1991) discusses the potential pathways in the evolution of the life-history traits, including egg-spreading and low population densities, that characterize the pollen-feeding heliconiine butterflies of the New World tropics. Pollen feeding, in having shifted the burden of resource acquisition to the adult, may have allowed larvae to feed on smaller plant parts such as meristems, which were, in turn, more effectively exploited by the deposition of single eggs. Both the territoriality of the pollen-feeding adults and the utilization of meristems by the larvae may limit population size in these species. Similarly, correlates of aggregation such as size (Hunter, Chapter 3, this volume) or the ability to tolerate variable plant quality (e.g., Hanski and Otronen, 1985) could increase the probability of outbreaks.

At the same time, attempts to bridge the gap between small-scale spatial patterns and large-scale temporal patterns require making assumptions about processes at the larger scale that are difficult to study. For example, when *U. nigrotuberculatum* feeds in aggregations on stems, individuals on some stems enjoy a safety in numbers because predators do not disproportionately accumulate in patches of high prey density. What would happen, however, if the density of the whole aphid population increased? It is entirely possible that at this level predators have a numerical response; they may reproduce more, for example, in fields with high overall aphid density. This could have a regulatory effect on overall aphid density. By studying spatial density dependence at the per-stem scale, we would not see this potentially regulatory temporal response.

The most appropriate experiment to test the influence of spatial behavior on dynamics would be one in which entire replicate populations were submitted to spatial-strategy treatments and the resultant dynamics recorded. The spatial scale would need to encompass the population of the most widely ranging species in the interaction—usually this is the predator or parasite. This sort of experiment, difficult for large animals, may prove feasible for small organisms with short generation times such as mites in greenhouses, although the difficulty may lie in performing the manipulations on such small species.

Some may view space as the final frontier (Kareiva, 1994), but we would argue that it is at the crossroads of space and time that population biologists will meet their greatest challenge. Forging the link between spatial patterns and temporal dynamics has been and continues to be a major impasse for population biologists. This is best illustrated by the often heated debate over if and how spatial patterns in predation and parasitism (i.e., whether mortality is density dependent, inversely density dependent, or simply variable in space) can lead to regulatory temporal density dependence (Strong, 1988; Taylor, 1993). How the spacing behaviors of organisms translate into temporal stability or fluctuations will provide the same frustrations for ecologists, unless we take bold steps to resolve the issue. What we have tried to do here is illustrate some of the experimental approaches that will allow us to take the final step toward understanding how spatial behavior influences population dynamics.

Acknowledgments

We thank A. M. Farmakis Velderman, S. Goodacre, D. Lavertu, I. Tessier, and L. Williams for help in the field. A. Hunter and T. Ohgushi provided thoughtful comments that helped greatly in revising earlier versions of the manuscript. This work was supported by a Natural Sciences and Engineering Council of Canada (NSERC) operating grant to H. D., an NSERC Women's Faculty Award to N. C., and a fellowship from the Ministère de l'enseignement supérieure et de la science (Québec) to J-F. D.

References

Bergelson, J., and Lawton, J. H. (1988). Does foliage damage influence predation on the insect herbivores of birch? *Ecology* **69**, 434–445.

Berryman, A. A. (1972). Resistance of conifers to invasion by bark-beetle–fungus associations. *BioScience* **22**, 598–602.

Brower, L. P. (1958). Bird predation and foodplant specificity in closely related procryptic insects. *Am. Nat.* **92**, 183–187.

Cain, M. L., Eccleston, J., and Kareiva, P. M. (1985). The influence of food plant dispersion on caterpillar searching success. *Ecol. Entomol.* **10**, 1–7.

Cappuccino, N. (1987). Comparative population dynamics of two goldenrod aphids: Spatial patterns and temporal constancy. *Ecology* **68**, 1634–1646.

Cappuccino, N. (1988). Spatial patterns of goldenrod aphids and the response of enemies to aphid patch density. *Oecologia* **76**, 607–610.

Cappuccino, N. (1991). Mortality of *Microrhopala vittata* (Coleoptera: Chrysomelidae) in outbreak and non-outbreak sites. *Environ. Entomol.* **20**, 865–871.

Cappuccino, N. (1993). Mutual use of leaf-shelters by lepidopteran larvae on paper birch. *Ecol. Entomol.* **18**, 287–292.

Codella, S. G., and Raffa, K. F. (1993). Defense strategies of folivorous sawflies. *In* "Sawfly Life History Adaptations to Woody Plants" (M. R. Wagner and K. F. Raffa, eds.), pp. 261–294. Academic Press, San Diego, CA.

Courtney, S. P. (1984). The evolution of egg clustering by butterflies and other insects. *Am. Nat.* **123**, 276–281.

Damman, H. (1991). Oviposition behaviour and clutch size in a group-feeding pyralid moth, *Omphalocera munroei*. *J. Anim. Ecol.* **60**, 193–204.

Damman, H. (1994). Defense and development in a gregarious leaf-mining beetle. *Ecol. Entomol.* **19**, 335–343.

Damman, H. (1995). Submitted for publication.

Damman, H., and Cappuccino, N. (1991). Two forms of egg defence in a chrysomelid beetle: Egg clumping and excrement cover. *Ecol. Entomol.* **16**, 163–167.

den Boer, P. J. (1968). Spreading of risk and stabilization of animal numbers. *Acta Biotheor.* **18**, 165–194.

Dethier, V. G. (1959). Food-plant distribution and density and larval dispersal as factors affecting insect populations. *Can. Entomol.* **88**, 581–596.

Dwyer, G. (1992). On the spatial spread of insect pathogens: Theory and experiment. *Ecology* **73**, 479–494.

Edson, J. L. (1985). The influences of predation and resource subdivision on the coexistence of goldenrod aphids. *Ecology* **66**, 1736–1743.

Gaston, K. J., and Lawton, J. H. (1988a). Patterns in the distribution and abundance of insect populations. *Nature (London)* **331**, 709–712.

Gaston, K. J., and Lawton, J. H. (1988b). Patterns in body size, population dynamics, and regional distribution of bracken herbivores. *Am. Nat.* **132**, 662–680.

Gilbert, L. E. (1991). Biodiversity of a Central American *Heliconius* community: Pattern, process, and problems. *In* "Plant–Animal Interactions: Evolutionary Ecology in Tropical and Temperate Regions" (P. W. Price, T. M. Lewinsohn, G. W. Fernandes, and W. W. Benson, eds.), pp. 403–427. Wiley, New York.

Haack, R. A., and Mattson, W. J. (1993). Life history patterns of North American tree-feeding sawflies. *In* "Sawfly Life History Adaptations to Woody Plants" (M. R. Wagner and K. F. Raffa, eds.), pp. 504–546. Academic Press, San Diego, CA.

Hamilton, W. D. (1971). Geometry for the selfish herd. *J. Theor. Biol.* **31**, 295–311.

Hanski, I. (1987). Pine sawfly population dynamics: Patterns, processes, problems. *Oikos* **50**, 327–335.

Hanski, I., and Otronen, M. (1985). Food quality induced variance in larval performance: Comparison between rare and common pine-feeding sawflies (Diprionidae). *Oikos* **44**, 165–174.

Hawkins, B. A. (1988). Foliar damage, parasitoids, and indirect competition: A test using herbivores of birch. *Ecol. Entomol.* **13**, 301–308.

Heinrich, B. (1979). Foraging strategies of caterpillars. *Oecologia* **42**, 325–337.

Heinrich, B., and Collins, S. L. (1983). Caterpillar leaf damage and the game of hide-and-seek with birds. *Ecology* **64**, 592–602.

Hunter, A. F. (1991). Traits that distinguish outbreaking and nonoutbreaking Macro-lepidoptera feeding on northern hardwood trees. *Oikos* **60**, 275–282.

Kareiva, P. (1984). Predator–prey dynamics in spatially-structured populations: Manipulating dispersal in a coccinellid–aphid interaction. *Lect. Notes Biomath.* **54**, 368–389.

Kareiva, P. (1994). Space: The final frontier for ecological theory. *Ecology* **75**, 1.

Knerer, G., and Atwood, C. E. (1973). Diprionid sawflies: Polymorphism and speciation. *Science* **179**, 1090–1099.

Larsson, S., Björkman, C., and Kidd, N. A. C. (1993). Outbreaks of diprionid sawflies: Why some species and not others. *In* "Sawfly Life History Adaptations to Woody Plants" (M. R. Wagner and K. F. Raffa, eds.), pp. 453–484. Academic Press, San Diego, CA.

Martineau, R. (1985). "Insectes Nuisibles des Forêts de l'Est du Canada." Éditions Marcel Broquet, LaPrairie, Québec.

Nothnagle, P. J., and Schultz, J. C. (1987). What is a forest pest? *In* "Insect Outbreaks" (P. Barbosa and J. C. Schultz, eds.), pp. 59–80. Academic Press, San Diego, CA.

Pilson, D. (1989). Aphid distribution and the evolution of goldenrod resistance. *Evolution (Lawrence, Kans.)* **46**, 1358–1372.

Pimm, S. L. (1991). "The Balance of Nature? Ecological Issues in the Conservation of Species and Communities." Univ. of Chicago Press, Chicago.

Redfearn, A., and Pimm, S. L. (1988). Population variability and polyphagy in herbivorous insect communities. *Ecol. Monogr.* **58**, 39–55.

Rhoades, D. F. (1985). Offensive–defensive interactions between herbivores and plants: Their relevance in herbivore population dynamics and ecological theory. *Am. Nat.* **125**, 205–238.

Root, R. B., and Cappuccino, N. (1992). Patterns in population change and the organization of the insect community associated with goldenrod. *Ecol. Monogr.* **62**, 393–420.

Root, R. B., and Kareiva, P. M. (1984). The search for resources by cabbage butterflies (*Pieris rapae*): Ecological consequences of Markovian movements in a patchy environment. *Ecology* **65**, 147–165.

Root, R. B., and Kareiva, P. M. (1986). Is risk-spreading so unrealistic? *Oikos* **47**, 114–116.

Root, R. B., and Messina, F. J. (1983). Defensive adaptations and natural enemies of a case-bearing beetle, *Exema canadensis* (Coleoptera: Chrysomelidae). *Psyche* **90**, 67–80.

Rose, A. H., and Lindquist, O. H. (1982). Insects of eastern hardwood trees. *Can. For. Serv. Tech. Rep.* **29F,** 304.

Spitzer, K., Rejmánek, M., and Soldán, T. (1984). The fecundity and long-term variability in abundance of noctuid moths (Lepidoptera, Noctuidae). *Oecologia* **62,** 91–93.

Stamp, N. E. (1980). Egg deposition patterns in butterflies: Why do some species cluster their eggs rather than deposit them singly? *Am. Nat.* **115,** 367–380.

Stiling, P. D. (1987). Density-dependent processes and key-factors in insect populations. *J. Anim. Ecol.* **57,** 581–593.

Strathmann, R. (1974). The spread of sibling larvae of sedentary marine invertebrates. *Am. Nat.* **108,** 29–44.

Strong, D. R. (1988). Parasitoid theory: From aggregation to dispersal. *Trends Ecol. Evol.* **3,** 277–280.

Taylor, A. (1993). Heterogeneity in host–parasitoid interactions: "Aggregation of risk" and the "$CV^2 > 1$ rule." *Trends Ecol. Evol.* **8,** 400–405.

Tsubaki, Y., and Shiotsu, Y. (1982). Group feeding as a strategy for exploiting food resources in the burnet moth *Pyeria sinica. Oecologia* **55,** 12–20.

Turchin, P., and Kareiva, P. (1989). Aggregation in *Aphis varians:* An effective strategy for reducing predation risk. *Ecology* **70,** 1008–1016.

Vulinec, K. (1990). Collective security: Aggregation by insects as a defense. *In* "Insect Defenses: Adaptive Mechanisms and Strategies of Prey and Predators" (D. L. Evans and J. O. Schmidt, eds.), pp. 251–288. State University of New York Press, Albany.

Walde, S., and Murdoch, W. W. (1988). Spatial density dependence in parasitoids. *Annu. Rev. Entomol.* **33,** 441–466.

Wallner, W. E. (1987). Factors affecting insect population dynamics: Differences between outbreak and non-outbreak species. *Annu. Rev. Entomol.* **32,** 317–340.

Watt, K. E. F. (1965). Community stability and the strategy of biological control. *Can. Entomol.* **97,** 887–895.

Chapter 5

Minor Miners and Major Miners: Population Dynamics of Leaf-Mining Insects

Michael J. Auerbach, Edward F. Connor, and Susan Mopper

I. Introduction

Leaf mining is a means by which some insects consume foliage while simultaneously dwelling inside it. In all leaf-mining taxa, only larvae feed within leaf mines. However, in many taxa, pupation occurs within the mine and the adult must break through the leaf epidermis to emerge. The duration of mining varies widely, from a single instar to the entire larval period. Most species remain in the natal leaf mine, but larvae of some species initiate new mines on the same or a different leaf. The number of generations per year also varies widely from univoltine species to facultative and obligate multivoltine ones.

Leaf mining has evolved independently numerous times and is found among Coleoptera, Diptera, Lepidoptera, and Hymenoptera (Needham *et al.*, 1928; Hering, 1951; Powell, 1980; Hespenheide, 1991). In the Coleoptera, the greatest number of leaf-mining species occurs within the Chrysomelidae. In the Diptera, the leaf-mining habit is most widely developed in the Agromyzidae and Ephydridae. In the Lepidoptera, leaf mining occurs in 34 families, with the greatest number of species in the Gracillariidae. Among Hymenoptera, fewer than 100 species worldwide are leaf miners, with most species in the Tenthredinidae (Smith, 1976, 1993).

Leaf-mining insects attack herbaceous and woody plants in both terrestrial and aquatic habitats. Leaf miners are important pests of agricultural crops, greenhouse plants, and orchard trees. However, a survey of the literature of eleven entomological and ecological journals for the years 1981–1991 indicated that only 11% of the literature on leaf-feeding insects in the orders Coleoptera, Diptera, Lepidoptera, and Hymenoptera dealt with leaf-mining species (E. F. Connor and M. P. Taverner, personal observation). Furthermore, 80% of the literature on leaf-mining insects deals with a few species of economically important pests such as flies in the genera *Liriomyza* and *Agromyza* that feed on various crops, and moths in the genus *Phyllonorycter* that attack apple. In this review we focus primarily on native species feeding on native host plants.

The central issues we address are (1) what are the dominant mortality sources for leaf miners, (2) do these sources impose spatially or temporally density-dependent mortality, (3) does variation in natality affect the dynamics of leaf miners, and (4) do mortality and natality vary between latent and eruptive species? We will examine these issues by summarizing the literature on leaf-mining insects in general, and by presenting case studies of both latent and eruptive species of leaf miners.

II. Abundance Patterns

Worldwide approximately 10,000 species of insects are leaf miners (Faeth, 1991a), but density estimates and demographic data are available for only 1% of these species. Approximately 75% of leaf miners with published density estimates occur at low, latent ($<<1$ mine/leaf) levels of abundance (Table 1). Given the inconspicuousness of most leaf-mining larvae, species that attain high densities are probably overrepresented.

Approximately 25% of leaf-mining species have been observed at high, eruptive densities ($>>1$ mine/leaf; Table 1). Some of these species typically occur at low densities, but in certain locations within their native ranges, or in particular years, eruptions occur. However, some species persist at chronically high densities. For example, *Odontota dorsalis* (Coleoptera: Chrysomelidae) inflicts obvious damage to its host tree *Robinia pseudoacacia* every year, although the extent of damage varies both spatially and temporally (Athey and Connor, 1989). We will refer to both chronic and occasional outbreaks as eruptive, high-lighting their variable population dynamics across sites and years.

III. Sources of Mortality

Both vertical and horizontal interactions can represent dominant causes of death. By vertical interactions we refer to those between miners and the trophic level below them (host plants), as well as the one above them (natural enemies). Horizontal interactions include inter- and intraspecific interactions between leaf miners and other herbivores. We caution that most studies of leaf-miner mortality have focused on the conspicuous larval stage. Egg mortality has been estimated for only a few species, largely because most eggs are minute. Pupal mortality sources are frequently documented for species that pupate within the leaf mine, but not for those that leave the mine to pupate. Similarly, causes of adult death are rarely recorded.

TABLE 1

Number of Species of Latent and Eruptive Leaf-Mining Insects on Various Genera of Host Plants in Different Geographic Regions

Host genus	Number of host taxa studied	Number of latent species	Number of eruptive species
Abies			
North America	1	0	1
Europe	1	0	1
Alnus	1	2	0
Betula			
North America	1	0	1
Europe	2	20	0
Carya	1	3	1
Ceanothus	1	0	1
Coffea	1	0	2
Commelina	1	1	0
Cyperus	1	1	0
Eucalyptus	1	0	1
Fagus	1	2	0
Ilex			
North America	1	0	1
Europe	1	1	0
Malus			
North America	1	0	2
Europe	1	0	1
Asia	1	1	1
Nyssa	1	0	1
Physalis	1	1	0
Populus			
North America	3	0	3
Asia	1	1	0
Prunus and *Crateagus*	4	2	0
Quercus			
North America	13	40	5
Europe	4	4	2
Middle East	1	0	1
Robinia	1	3	2
Salix	1	1	0
Solanum	1	1	0
Solidago	1	5	0
Spartina	1	0	1
Vaccinium	1	0	1
Total		89	29

A. Vertical Sources of Mortality

1. Natural Enemies

Most leaf-mining species apparently suffer little egg mortality from natural enemies (e.g., Delucchi, 1958; Auerbach, 1991; Mopper and Simberloff, 1995). This is because few species of egg parasitoids are reported to attack the minute eggs of leaf miners.

Virtually every study examining mortality of leaf miners, whether at latent or eruptive densities, has identified parasitism of larvae, and sometimes pupae, as an important source of mortality. This is hardly surprising since for most species the larval stage is the most exposed life-history stage. The parasitoid communities associated with leaf-mining insects are more species rich and comprise a greater number of generalist parasitoid species than are the parasitoid communities of any other phytophagous insect guild (Askew, 1975; Askew and Shaw, 1986; Hawkins and Lawton, 1987; Hawkins, 1988a, 1990; Hawkins *et al.*, 1990, 1992). Several studies have also documented host feeding by adult parasitoids as a major source of larval mortality (Askew and Shaw, 1979; Connor and Beck, 1993), although host feeding is difficult to assess and has probably been overlooked in many instances.

Both vertebrate and invertebrate predators of leaf-miner larvae and pupae have been reported as important mortality agents (Faeth, 1980; Munster-Swendsen, 1989; Itamies and Ojanen, 1977; Owen, 1975; Heads and Lawton, 1983a,b). These include various species of birds (particularly members of the genus *Parus*), ants, and other carnivorous insects. Accidental predation, the inadvertent consumption of leaf-miner eggs, larvae, or pupae by ectophagous insects, can also be responsible for considerable mortality (Auerbach, 1991).

Leaf miners seldom suffer mortality from diseases even when density per leaf is extremely high (E. F. Connor and M. P. Taverner, personal observation). However, Munster-Swendsen (1989) reports that two species of needle-mining *Epinotia* (Lepidoptera: Tortricidae) suffer substantial pupal mortality from a pathogenic fungus while pupating in the soil, and that *Epinotia tedella* also suffers a reduction in fecundity from a sublethal protozoan infection while pupating in the soil (Munster-Swendsen, 1991).

2. Host-Plant Attributes

Several studies have noted that eggs can be displaced by some host-plant species if they are placed on young, rapidly expanding leaves (Mazanec, 1985; M. J. Auerbach, personal observation). Leaf toughness and the chemical composition of leaves, in terms of both nutritional value and constitutive and induced phytochemical defenses, can be responsible for significant larval mortality (Faeth, 1986; Potter and Kimmerer, 1986; Auerbach and Alberts, 1992; Marino *et al.*, 1993). Death from host-plant resistance has generally been inferred after all larval mortality that can be readily ascribed to an obvious cause of death, such as parasitism or predation, has been documented. The remaining "unknown mortality" often varies among leaf miners on conspecific hosts, lending credence to its classification as host-plant resistance. Several studies found correlations (some negative) between leaf-miner densities or larval performance and foliar nitrogen content of host plants (Faeth *et al.*, 1981a; Stiling *et al.*, 1982; Auerbach and Simberloff, 1988; Marino *et al.*, 1993).

Numerous studies have also identified host-plant phenology as a dominant source of mortality. Although some leaf-mining larvae can complete development in abscised leaves (Kahn and Cornell, 1983, 1989), most cannot unless they are in the terminal instar when abscission occurs (Faeth *et al.*, 1981b; Pritchard and James, 1984; Stiling and Simberloff, 1989; Stiling *et al.*, 1991). The likelihood of premature leaf abscission correlates with density or position of mines in some species (Faeth *et al.*, 1981b; Mopper and Simberloff, 1995). Premature abscission resulting from abiotic factors such as drought can also significantly reduce leaf-miner survival (Auerbach, 1991). Similarly, annual variation in the timing of autumn leaf fall in temperate, deciduous hosts can greatly influence mortality, particularly for multivoltine species with facultative late summer or early autumn generations (Pottinger and LeRoux, 1971; Miller, 1973; Barrett and Brunner, 1990; Connor *et al.*, 1994).

B. Horizontal Sources of Mortality

Interactions with conspecifics and other phytophagous insect species have generally been viewed as having little effect on population dynamics, largely because the characteristically low densities of most phytophages were presumed to result in little competition for limiting resources (Strong *et al.*, 1984). However, many species of leaf miners are seriously affected by both direct interference competition or indirect exploitative competition with conspecifics. This is particularly true for the high-density phase of eruptive species. Interference competition, including cannibalism or at least the killing of larvae by conspecifics on multiply mined leaves, can be a dominant source of mortality (Martin, 1956; Auerbach, 1991). The frequency of conspecific encounters, and their impact on survival, is often a product of both mine density and the dispersion pattern of miners within and among leaves (Faeth, 1990; Auerbach, 1991). Miners are commonly aggregated on leaves of individual host plants, although random and hyperdispersed patterns have been observed (Auerbach and Simberloff, 1989; Faeth, 1990). Though aggregation can increase the occurrence of intraspecific interference, within-leaf spatial arrangement can ameliorate the potential negative effects. For example, several species of leaf miners tend to place eggs on opposite sides of the leaf midvein, thus reducing the likelihood of intraspecific encounters (Stiling *et al.*, 1987; Auerbach and Simberloff, 1989).

Several types of exploitative competition have been observed. Some species use egg-marking pheromones or shed abdominal scales to deter oviposition by conspecifics (McNeil and Quiring, 1983). Ovipositing females of some multi-voltine species avoid placing eggs on previously mined leaves. Avoidance of conspecific eggs, previously mined leaves, or leaves damaged by other herbivores may result in use of inferior leaves (Bultman and Faeth, 1986a; Faeth, 1986). In some species, pupal mass correlates inversely with density of mines on a leaf (Quiring and McNeil, 1984; Stiling *et al.*, 1984; Potter, 1985).

C. Abiotic Sources of Mortality

Direct effects of abiotic factors on leaf-miner survival have seldom been quantified, except for a few studies examining causes of overwintering mortality in several temperate-zone species (Pottinger and LeRoux, 1971; Connor, 1984; Connor *et al.*, 1994). Wind storms, frosts, and hard rain can tear or dislodge leaf mines, resulting in larvae being tossed from mines or desiccation (Delucchi, 1958). Cool weather may also prolong development, thereby increasing mortality from parasitism, predation, leaf abscission, or physical changes in leaves (Blais and Pilon, 1968; Nielsen, 1968).

IV. Regulatory Effects of Variation in Mortality

Tests of the regulatory role of dominant mortality sources, usually with regression-based tests for spatial and or temporal density dependence, have been equivocal. Among species with low population densities, mortality from predation, or larval or pupal parasitism, was spatially density dependent in only 36% of the species examined (Table 2). During the high-density phase of species with eruptive population dynamics, parasitism was always spatially density independent. Mortality from host-plant resistance was generally independent of density, whereas death from premature leaf abscission was usually density dependent. The only sources of mortality that were consistently spatially density dependent were intraspecific exploitative and interference competition.

Fewer studies have compiled data over sufficient periods of time for adequate tests of temporal density dependence, particularly for species with low population densities. When conducted, such tests have generally found density independence or inverse density dependence among dominant mortality sources. Again, the lone exception tends to be larval interference, which sometimes imposed mortality in a temporally density-dependent fashion (Table 2).

The general conclusion emerging from these studies is that although vertical sources of mortality are usually responsible for most egg, larval, and sometimes pupal mortality, they generally cannot "regulate" population size in the strict

TABLE 2

**Results of Tests for Spatial or Temporal Density Dependence among Leading
Mortality Sources of Leaf-Mining Insects**[a]

Order	Density	Source of mortality	Spatial/ temporal	Density relationship	Reference
Lep	H	Egg	S	DI	Auerbach and Simberloff (1989)
		Int	S	DD	
		HR	S	DI	
		Pred	S	DI	
		Para	S	DI	
		Abs	S	DD	
Lep	H	Egg	S	DI	Auerbach (1991)
		Int	S	DD	
		HR	S	DI	
		Abs	S	DI	
		Para	S	DI	
Lep	H	Egg	T	DI	M. J. Auerbach (personal observation)
		Int	T	DD	
		HR	T	DI	
		Para	T	IDD	
Lep	H	HR	S	DD	Connor and Beck (1993)
		Comp	S	DD	
		Para	S	DI	
		Pred	S	IDD	
	L & H	Para	T	DI	
		Pred	T	IDD	
		Comp	T	DI	
Lep	L	Para	S	DD	Miller (1973)
		Para	T	DD	
Lep	L	Pred	S	DD	Itamies and Ojanen (1977)
Lep	?	Para	?	DI	Sekita and Yamada (1979)
Lep	L	Para	S	DD	Mopper et al. (1984)
		Pred	S	IDD	
Lep	L	Para	S	DD	Bultman and Faeth (1985)
		Pred	S	DD	
Lep	L	Comp	S	DD	Bultman and Faeth (1986b)
Lep	L	Pred	S	DI	Simberloff and Stiling (1987)
		Para	S	DI	
		Abs	S	DD	
Lep	L	Abs	S	DD	Stiling et al. (1987)
Lep	L	Para	S	DI	Hawkins (1988b)
Lep	L	Abs	S	DD	Stiling and Simberloff (1989)
Lep	L	Egg	S	DI	Alberts (1989)
		Int	S	DD	
		HR	S	DI	
		Para	S	DI	

(continues)

TABLE 2   (*continued*)

Order	Density	Source of mortality	Spatial/ temporal	Density relationship	Reference
Lep	L	Int	S	DD	Faeth (1990)
		Abs	S	DD	
		Para	S	DI/IDD	
		Pred	S	DI	
Lep	L	Para	S	DD	Fritz (1990)
Lep	L	Pred	S	DI	Stiling *et al.* (1991)
		Para	S	DI	
		HR	S	DI	
		Abs	S	DD	
Lep	L	TLM	S	DI/IDD	Faeth (1991b)
Lep	L	Pred	S	DI	Cappuccino (1991)
		Para	S	DI/DD	
		HR	S	DI	
Lep	L	Para	S	DI	Koricheva (1994)
		Para	T	DI	
Lep	L	Para	S	DD	Preszler and Boeckle (1994)
Lep	L	Abs	S	DD	Mopper and Simberloff (1995)
		HR	S	DI	
		Pred	S	DI	
		Para	S	DI	
Col	L	Para	T	DI	Day and Watt (1989)
Col	L	Pred	S	DI/DD	Cappuccino (1991)
		Para	S	DI	
		HR	S	DI	
Dip	L	Pred	S	IDD	Heads and Lawton (1983a)
		Para (l)	S	DD	
		Para (p)	S	DI	
Dip	L & H	Int	S	DD	Quiring and McNeil (1984)
		Comp	S	DD	
Dip	L & H	Int	S	DD	Potter (1985)
		Para	S	DI	
Dip	L	Pred	S	DI	Freeman and Smith (1990)
		Para	S	DI/DD	
		Comp	S	DD	
Dip	L	Pred	S	IDD	Valladares and Lawton (1991)
		HR	S	DI	
		Para	S	DD	
Dip	L	Pred	S	DI	Cappuccino (1991)
		Para	S	DI	
Dip	L	Pred	S	DI	Cappuccino (1991)
Dip	L	Pred	S	DI	Cappuccino (1991)
		Para	S	DI	
		HR	S	DI	

(*continues*)

TABLE 2 (*continued*)

Order	Density	Source of mortality	Spatial/ temporal	Density relationship	Reference
Dip	L	Comp	S	DD	Kato (1994)
		Para (l)	S	DI	
		Para (p)	S	DD	

[a]Order: Lep = Lepidoptera, Dip = Diptera, Col = Coleoptera. Density: H = high (>1 mine/leaf), L = low (<1 mine/leaf). Source of mortality: Egg = egg death, Para = parasitism, Pred = predation, Int = larval interference, HR = host resistance, Comp = larval exploitative competition, Abs = premature leaf abscission, TLM = total larval mortality. Density relationship: DD = positive density dependence, DI = density independence, IDD = inverse density dependence. l = larval, p = pupal.

sense. This is particularly true during population eruptions when parasitism and host-plant resistance are generally density independent. Premature leaf abscission is often density dependent, but most species suffer little mortality from this source.

In contrast, horizontal interactions, particularly those with conspecifics, impose both spatially and temporally density-dependent mortality and could play an important role in regulating leaf-miner population dynamics. Whether horizontal interactions are capable of terminating population eruptions may depend on several factors, including leaf size and larval dispersal patterns among and within leaves (Faeth, 1990; Auerbach, 1991).

V. Variation in Natality

One additional caveat must be added to any discussion of population dynamics of leaf-mining insects. Very limited data are available on variation in fecundity among individuals, species, locations, or years. Thus, discussions of population dynamics have focused on mortality sources rather than reproductive output. However, two important influences on natality and oviposition success have been identified. First, as noted earlier, inter- and intraspecific exploitative competition can result in reduced pupal mass and, presumably, adult fecundity. Second, numerous species of leaf miners feed preferentially or exclusively on young leaves of their host plants (Auerbach and Simberloff, 1984; Hespenheide, 1991), and fecundity can vary with use of differently aged leaves (Bale, 1984). Stochastic abiotic events, such as late frosts or droughts, can alter the synchrony between oviposition and availability of young foliage, which in turn affects oviposition success and dispersal patterns of eggs, and therefore the survival and eventual fecundity of offspring (Wallace, 1970; Auerbach and Simberloff, 1984; Potter and Redmond, 1989).

VI. Case Studies

In this section we detail the population dynamics of three species of leaf miners that we have investigated extensively. Unlike most studies, we have compiled fairly long-term data on densities and sources of mortality. One species is latent and two are eruptive. In each instance, natural enemies are a dominant source of mortality, but their effects are not spatially or temporally density dependent. Intraspecific competition is an important mortality source for two of the species and all three suffer mortality from host-plant resistance. The common theme emerging from a comparison of these three cases is that host-plant phenology, and consequently resource availability, play a critical role in determining population dynamics. Variation in host-plant phenology also appears to be an important factor governing shifts between low- and high-density population phases of the eruptive species.

A. *Phyllonorycter tremuloidiella,* an Eruptive Species on *Populus tremuloides*

Larvae of *Phyllonorycter tremuloidiella* Braun (Lepidoptera: Gracillariidae) mine leaves of quaking aspen, *Populus tremuloides,* and occasionally other *Populus* species. This species has been previously identified as *P. salicifoliella* in several studies (Martin, 1956; Auerbach, 1990, 1991; Auerbach and Alberts, 1992). Although densities are generally low (<1 mine/leaf), eruptions occur at some sites and often persist for more than 5 years (Martin, 1956). One of us (M. J. A.) has been investigating a population eruption that has persisted for at least 15 years in north-central Minnesota at Lake Itasca State Park (Auerbach, 1991). Although densities have been consistently high at Itasca, they have varied severalfold among years (Fig. 1). Thus, densities do not appear to fluctuate around a high-density equilibrium as envisioned in some models of population dynamics with multiple stable states (e.g., Southwood and Comins, 1976).

Phyllonorycter tremuloidiella is univoltine at Itasca. In most years, adult females begin oviposition flights in mid-May, shortly after the peak budbreak in quaking aspen. Ovipositing females show a strong preference for small trees (<5 m in height). Females also preferentially oviposit on young, not quite fully expanded leaves. Within leaves, females place eggs on opposite sides of the leaf midvein more often than expected by chance (Alberts, 1989). Larger leaves receive more eggs than smaller ones, however, leaf area only explains 8–20% of the variation in egg density and there appears to be only slight preference for larger leaves (M. J. Auerbach, personal observation). Eggs are aggregated among leaves within a tree across a wide range of population densities, with the degree of aggregation generally increasing with density (Auerbach, 1990).

Larvae eclose in 5–28 days depending on temperature. During the first three

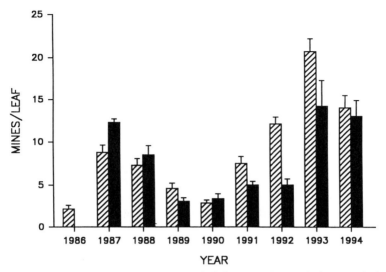

Figure 1. Average densities of *Phyllonorycter tremuloidiella* on *Populus tremuloides* at north (hatched bar) and south (solid bar) sites in Lake Itasca State Park, Minnesota, U.S.A. Values are means of estimates from 9–40 trees per site per year. Thin vertical bars represent one standard error.

of five larval instars, larvae expand the mine while feeding on tissue sap. When mines inadvertently coalesce during the early instars, one larva kills the other(s). We refer to this mortality as larval interference. Upon molting in the fourth instar, larvae begin tissue-feeding and excavate the mine without increasing its perimeter. Pupation occurs within the mine. Adults exit the mine and then over-winter under the bark scales of red pine (*Pinus resinosa*) and other coniferous trees.

1. Dominant Mortality Factors and Population Dynamics

An extensive 3-year study of egg, larval, and pupal mortality revealed that most mortality occurred during the larval stage (Auerbach, 1991). Parasitism, host-plant resistance, and larval interference were consistently the dominant sources of mortality, although their rank order varied among years. However, neither of the dominant vertical sources of mortality operated in a spatially density-dependent fashion across a wide range of spatial scales (Auerbach, 1991). Most surprising was the lack of density-dependent parasitism, given the presence of a diverse assemblage of at least 17 species of parasitic Hymenoptera. Only larval interference was consistently density dependent across all spatial scales. Despite this, the probability of a miner completing development was independent of mine density on a leaf, because winners of intraspecific encounters had higher probabilities of completing development than did larvae not encountering conspecifics. Vertical interactions also did not impose density-dependent mortality in a chronically low-density population of *P. tremuloidiella* (Alberts, 1989).

A recently completed parallel study extends these analyses to test for temporal density dependence by following miners on eight trees over a 6-year period (M. J. Auerbach, personal observation). Again, effects of vertical sources of mortality were either independent of density or declined with increased density. Mortality from larval interference was density dependent, but the relationship was generally weak and regression slopes varied considerably among trees and years.

Apparently, the only source of egg, larval, or pupal mortality that might regulate populations of *P. tremuloidiella* around an equilibrium value is larval interference, which is both spatially and temporally density dependent. Though the likelihood of larval interference is enhanced by aggregation of mines among leaves of host trees, interference cannot impose strict regulation because several factors ameliorate its impact. First, interference can never impose complete mortality on a leaf because at least one larva per leaf is immune to it. In addition, the tendency of females to place eggs on opposite sides of a midvein further reduces the chance of interference since larvae seldom cross a midvein. Finally, although there is a significant relationship between interference rate and mine density on a leaf, the slope of this relationship varies among years (Auerbach, 1991). In some years, miners on high-density leaves encounter less interference than do ones on lower-density leaves in other years because the size of a leaf with a given number of mines, and hence the likelihood of interference, varies among years. Linkage between the spatial and temporal density dependence of interference is further weakened by variation in the degree of aggregation resulting from annual differences in availability of young leaves during ovipiosition. Thus, changes in resource quality (leaf size) and availability make the coupling loose and strict regulation unlikely.

The lack of density-dependent vertical sources of mortality and the variable and non-linear influence of larval interference imply that eruptive populations of *P. tremuloidiella* are unregulated, at least with respect to any definable equilibrium level. Instead, during eruptions *P. tremuloidiella* occurs at or near population "ceilings" delimited by the availability of leaves, particularly young leaves preferred by ovipositing females (Auerbach, 1991). Thus, synchrony between termination of diapause and peak budbreak of host trees is a critical determinant of population density. Variable host-plant phenology also imposes many of the density patterns described here. For instance, trees producing leaves before the onset of oviposition recruit low densities of miners, whereas adjacent later-flushing trees are heavily infested. Aggregation of miners among leaves within trees also reflects staggered availability of leaves.

Host-plant phenology affects leaf-miner density and survival since synchrony between oviposition flights and availability of host leaves largely determines density per leaf, and influences mortality from larval interference. Patterns of dispersion of miners among leaves within trees also reflect budbreak patterns, which again affect the probability of larval interference and parasitism. At high densities or during years when synchrony between oviposition and availability of preferred leaves is poor, many females must oviposit on older, nonpreferred leaves. Studies testing the population consequences of oviposition on very young or mature leaves are currently in progress.

2. Initiation and Termination of Eruptions

Synchrony between leaf-miner and host-plant phenologies may be largely responsible for initiating eruptions of *P. tremuloidiella,* as in other leaf miners that prefer young leaves (Wallace, 1970; Auerbach and Simberloff, 1984; Potter and Redmond, 1989). Quaking aspen frequently grows in large, clonal stands and phenological variation within clones is generally much less than variation among clones. Thus, a fortuitous match between host-plant and leaf-miner phenologies involves many more trees (ramets) than would be the case for nonclonal host plants. Intensive logging can result in very dense stands of preferred juvenile trees. One such stand at Itasca maintained low densities of *P. tremuloidiella* for about 8 years until aspen reached 2–3 m in height. Leaf-miner densities then rose rapidly over several years, reaching eruptive levels in 1994.

Meteorological patterns also directly and indirectly influence population dynamics of *P. tremuloidiella.* We have been unable to quantify overwintering mortality, however, over the last decade, densities have climbed following unusually mild winters or ones with heavy snowfall. In addition, early spring weather patterns influence timing of host-tree budbreak and diapause termination in *P. tremuloidiella* along with the degree of synchrony between them.

It is still unclear exactly what forces terminate eruptions of *P. tremuloidiella,* but two factors appear to be important. Extreme disruption of synchrony of oviposition and leaf availability, such as that following a late frost resulting in leaf abscission, could drastically reduce population density. Alternatively, given that budbreak of large trees usually precedes oviposition flights, densities may decline with maturation of even-aged aspen stands. Trees surviving natural thinning and heavy leaf-miner infestations may simply outgrow the eruption.

B. *Cameraria hamadryadella,* an Eruptive Species on *Quercus alba*

Cameraria hamadryadella (Clemens) (Lepidoptera: Gracillariidae) is an upper-surface blotch leaf miner that feeds on oaks in the subgenus *Lepidobalanus.* It occurs most frequently on *Quercus alba* L. in eastern North America (Needham *et al.,* 1928; Hinckley, 1972; Maier and Davis, 1989; Connor, 1991).

Host and leaf selection are made by ovipositing females, which cement eggs singly to the upper leaf surface along the leaf midvein and lateral veins. Eggs hatch in 1–2 weeks, and larvae chew through the egg chorion and the leaf epidermis to commence mining. Development from egg to adult occurs in a single mine and larvae feed for 4–6 weeks before pupating within the mine. The first generation completes development, mates, and oviposits by late July or early August. The second generation appears in middle to late August and larvae feed until leaf senescence and abscission occur by the end of October. Larvae diapause within the abscised leaf and complete development, emerging as adults in the spring.

Cameraria hamadryadella has population eruptions that can cover broad geographic areas (Solomon *et al.,* 1980). One of us (E. F. C.) has observed extensive eruptions of *C. hamadryadella* on *Q. alba* in the piedmont of Georgia and the Carolinas in 1977 and in the coastal plain of New Jersey in 1989. However, most of our knowledge of the dynamics of eruptive and latent populations of *C. hamadryadella* derives from a 13-year study at Blandy Experimental Farm in northern Virginia (Connor, 1991; Connor and Beck, 1993; Connor *et al.,* 1994; Connor and Cargain, 1994).

Both eruptive and latent populations of *C. hamadryadella* have been present at Blandy Farm over the past 12 years. Latent populations (densities of less than 1 mine/leaf) persist in natural woodlots on the periphery of this 280-hectare farm. A single eruptive population, which has undergone two distinct eruptions, persists in a grove of planted oaks in the Orland E. White Arboretum located in the central portion of Blandy Farm. The eruptive population has fluctuated from a low of fewer than 0.1 individual/leaf in 1987 to a high of over 40/leaf in 1993–1994 on foliage within 3 m of the ground (Connor and Beck, 1993; Connor *et al.,* 1994). Densities in latent populations of *C. hamadryadella* are often as low as 0.03 individual/leaf (Fig. 2). Preliminary indications in the first generation of 1994 suggest that dispersal from the eruptive population has led to increased densities at the sites that had not previously experienced eruptions (Fig. 3).

1. Factors Leading to Eruptions of *C. hamadryadella*

Since 1982 the eruptive population of *C. hamadryadella* at Blandy Farm has experienced a decline from an eruption (1982–1984), a trough (1985–1990), and an eruption (1991–1994). The most detailed information about the population dynamics of *C. hamadryadella* has been collected since 1989. The evidence suggests that the eruption of *C. hamadryadella* that began in 1991 was caused by an extended growing season in 1990. The extended growing season resulted in late autumn leaf fall in 1990. Late leaf fall resulted in a significantly larger proportion of second-generation larvae having adequate time to gain sufficient mass, enter the larval diapause, and successfully emerge as adults the following spring. In fact, survival in the second generation of 1990 was an order of

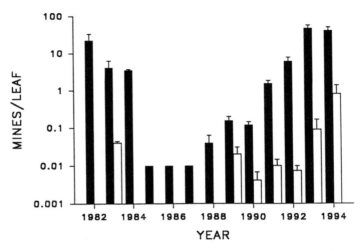

Figure 2. The density of first-generation *Cameraria hamadryadella* leaf mines (mines/leaf) on *Quercus alba* in an eruptive (arboretum, solid bar) and a noneruptive (southwest woodland, open bar) site from 1982 to 1994 at Blandy Experimental Farm, Boyce, Virginia, U.S.A. Thin vertical bars represent one standard error.

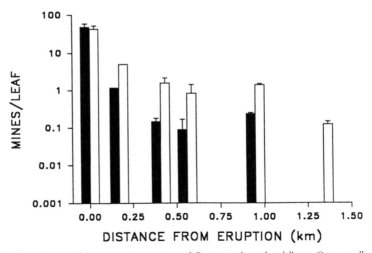

Figure 3. The density of first-generation mines of *Cameraria hamadryadella* on *Quercus alba* in the eruptive site and several noneruptive sites in 1993 (solid bar) and 1994 (open bar) at Blandy Experimental Farm in relation to distance from the eruptive site. The eruptive site is represented by density estimates at distance 0. Thin vertical bars represent one standard error.

magnitude higher than that in the same generation in 1989 (Connor *et al.*, 1994). The density of *C. hamadryadella* increased by an order of magnitude between 1990 and 1991, rose steadily through 1993, and leveled off in 1994.

One factor that distinguishes the eruptive site from the latent sites is the presence of *Q. macrocarpa* at the eruptive site. It foliates before *Q. alba* in the spring and retains its leaves longer in the fall (Connor *et al.*, 1994). Hence, even in years when the peak of leaf fall is early in the autumn, second-generation *C. hamadryadella* developing on *Q. macrocarpa* are able to complete development, emerge, and repopulate both *Q. alba* and *Q. macrocarpa* the following spring. The absence of *Q. macrocarpa* from the natural woodlands results in a smaller cohort of *C. hamadryadella* surviving to emerge the following spring and therefore lower-density populations.

We have some evidence pertaining to other possible explanations for the eruptions of *C. hamadryadella*. The host-plant quality hypothesis suggests that insect eruptions arise because of temporal variation in the nutritional quality of the host plant. However, offspring reared from adults captured from the eruptive population survive equally well on trees in the latent sites as on trees in the eruptive site (M. S. Nuckols and E. F. Connor, personal observation). Furthermore, measurements of foliage toughness, moisture content, leaf water potential, and phenology do not differ between *Q. alba* trees in eruptive and latent sites. If temporal variation in host quality results from temporal variation in the environment, then we would have expected eruptions of *C. hamadryadella* at all sites. Nevertheless, further direct observations on host quality over the course of an eruption would be valuable in evaluating this hypothesis.

Even though we have a limited number of generations of data on the effects of natural enemies, they do not support the idea that eruptions arise because of escape from the temporally or spatially density-dependent effects of natural enemies (Connor and Beck, 1993). Predation appears to be both spatially and temporally inversely density dependent, and parasitism is spatially density independent in years of moderate leaf-miner density and inversely density dependent in years of high leaf-miner density (Connor and Beck, 1993; Green, 1994). Furthermore, behavioral studies of the major parasitoid of *C. hamadryadella*, *Closterocerus tricinctus* (Hymenoptera: Eulophidae), in the eruptive population show that it forages in a spatially inversely density-dependent manner (Connor and Cargain, 1994). Natural enemies are never spatially density dependent, and their tendency to be inversely density dependent in high-density populations might amplify rather than reduce population eruptions.

Although a combination of host-plant resistance and intraspecific competition accounts for the largest proportion of mortality and these factors are spatially density dependent within leaves (Connor and Beck, 1993), it is unlikely that release from competition or induced host-plant defenses are the cause of the observed eruption of *C. hamadryadella*. If release from the effects of competition or induced host defenses were important, then the populations in the latent sites should also have experienced eruptions, since with population densities below 0.1 mine/leaf, they are largely free from the effects of induced host-plant defenses and intraspecific competitive interactions.

2. Factors Leading to Population Crashes in *C. hamadryadella*

In 1984 when population densities were 3.59 mines/leaf, intraspecific competition accounted for substantial mortality and was spatially density dependent. Similar results were obtained in 1991 and in 1993, when densities were 1.5 and 49.3 mines/leaf, respectively. This suggests that intraspecific competition could limit population growth and could initiate a population decline.

We also have casual observations that the abundance of parasitoids has increased since 1991. However, whether the impact of parasitoids can or will result in a population decline is not clear. Parasitoids have a spatially inversely density-dependent impact in high-density populations, partly because the most abundant species forages in an inversely density-dependent manner (Connor and Cargain, 1994), and partly because parasitized larvae and their parasitoids die because of intraspecific competition on high-density leaves (Green, 1994).

The most likely reason for population declines in *C. hamadryadella* is similar to the cause of the eruption, a phenological event that leads to low rates of successful development of the final autumn generation and low emergence rates the following spring. This could arise because of early autumn leaf fall, or because of the tendency of *C. hamadryadella* to attempt to complete additional generations. In 2 of 13 years, we have observed early second-generation larvae pupate, emerge as adults, mate, and oviposit rather than enter a larval diapause to overwinter. If a substantial fraction of the second generation attempts a third generation, and does not gain sufficient mass to enter diapause, then even a year with an average time of leaf fall could result in a population crash.

C. *Stilbosis quadricustatella*, a Latent Species on *Quercus geminata*

Stilbosis quadricustatella (Cham.) is a univoltine leaf miner (Lepidoptera: Cosmopterigidae) commonly found on *Quercus geminata* (sand-live oak) in northern Florida. *Quercus geminata* budbreak occurs in late April and most leaves are fully expanded when adult leaf miners emerge from pupation in May. Adults fly to the canopy, where they mate and lay single eggs on the upper leaf surface. Eggs are 0.2 mm in diameter and usually occur in a clumped distribution on large, peripheral, and undamaged leaves (Simberloff and Stiling, 1987). Observations of as many as 5 mines per leaf are common, and leaves of heavily attacked trees may contain as many as 13 mines. Larvae hatch within a week of oviposition and excavate a full-depth mine, which gradually expands to a maximum size of 2 cm². In late September, larvae quit leaves and drop to the ground, where they pupate and overwinter (Mopper *et al.,* 1984).

1. Sources of *S. quadricustatella* Mortality

Sources of leaf-miner mortality are readily identified by egg coloration and characteristic scars on the mine (Mopper *et al.,* 1984, 1995). Egg parasitism is rare; only one wasp species (*Trichogramma* sp.) has been reared from eggs (S. Mopper, personal observation). There are no data on egg predation. Most early larval mortality results from host-plant resistance, but in later instars both the host plant and natural enemies inflict substantial mortality (Mopper *et al.,* 1995). Little is known about mortality suffered in the pupal and adult stages.

Stilbosis quadricustatella is classified as latent since densities rarely exceed 0.5 mine/leaf. There is considerable variation in leaf-miner density among trees (Mopper *et al.,* 1995). Some *Q. geminata* are repeatedly infested, whereas adjacent trees remain relatively free of miners (Fig. 4). This variation in density appears to be unrelated to leaf-miner performance on individual trees. Transfer experiments revealed no difference in plant- and enemy-mediated larval mortality on heavily and lightly attacked trees (Mopper and Simberloff, 1995). Nor were there significant differences between years in larval survival. However, heavily and lightly infested trees differed significantly in leaf-expansion phenology and leaf area, factors that may influence adult oviposition behavior (Fig. 5).

2. Density and *S. quadricustatella* Fitness Components

Unlike *Phyllonorycter tremuloidiella* and *Cameraria hamadryadella, S. quadricustatella* mines never coalesce, and there is no evidence of direct interactions among larvae inhabiting the same leaf. Of the three studies that examined the impact of conspecifics on pupal mass, two detected no difference in the pupal mass of miners collected from leaves with one to three mines (P. D. Stiling, personal observation; Mopper and Simberloff, 1995), and the third study revealed a significant negative correlation between pupal mass and leaf-miner density when densities exceeded three mines per leaf (S. Mopper, personal observation). We conclude that at the leaf spatial scale, there are no negative intraspecific effects on pupal mass at low densities. However, on leaves with more than three mines, pupal mass and, ostensibly, fecundity are reduced.

In general, there is a positive relationship between larval mortality and density within leaves. Both plant-mediated and enemy-mediated mortality rise when several larvae share the same leaf (S. Mopper, personal observation; Simberloff and Stiling, 1987), although one study observed an inverse correlation between density and predation at high leaf-miner densities (Mopper *et al.,* 1984). There is a strong positive correlation between premature leaf abscission and the number of miners that occupy the same leaf (Simberloff and Stiling, 1987). We conclude that sharing a leaf with other conspecifics is generally detrimental

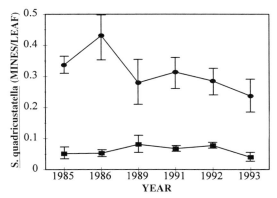

Figure 4. *Stilbosis quadricustatella* density on six lightly attacked (solid square) and six heavily attacked (solid circle) *Q. geminata* trees at Lost Lake in northern Florida from 1985 through 1993. Peter Stiling collected the 1985, 1986, and 1989 density data. After Mopper and Simberloff (1995).

to *S. quadricustatella* because of increased plant- and enemy-mediated mortality.

Density of leaf miners on a tree has little impact on pupal mass. P. D. Stiling (personal observation) observed no correlation between pupal mass and density at the tree spatial scale. Mopper and Simberloff (1995) detected a weak but significant positive correlation between pupal mass and leaf-miner density within

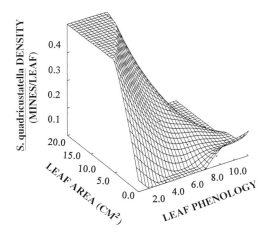

Figure 5. The relationship between leaf phenology, leaf area, and *S. quadricustatella* densities [asine $Y = 0.485 + 0.022$ (area) $- 0.055$ (phenology)]. Phenology explained 61% and leaf area explained 13% of the variation in leaf-miner density ($F_{2,7} = 9.8$, $P = 0.009$). After Mopper and Simberloff (1995).

trees. These two studies indicate that density of leaf miners is unrelated to pupal mass, and probably fecundity, at the tree spatial scale.

Leaf-miner density at the tree spatial scale has little effect on rates of natural enemy attack (Stiling *et al.*, 1991; Mopper and Simberloff, 1995). In contrast, plant-mediated mortality increases with density. There is a positive nonlinear correlation between premature leaf abscission and leaf-miner density at the tree spatial scale (Stiling and Simberloff, 1989; Stiling *et al.*, 1991; Mopper and Simberloff, 1995). Leaf abscission is usually lethal because larvae in abscised leaves suffer substantial plant- and enemy-mediated mortality (Simberloff and Stiling, 1987; Stiling *et al.*, 1991; Mopper and Simberloff, 1995).

Timing of *Q. geminata* spring leaf production is a reliable predictor of leaf-miner densities. Trees that produced leaves early in the season suffered substantially less herbivory than trees that produced leaves later (Mopper and Simberloff, 1995).

3. Regulation of *S. quadricustatella* Populations

The host plant has the greatest impact on *S. quadricustatella* survival, which can be affected by three different mechanisms: timing of spring foliation, host-plant resistance, and premature leaf abscission. However, only premature leaf abscission is density dependent. Natural enemies also respond positively to leaf-miner density, although their impact on *S. quadricustatella* fitness is much weaker than that of the host plant. Density-dependent forces do not seem to be important determinants of variation in leaf-miner densities among individual trees. Whether they maintain *S. quadricustatella* populations at consistently latent levels remains to be determined.

VII. Conclusions

A. Comparison of Latent and Eruptive Leaf Miners

We have identified no obvious differences between latent and eruptive species in either life-history attributes or sources of mortality that might account for why some species reach epidemic densities. In compiling Table 1, we noted no tendency for eruptive species to belong to particular taxa, to feed on deciduous versus evergreen host plants, or to have particular voltinism patterns. There were few differences in dominant sources of mortality, or in those sources of mortality that are density dependent, between species in these two classes of population dynamics, except that mortality from interference and exploitative competition is more prevalent in eruptive species.

B. Are Leaf-Miner Populations Regulated?

Results of our case studies and those studies listed in Table 2 indicate that among the leading causes of death only intraspecific competition and leaf abscission impose density-dependent mortality. Thus, horizontal and bottom-up forces generally appear to have a greater impact on leaf-miner population dynamics than do top-down effects. In his review of insects in general, Stiling (1988) also found that density dependence was more common among bottom-up than top-down sources of mortality. We are not suggesting, however, that either leaf abscission or competition is universally important in regulating leaf-miner populations. The prevalence of premature leaf abscission among different taxa of host plants is unknown and not all premature abscission imposes density-dependent mortality (Auerbach, 1991; Kahn and Cornell, 1989). Similarly, many leaf miners do not engage in interference competition, and the intensity of exploitative competition has been tested in only a few species. Even when it does operate in a density-dependent fashion, competition need not be regulatory. Whether it is depends on attributes of both the leaf miner (e.g., larval dispersion and mode of competition) and their host plants (e.g., phenology and leaf size) (Faeth, 1990; Auerbach, 1991).

Studies employing cages to exclude parasitoids and predators (Faeth and Simberloff, 1981) and successful biological control of leaf miners (Dharmad-hikari *et al.,* 1977; Drea and Hendrickson, 1986) demonstrate that natural enemies can profoundly influence the population dynamics of leaf miners. However, mortality from natural enemies was density dependent in only 36% of the latent species examined (Table 2), suggesting that parasitoids and/or predators are generally not responsible for regulation around an equilibrium density. These results are in accord with several reviews that found parasitism to be spatially density dependent in only about 25% of the cases examined (Lessells, 1985; Stiling, 1987; Walde and Murdoch, 1988).

The role of natural enemies in regulating populations of eruptive species is even more suspect since parasitism and predation do not impose density-dependent mortality during the high-density phase of these species (Table 2). No studies have simultaneously examined latent and eruptive populations of the same species over several years, so it is presently impossible to determine if these species have multiple stable states with low- and high-density populations regulated around different equilibrium levels (Southwood and Comins, 1976). However, this seems unlikely either for *Phyllonorycter tremuloidiella,* which suffered no density-dependent mortality from vertical sources at either latent or eruptive densities over one growing season (Alberts, 1989; Auerbach, 1991), or for *Cameraria hamadryadella,* which shows no density-dependent effects of natural enemies in low-density populations and strong inversely density-dependent effects in high-density populations (Connor and Beck, 1993; Green, 1994).

The frequent lack of density-dependent mortality from leading causes of death, particularly among species at eruptive densities, points to an important question: Are leaf miner populations regulated around equilibrium densities? Or is wide annual variation in density common? Some multiple-year studies involving latent species have noted that although densities remain low, annual variation can be quite large (e.g., Hinckley, 1972; Faeth *et al.*, 1981a; Kato, 1994). The same is true for intergenerational changes in density over a growing season among multivoltine temperate-zone species (Faeth *et al.*, 1981a; Connor *et al.*, 1983), which may be paralleled by changes in dominant sources of mortality (Kato, 1994). In addition, studies with species at eruptive densities have generally found wide annual variation in density (Auerbach, 1991; Connor and Beck, 1993). Thus, many leaf-miner populations do not appear to reside near equilibrium densities, but rather fluctuate widely as has been suggested for other phytophages (Dempster and Pollard, 1981; Stiling, 1988).

C. Host-Plant Phenology and Leaf-Miner Population Dynamics

A common theme emerging from many studies of leaf-miner populations is that host-plant phenology can have profound effects on population dynamics. Our case studies illustrate some of the varied ways in which patterns of leaf production and abscission can directly and indirectly affect leaf-miner mortality or natality and thus be a critical determinant of population density.

Direct effects of host-plant phenology on mortality rates generally occur toward the end of the mining phase of development. Premature leaf abscission is an important source of mortality for many species, both latent and eruptive. Similarly, as we described for *Cameraria hamadryadella,* annual variation in the timing of autumn leaf fall can greatly affect mortality of some multivoltine temperate-zone species, particularly species with the ability to impose additional facultative generations (Connor *et al.*, 1994).

Variation in host-plant phenology can also profoundly influence oviposition success and natality. Many species of leaf-mining insects feed exclusively or preferentially on young leaves produced either at budbreak or in secondary leaf flushes. Limited temporal or spatial availability of leaves appears to be largely responsible for density patterns of both latent and eruptive species, as we have detailed for *Stilbosis quadricustatella* and *Phyllonorycter tremuloidiella.* Indeed, close matching of leaf-miner and host-plant phenologies or unusually high availability of preferred leaves can initiate eruptions of some species of leaf miners (Wallace, 1970; Auerbach and Simberloff, 1984; Potter and Redmond, 1989; Auerbach, 1991). Variation in the quality and availability of leaves during critical life-history stages appears to be a major force driving the population dynamics of many species of leaf-mining insects.

Acknowledgments

We would like to thank M. W. Beck, T. C. Carr, H. V. Cornell, H. Damman, P. W. Price, and P. Stiling for critically reading the manuscript. Special thanks to Peter Stiling, who generously provided density data on *S. quadricustatella*. We owe a debt to Mike Beck for his thoughtful insights and assistance with many experiments described herein. Financial support was provided by NSF Grants RII-8610675, DEB-9306657 (M. J. A.), BSR 90-07144 (S. M.), DEB-8021779, BSR-8812989, BSR-9112013, DIR-9115117, and BBS-9200136 (E. F. C.).

References

Alberts, J. D. (1989). Success of the aspen blotch miner, *Phyllonorycter salicifoliella,* on three host-tree species. M.S. Thesis, University of North Dakota, Grand Forks.

Askew, R. R. (1975). The organisation of chalcid-dominated parasitoid communities centered upon endophytic hosts. *In* "Evolutionary Strategies of Parasitic Insects and Mites" (P. W. Price, ed.), pp. 130–153. Plenum, London.

Askew, R. R., and Shaw, M. R. (1979). Mortality factors affecting the leaf-mining stages of *Phyllonorycter* (Lepidoptera: Gracillariidae) on oak and birch. 1. Analysis of mortality factors. *Zool. J. Linn. Soc.* **67,** 31–49.

Askew, R. R., and Shaw, M. R. (1986). Parasitoid communities: Their size, structure, and development. *In* "Insect Parasitoids" (J. Waage and D. Greathead, eds.), pp. 225–264. Academic Press, London.

Athey, L. A., and Connor, E. F. (1989). The relationship between foliar nitrogen content and feeding by *Odontota dorsalis* Thun. on *Robinia pseudoacacia. Oecologia* **79,** 390–394.

Auerbach, M. J. (1990). Population dynamics of a leaf-mining insect at endemic and epidemic densities. *Symp. Biol. Hung.* **39,** 435–436.

Auerbach, M. J. (1991). Relative impact of interactions within and between trophic levels during an insect outbreak. *Ecology* **72,** 1599–1608.

Auerbach, M. J., and Alberts, J. D. (1992). Occurrence and performance of the aspen blotch miner, *Phyllonorycter salicifoliella,* on three host-tree species. *Oecologia* **89,** 1–9.

Auerbach, M. J., and Simberloff, D. (1984). Responses of leaf miners to atypical leaf production patterns. *Ecol. Entomol.* **9,** 361–367.

Auerbach, M. J., and Simberloff, D. (1988). Rapid leaf-miner colonization of introduced trees and shifts in sources of herbivore mortality. *Oikos* **52,** 41–50.

Auerbach, M. J., and Simberloff, D. (1989). Oviposition site preference and larval mortality in a leaf-mining moth. *Ecol. Entomol.* **14,** 131–140.

Bale, J. S. (1984). Budburst and success of the beech weevil, *Rhynchaenus fagi* L.: Feeding and oviposition. *Ecol. Entomol.* **9,** 139–140.

Barrett, B. A., and Brunner, J. F. (1990). Temporal distribution of *Phyllonorycter elmaella* (Lepidoptera: Gracillariidae) and its major parasitoid, *Pnigalio flavipes* (Hymenoptera: Eulophidae) in Washington apple orchards. *Environ. Entomol.* **19,** 362–369.

Blais, J. R., and Pilon, J. G. (1968). Influence of temperature and moisture on the survival of cocoons, and on adult emergence of *Bucculatrix canadensisella. Can. Entomol.* **100,** 742–749.

Bultman, T. L., and Faeth, S. H. (1985). Patterns of intra- and interspecific association in leaf-mining insects on three oak host species. *Ecol. Entomol.* **10,** 121–129.

Bultman, T. L., and Faeth, S. H. (1986a). Experimental evidence for intraspecific competition in a lepidopteran leaf miner. *Ecology* **67**, 442–448.

Bultman, T. L., and Faeth, S. H. (1986b). Effects of within-leaf density and leaf size on pupal weight of a leaf-miner, *Cameraria* (Lepidoptera: Gracillariidae). *Southwest. Nat.* **51**, 201–206.

Cappuccino, N. (1991). Density dependent mortality of phytophagous insects on goldenrod. *Environ. Entomol.* **20**, 1121–1128.

Connor, E. F. (1984). The causes of overwintering mortality in *Phyllonorycter* on *Quercus robur*. *Ecol. Entomol.* **9**, 23–28.

Connor, E. F. (1991). Colonization, survival, and causes of mortality of *Cameraria hamadryadella* (Lepidoptera: Gracillariidae) on four species of host plants. *Ecol. Entomol.* **16**, 315–322.

Connor, E. F., and Beck, M. W. (1993). Density-related mortality in *Cameraria hamadryadella* (Lepidoptera: Gracillariidae) at epidemic and endemic densities. *Oikos* **66**, 515–525.

Connor, E. F., and Cargain, M. J. (1994). Density-related foraging behaviour in *Closterocerus tricinctus*, a parasitoid of the leaf-mining moth, *Cameraria hamadryadella*. *Ecol. Entomol.* **19**, 327–334.

Connor, E. F., Faeth, S. H., and Simberloff, D. (1983). Leaf miners on oak: The role of immigration and *in situ* reproductive recruitment. *Ecology* **64**, 191–204.

Connor, E. F., Adams-Manson, R. J., Carr, T. G., and Beck, M. W. (1994). The effects of host plant phenology on the demography and population dynamics of the leaf-mining moth, *Cameraria hamadryadella* (Lepidoptera: Gracillariidae). *Ecol. Entomol.* **19**, 111–120.

Day, K. R., and Watt, A. D. (1989). Population studies of the beech mining weevil (*Rhynchaenus fagi*) in Ireland and Scotland. *Ecol. Entomol.* **14**, 23–30.

Delucchi, V. L. (1958). *Lithocolletis messaniella* Zeller (Lep. Gracillariidae): Analysis of some mortality factors with particular reference to its parasite complex. *Entomophaga* **3**, 203–270.

Dempster, J. P., and Pollard, E. (1981). Fluctuations in resource availability and insect populations. *Oecologia* **50**, 412–416.

Dharmadhikari, P. P., Perera, P. A. C., and Hassen, T. M. F. (1977). A short account of the biological control of *Promecotheca cumingi* (Col.: Hispidae) the coconut leaf-miner, in Sri Lanka. *Entomophaga* **22**, 3–18.

Drea, J. J., Jr., and Hendrickson, R. M., Jr. (1986). Analysis of a successful biological control project: The alfalfa blotch leafminer (Diptera: Agromyzidae) in the northeastern United States. *Environ. Entomol.* **15**, 448–455.

Faeth, S. H. (1980). Invertebrate predation of leaf-miners at low densities. *Ecol. Entomol.* **5**, 111–114.

Faeth, S. H. (1986). Indirect interactions between temporally separated herbivores mediated by the host plant. *Ecology* **67**, 479–494.

Faeth, S. H. (1990). Aggregation of a leafminer, *Cameraria* sp. nov. (Davis): Consequences and causes. *J. Anim. Ecol.* **59**, 569–586.

Faeth, S. H. (1991a). Novel aspects of host tree resistance to leafminers. *USDA For. Serv. Gen. Tech. Rep.* **NE-153**, 219–239.

Faeth, S. H. (1991b). Effect of oak leaf size on abundance, dispersion, and survival of the leafminer *Cameraria* sp. (Lepidoptera: Gracillariidae). *Environ. Entomol.* **20**, 196–204.

Faeth, S. H., and Simberloff, D. (1981). Population regulation in a leaf-mining insect, *Cameraria* sp. nov., at increased field densities. *Ecology* **62**, 620–624.

Faeth, S. H., Mopper, S., and Simberloff, D. (1981a). Abundances and diversity of leaf-mining insects on three oak host species: Effects of host–plant phenology and nitrogen content of leaves. *Oecologia* **37**, 238–251.

Faeth, S. H., Connor, E. F., and Simberloff, D. (1981b). Early leaf abscission: A neglected source of mortality for folivores. *Am. Nat.* **117**, 409–415.

Freeman, B. E., and Smith, D. C. (1990). Variation of density-dependence with spatial scale in the leaf-mining fly *Liriomyza commelinae* (Diptera: Agromyzidae). *Ecol. Entomol.* **15**, 265–274.

Fritz, R. (1990). Variable survival and impact of natural enemies on two herbivores among willow genotypes. *Bull. Ecol. Soc. Am.* **71**, 160.

Green, E. V. (1994). The effect of competing sources of mortality on the detection of density dependence in insect parasitism. B.A. Honors Thesis, University of Virginia, Charlottesville.

Hawkins, B. A. (1988a). Species diversity in the third and fourth trophic levels: Patterns and mechanism. *J. Anim. Ecol.* **57**, 137–162.

Hawkins, B. A. (1988b). Foliar damage, parasites, and indirect competition: A test using herbivores of birch. *Ecol. Entomol.* **13**, 301–308.

Hawkins, B. A. (1990). Global patterns of parasitoid assemblage size. *J. Anim. Ecol.* **59**, 57–72.

Hawkins, B. A., and Lawton, J. H. (1987). Species richness for parasitoids of British phytophagous insects. *Nature (London)* **326**, 788–790.

Hawkins, B. A., Askew, R. R., and Shaw, M. R. (1990). Influences of host feeding-niche and foodplant type on generalist and specialist parasitoids. *Ecol. Entomol.* **15**, 275–280.

Hawkins, B. A., Shaw, M. R., and Askew, R. R. (1992). Relations among assemblage size, host specialization, and climatic variability in North American parasitoid communities. *Amer. Nat.* **139**, 58–79.

Heads, P. A., and Lawton, J. H. (1983a). Studies on the natural enemy complex of the holly leafminer: The effects of scale on the detection of aggregative responses and the implications for biological control. *Oikos* **40**, 267–276.

Heads, P. A., and Lawton, J. H. (1983b). Tit predation on the holly leaf-miner: The effect of prickly leaves. *Oikos* **41**, 161–164.

Hering, E. M. (1951). "Biology of the Leafminers." Dr. W. Junk, s'Gravenhage, The Netherlands.

Hespenheide, H. A. (1991). Bionomics of leaf-mining insects. *Annu. Rev. Entomol.* **36**, 535–560.

Hinckley, A. D. (1972). Comparative ecology of two leafminers on white oak. *Environ. Entomol.* **1**, 258–261.

Itamies, J., and Ojanen, M. (1977). Autumn predation of *Parus major* and *P. montanus* upon two leaf-mining species of *Lithocolletis* (Lepidoptera, Lithocolletidae). *Ann. Zool. Fenn.* **14**, 235–241.

Kahn, D. M., and Cornell, H. V. (1983). Early leaf abscission and folivores: Comments and considerations. *Am. Nat.* **122,** 428–432.

Kahn, D. M., and Cornell, H. V. (1989). Leafminers, early leaf abscission, and parasitoids: A tritrophic interaction. *Ecology* **70,** 1219–1226.

Kato, M. (1994). Alternation of bottom-up and top-down regulation in a natural population of an agromyzid leafminer, *Chromatomyia suikazurae. Oecologia* **97,** 9–16.

Koricheva, J. (1994). Air pollution and Eriocrania miners: Observed interactions and possible mechanisms. Ph.D. Thesis, University of Turku.

Lessells, C. M. (1985). Parasitoid foraging: Should parasitism be density dependent? *J. Anim. Ecol.* **54,** 27–41.

Maier, C. T., and Davis, D. R. (1989). Southern New England host and distributional records of Lithocolletine Gracillariidae (Lepidoptera). *Misc. Publ. Entomol. Soc. Am.* **70.**

Marino, P. C., Cornell, H. V., and Kahn, D. H. (1993). Environmental and clonal influences on host choice and larval survival in a leafmining insect. *J. Anim. Ecol.* **62,** 503–510.

Martin, J. L. (1956). The bionomics of the aspen blotch miner, *Lithocolletis salicifoliella* Cham. (Lepidoptera: Gracillariidae). *Can. Entomol.* **88,** 155–168.

Mazanec, Z. (1985). Resistance of *Eucalyptus marginata* to *Perthida glyphopa* (Lepidoptera: Incurvariidae). *J. Aust. Entomol. Soc.* **24,** 209–221.

McNeil, J. N., and Quiring, D. T. (1983). Evidence of an oviposition-deterring pheromone in the alfalfa blotch leaf-miner, *Agromyza frontella* (Rond.) (Diptera: Agromyzidae). *Environ. Entomol.* **12,** 990–992.

Miller, P. F. (1973). The biology of some *Phyllonorycter* species (Lepidoptera: Gracillariidae) mining leaves of oak and beech. *J. Nat. Hist.* **7,** 391–409.

Mopper, S., and Simberloff, D. (1995). Differential herbivory in an oak population: The role of plant phenology and insect performance. *Ecology* **76,** 1233–1241.

Mopper, S., Faeth, S. H., Boecklen, W. J., and Simberloff, D. S. (1984). Host-specific variation in leafminer population dynamics: Effects on density, natural enemies and behaviour of *Stilbosis quadricustatella* (Lepidoptera: Cosmopterigidae). *Ecol. Entomol.* **9,** 169–177.

Mopper, S., Beck, M., Simberloff, D., and Stiling, P. (1995). Local adaptation and agents of mortality in a mobile insect. *Evolution,* (Lawrence, Kans.) (in press).

Munster-Swendsen, M. (1989). Phenology and natural mortalities of the fir needleminer, *Epinotia fraterna* (Hw.) (Lepidoptera, Tortricidae). *Entomologiske Meddelelser.* **57,** 111–120.

Munster-Swendsen, M. (1991). The effect of sublethal neogregarine infections in the spruce needleminer, *Epinotia tedella* (Lepidoptera: Tortricidae). *Ecol. Entomol.* **16,** 211–219.

Needham, J. G., Frost, S. W., and Tothill, B. H. (1928). "Leaf-Mining Insects." Williams & Wilkins, Baltimore.

Nielsen, B. O. (1968). Studies of the fauna of beech foliage. 2. Observations on the mortality and mortality factors of the beech weevil [*Rhynchaenus (Orchestes) fagi* L.] (Coleoptera: Curculionidae). *Natura Jutlandica* **14,** 99–125.

Owen, D. F. (1975). The efficiency of blue-tits *Parus caeruleus* preying on larvae of *Phytomyza ilicis. Ibis* **117,** 515–516.

Potter, D. A. (1985). Population regulation of the native holly leafminer, *Phytomyza ilicicola* Loew (Diptera: Agromyzidae), on American holly. *Oecologia* **66**, 499–505.

Potter, D. A., and Kimmerer, T. W. (1986). Seasonal allocation of defensive investment in *Ilex opaca* Aiton and constraints on a specialist leaf-miner. *Oecologia* **69**, 217–224.

Potter, D. A., and Redmond, C. T. (1989). Early spring defoliation, secondary leaf flush, and leafminer outbreaks on American holly. *Oecologia* **81**, 192–197.

Pottinger, R. P., and LeRoux, E. J. (1971). The biology and dynamics of *Lithocolletis blancardella* (Lepidoptera: Gracillariidae) on apple in Quebec. *Mem. Entomol. Soc. Can.* **77**, 1–437.

Powell, J. A. (1980). Evolution of larval food preferences in microlepidoptera. *Annu. Rev. Entomol.* **25**, 133–159.

Preszler, R. W., and Boecklen, W. J. (1994). A three-trophic-level analysis of the effects of plant hybridization on a leaf-mining moth. *Oecologia* **100**, 66–73.

Pritchard, I. M., and James, R. (1984). Leaf fall as a source of leaf miner mortality. *Oecologia* **64**, 140–141.

Quiring, D. T., and McNeil, J. N. (1984). Exploitation and interference intraspecific larval competition in the dipteran leaf-miner, *Agromyza frontella* (Rondani). *Can. J. Zool.* **62**, 421–427.

Sekita, N., and Yamada, M. (1979). Studies on the population of the apple leaf miner *Phyllonorycter ringoniella* Matsumura (Lepidoptera: Lithocolletidae). III. Some analyses of the mortality factors operating on the population. *Appl. Entomol. Zool.* **14**, 137–148.

Simberloff, D. S., and Stiling, P. (1987). Larval dispersion and survivorship in a leaf-mining moth. *Ecology* **68**, 1647–1657.

Smith, D. R. (1976). World genera of the leaf-mining sawfly tribe Fenusini (Hymenoptera: Tenthredinidae). *Entomol. Scand.* **7**, 253–260.

Smith, D. R. (1993). Systematics, life history, and distribution of sawflies. *In* "Sawfly Life History Adaptations to Woody Plants" (M. P. Wagner and K. F. Raffa, eds.), pp. 3–32. Academic Press, San Diego, CA.

Solomon, J. D., McCracken, F. I., Anderson, R. L., Lewis, R., Jr., Olivera, F. L., Flier, T. H., and Barry, P. J. (1980). Oak pests: A guide to major insects, diseases, air pollution, and chemical injury. *USDA For. Serv. Gen. Rep.* **SA-GR-11.**

Southwood, T. R. E., and Comins, H. N. (1976). A synoptic population model. *J. Anim. Ecol.* **65**, 949–965.

Stiling, P. D. (1987). The frequency of density dependence in insect host–parasitoid systems. *Ecology* **68**, 844–856.

Stiling, P. D. (1988). Density-dependent processes and key factors in insect populations. *J. Anim. Ecol.* **57**, 581–593.

Stiling, P. D., and Simberloff, D. S. (1989). Leaf abscission: Induced defense against pests or response to damage? *Oikos* **55**, 44–49.

Stiling, P. D., Brodbeck, B. V., and Strong, D. R. (1982). Foliar nitrogen and larval parasitism as determinants of leafminer distribution patterns on *Spartina alterniflora*. *Ecol. Entomol.* **7**, 447–452.

Stiling, P. D., Brodbeck, B. V., and Strong, D. R. (1984). Intraspecific competition in *Hydrellia valida* (Diptera: Ephydridae), a leaf miner of *Spartina alterniflora*. *Ecology* **65**, 660–662.

Stiling, P. D., Simberloff, D., and Anderson, L. C. (1987). Non-random distribution patterns of leaf miners on oak trees. *Oecologia* **74,** 102–105.

Stiling, P. D., Simberloff, D. S., and Brodbeck, B. V. (1991). Variation in rates of leaf abscission between plants may affect the distribution patterns of sessile insects. *Oecologia* **88,** 367–370.

Strong, D. R., Lawton, J. H., and Southwood, T. R. E. (1984). "Insects on Plants." Harvard Univ. Press, Cambridge, MA.

Valladares, G., and Lawton, J. H. (1991). Host–plant selection in the holly leaf-miner: Does mother know best? *J. Anim. Ecol.* **60,** 227–240.

Walde, S. J., and Murdoch, W. W. (1988). Spatial density dependence in parasitoids. *Annu. Rev. Entomol.* **33,** 441–466.

Wallace, M. M. (1970). The biology of the jarrah leaf miner, *Perthida glyphopa* Common (Lepidoptera: Incurvariidae). *Aust. J. Zool.* **18,** 91–104.

PART III

MECHANISMS AND PROCESSES OF POPULATION DYNAMICS

Chapter 6

Density-Dependent Dispersal and Its Consequences for Population Dynamics

Robert F. Denno and Merrill A. Peterson

I. Introduction

A fundamental concern in insect ecology has been to determine the forces that dictate the temporal and spatial fluctuations of populations. This endeavor has engendered heated debate over numerous issues, perhaps the most divisive of which has been the role of density-dependent processes in governing population dynamics (Dempster, 1983; Hassell, 1985; Dempster and Pollard, 1986; Murdoch and Reeve, 1987). In an attempt to resolve this question, researchers have scrutinized extrinsic factors such as predation and parasitism, establishing that natural enemies act in a density-dependent manner in some cases, but not others (Stiling, 1987, 1988; Walde and Murdoch, 1988). At the same time, relatively little attention has been given to how the intrinsic process of density-dependent dispersal influences local population dynamics (but see Stiling, 1988).

Density-dependent dispersal can have a significant suppressing and stabilizing influence on population growth (McClure, 1991; Denno *et al.*, 1994; Fujisaki, 1994). For example, by minimizing losses due to emigration, density-dependent dispersal may allow local populations to build from low densities. Furthermore, increased dispersal at high densities may suppress populations and limit their potential for outbreak (Way, 1968; Dixon, 1969; Denno *et al.*, 1994). At a larger metapopulation scale, the redistribution of individuals among habitat patches via density-dependent dispersal can theoretically influence the relative densities of local populations (Gadgil, 1971; Hanski, 1991). Without a thorough review to date, we are left without a general understanding of the extent to which and the means by which density-dependent dispersal influences local population dynamics.

The primary goal in this chapter is to examine published field studies in order to answer several questions concerning the impact of dispersal on the local population dynamics of phytophagous insects. Specific objectives are to

(1) review the evidence in support of density-dependent dispersal (emigration or the production of migratory forms), (2) determine the spatial scale (intraplant, interplant, or interhabitat) at which dispersal occurs, (3) identify what mechanisms (host plant quality and quantity, feeding site preemption, or aggression) mediate emigration, and (4) determine if and how density-dependent emigration contributes importantly to local population suppression.

To address these questions, we limited our examination to the literature on the hemipteroid orders of insects, including the Thysanoptera, Homoptera, and Heteroptera. All studies included in this review directly examined the relationship between population density and either emigration or the production of migratory forms (alates, macropters, or sexuparae). Most of the studies we surveyed employed either a rigorous experimental approach or a life table/key factor analysis of density dependence.

We selected sap-feeding insects (including haustellate seed-feeders) for this review because they frequently display striking dimorphisms in flight capability, facilitating the identification of migrants for studies of density-dependent dispersal. As a final objective, we considered whether the impact of density-dependent emigration on the suppression of populations of sap-feeders is representative of its importance to the population dynamics of phytophagous insects as a whole.

II. Evidence for Density-Dependent Dispersal

Our review of the literature found widespread evidence for density-dependent dispersal in the Thysanoptera, Homoptera, and Heteroptera. In all, there was a clear demonstration of density-dependent emigration or the expression of the migratory form in 72 species distributed in 12 families (Table 1). Examples included the density-dependent production of migratory forms in wing-dimorphic insects such as the winged macropters of planthoppers (Fig. 1A), seed bugs (Fig. 1B), and thrips, and the alates (Fig. 1C) and sexuparae (Fig. 1D) of aphids. For wing-dimorphic sap-feeders in general, wing form is controlled by a hormonally mediated developmental switch that responds to environmental cues such as crowding (Dixon, 1985; Denno, 1994). The sensitivity of the switch to population density, however, is heritable and often under polygenic control (Denno *et al.*, 1991). Thus, as populations grow in size, the fraction of individuals molting into flight-capable adults increases; at low densities most individuals molt into flightless adults with reduced wings (brachypters) or no wings (apterae) (Dixon, 1985; Denno, 1994). In all cases where they have been explicitly compared, the volant forms of these wing-dimorphic sap-feeders emigrate more readily and disperse much farther than their flightless counterparts (e.g., Murai, 1977; Waloff, 1980; Roderick, 1987). In addition to evidence that the production of these migratory forms is density dependent, several studies provided evidence that the

TABLE 1
**Number of Sap-Feeding Insect Species Exhibiting Density-Dependent Dispersal (Emigration)
or the Density-Dependent Expression of the Migratory Wing Morph[a]**

Taxon	Dispersal by winged adult	Dispersal by wingless adult	Dispersal by nymph
THYSANOPTERA (Thrips)			
Tubulifera			
Phlaeothripidae	1 (macropter)		
Terebrantia			
Thripidae	1 (macropter)		
HOMOPTERA			
Sternorrhyncha			
Aphidoidea			
Adelgidae	3 (sexuparae)		3 (early instars)
(conifer woolly aphids)			
Aphididae (aphids)	23 (alatae)	11 (apterae)	2 (several instars)
Psylloidea			
Psyllidae (psyllids)	2 (monomorphic)		2 (several instars)
Aleyrodoidea			
Aleyrodidae (whiteflies)	1 (monomorphic)		
Coccoidea			
Diaspididae (hard scales)			3 (1st-instar crawler)
Margarodidae			1 (1st-instar crawler)
(margarodid scales)			
Auchenorrhyncha			
Cicadellidae (leafhoppers)	2 (monomorphic)		
Delphacidae (planthoppers)	12 (macropter)		
HETEROPTERA			
Lygaeidae (seed bugs)	2 (monomorphic)		
	1 (macropter)		
Miridae (mirid bugs)	2 (monomorphic)		
Total	50	11	11 (72)

[a]References: THYSANOPTERA: Phlaeothripidae (Crespi, 1988); Thripidae (Kamm, 1972). HO-
MOPTERA: Adeligidae (Varty, 1956; Eichhorn, 1969; McClure, 1984, 1989, 1991); Aphididae (Itô,
1960; Johnson, 1965; Toba et al., 1967; Waloff, 1968; Way, 1968; Dixon, 1969, 1972, 1985;
Tamaki and Allen, 1969; Dixon and Glen, 1971; Judge and Schaefers, 1971; Shaw, 1973; Dixon and
Barlow, 1979; Watt and Dixon, 1981; Delisle et al., 1983; Kidd and Tozier, 1984; Edson, 1985;
Kidd and Cleaver, 1986; Moran, 1986; Whitham, 1986; Cappuccino, 1987; Lamb and MacKay,
1987; Antolin and Addicott, 1988; Wright and Cone, 1988; Kidd, 1988; Vehrs et al., 1992);
Psyllidae (Waloff, 1968); Aleyrodidae (Baumgärtner and Yano, 1990); Diaspididae (Willard, 1973;
McClure, 1979, 1980); Margarodidae (McClure, 1977, 1990); Cicadellidae (Kuno, 1984; Widiarta et
al., 1992; Aryawan et al., 1993); Delphacidae (Kisimoto, 1965; Raatikainen, 1967; Everett, 1969;
Kuno and Hokyo, 1970; Metcalfe, 1972; Mochida, 1973; May, 1975; Denno, 1976; Drosopoulos,
1977; Bull, 1981; Fisk et al., 1981; Denno et al., 1985, 1991; Iwanaga et al., 1987; Mori and
Nakasuji, 1991; Denno and Roderick, 1992; Matsumura, 1994). HETEROPTERA: Miridae (Gibson,
1980; Gibson and Visser, 1982); Lygaeidae (Murai, 1977; Solbreck, 1978, 1985; Fujisaki, 1985,
1994; McLain and Shure, 1990; Solbreck and Sillén-Tullberg, 1990).

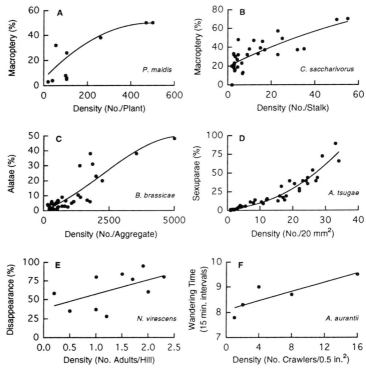

Figure 1. Evidence for the density-dependent expression of migratory forms and density-dependent dispersal in sap-feeding insects. Density-dependent production of (A) macropters in the corn plant-hopper *Peregrinus maidis* (adapted from Fisk *et al.*, 1981), (B) macropters of the oriental chinch bug *Cavelerius saccharivorus* (adapted from Fujisaki, 1994), (C) alatae of the cabbage aphid *Brevicoryne brassicae* (adapted from Way, 1968), and (D) winged sexuparae of the hemlock woolly adelgid *Adelges tsugae* (adapted from McClure, 1991). (E) Density-dependent disappearance (emigration) by flight of the green rice leafhopper *Nephotettix virescens* (adapted from Widiarta *et al.*, 1992). (F) Density-dependent increase in wandering time of the first-instar crawlers of the California red scale *Aonidiella aurantii*. Increased wandering time results in crawlers settling farther from their site of hatching (adapted from Willard, 1973).

emigration of migratory forms from the immediate habitat is also density dependent (e.g., Way, 1968; Murai, 1977).

For monomorphic winged species such as leafhoppers, psyllids, whiteflies, and some heteropterans, there were several studies suggesting that emigration by flight is density dependent (e.g., Fig. 1E). In all, 50 species exhibited density-dependent dispersal as winged adults (Table 1). Similarly, small-scale, density-dependent dispersal by walking within and among plants was noted for the apterous adults of aphids (11 species), the nymphs of aphids, adelgids, and psyllids (7 species), and first-instar crawlers of scale insects (4 species) (Fig. 1F and Table 1). For most of the dispersal responses, the relationship between population size and dispersal capability (e.g., percentage macropterous form) or emigration rate was often quite tight; there was little evidence that the relationship exhibited features of being density vague (Fig. 1). For instance, the fraction of adults molting to migratory forms or the proportion of emigrants rose continuously with an increase in crowding.

Our review also revealed that density-dependent dispersal occurred across a range of spatial scales. Intraplant dispersal was found for 20 species of sap-feeders, mostly the rather immobile nymphs of aphids, psyllids, and scale insects (e.g., Waloff, 1968; Willard, 1973; McClure, 1980, 1984), and the apterous adults of aphids (Itô, 1960; Whitham, 1986). At the interplant scale, plant contact typically facilitated the density-dependent dispersal (mostly by walking) of 14 species of aphids, whiteflies, and mirid bugs (e.g., Gibson, 1980; Moran, 1986; Cappuccino, 1987; Baumgärtner and Yano, 1990). Interhabitat dispersal occurred in 25 species and resulted from the density-dependent emigration of the winged macropters and alates of thrips, homopterans, and lygaeid bugs (e.g., Kamm, 1972; Denno *et al.,* 1994; Fujisaki, 1994) and the airborne dispersal of scale crawlers (McClure, 1977).

The mechanisms mediating density-dependent dispersal were diverse, but were dominated by host-plant-related factors (Table 2). For example, for 28 of 40 species, host-plant condition influenced either the density-dependent production of migratory forms or the emigration process itself. For the most part, density-dependent dispersal was intensified on poor-quality or nutrient-deficient plants (e.g., Kisimoto, 1965; Fisk *et al.,* 1981; Watt and Dixon, 1981; Edson, 1985; Denno *et al.,* 1985). For 15 of these species, the deteriorated plant condition prompting the dispersal response was feeding induced by the insects themselves. For 14 of these species, a decline in plant vigor or nutritional quality at high sap-feeder densities triggered an increase in dispersal (e.g., Kisimoto, 1965; Fisk *et al.,* 1981; Dixon, 1985). There was also one case in which gall aphids induced premature leaf abscission in a density-dependent fashion; some aphids escaped leaf drop by migrating from deteriorating galls to nearby galls in less danger of abscission (Williams and Whitham, 1986).

Another plant-mediated dispersal response occurred in two lygaeid species when feeding influenced the quantity of the seed resource. Subsequently, adult bugs emigrated from areas of low seed abundance (Solbreck, 1978, 1985; Solbreck and Sillén-Tullberg, 1990). For several other species of Heteroptera, food shortage stimulates flight muscle development and/or promotes migration (Dingle and Arora, 1973; Kehat and Wyndham, 1973; Solbreck, 1986), but the link between resource abundance and herbivore density has not been explicitly demonstrated.

TABLE 2
Mechanisms Mediating Density-Dependent Dispersal/Expression of the Dispersal Morph[a]

Taxon	Host-plant condition influences density effect (feeding induced)	Resource abundance (feeding influenced)	Feeding site preemption	Aggressive behavior/tactile stimulation
THYSANOPTERA				
Tubulifera				
Phlaeothripidae	1			
Terebrantia				
Thripidae	1 (1)			
HOMOPTERA				
Sternorrhyncha				
Aphidoidea				
Adelgidae	3 (3)		1	
Aphididae	16 (8)			2
Psylloidea				
Psyllidae			2	
Aleyrodoidea				
Aleyrodidae	1			
Coccoidea				
Diaspididae			3	
Margarodidae	1 (1)			
Auchenorrhyncha				
Delphacidae	5 (2)			
HETEROPTERA				
Lygaeidae		2 (1)		
Miridae				2
Total	28 (15)	2 (1)	6	4

[a]For references, see Table 1.

Two other less frequent dispersal-mediating mechanisms were feeding site preemption (six species) and aggressive behavior/tactile stimulation (four species). Feeding site preemption occurred mainly in scale insects and psyllids, when at high standing-room-only densities, nymphs were forced to emigrate farther on the plant to locate suitable feeding sites (Waloff, 1968; McClure, 1979, 1980, 1984, 1989, 1990). Also, aggressive and territorial behavior resulted in small-scale dispersal in pemphigine aphids and mirid bugs when direct encounters forced conspecifics and heterospecifics to occupy suboptimal sites elsewhere, or caused them to flee outright (Gibson, 1980; Gibson and Visser, 1982; Whitham, 1986). In addition, tactile stimulation among individuals independent of induced changes in plant condition triggered the production of alates in the aphid *Megoura viciae* (Dixon, 1985).

III. Consequences of Density-Dependent Dispersal for Local Population Dynamics

As our primary pursuit, we examined the literature for evidence that density-dependent emigration acts as an important suppressing or "stabilizing" force in local population dynamics. Support for density-dependent dispersal as a stabilizing force was provided by evidence either that high levels of dispersal contributed to population suppression or stabilization around some equilibrium level or that low levels of dispersal conferred little suppressing effect on the population. For this assessment, which was restricted to species exhibiting density-dependent dispersal, most studies fell into one of two groupings, either rigorous experimental demonstrations (e.g., Dixon, 1969; Cappuccino, 1987; McClure, 1991; Fujisaki, 1994) or life table and key factor analyses (e.g., Metcalfe, 1972; Aryawan *et al.*, 1993).

Our assessment showed that for 27 species representing a diversity of both wing-dimorphic and monomorphic sap-feeders, high levels of dispersal contributed to population suppression or "stabilization" (e.g., Way, 1968; Kuno and Hokyo, 1970; Dixon and Barlow, 1979; McClure, 1991). In 9 other species of relatively immobile aphids and scale insects, low levels of dispersal were inadequate to suppress populations and often encouraged outbreak, especially when other factors such as natural enemies were inconsequential in the system (e.g., McClure, 1979, 1980, 1989; Kidd, 1988).

To make a more quantitative assessment of the impact of density-dependent dispersal on local population dynamics, we restricted our analysis to wing-dimorphic species because the fraction of potential dispersers is easily quantified. Species were divided into two categories, one in which adult-stage dispersal was determined to be an important suppressing process in population dynamics, and another in which adult dispersal was considered to be comparatively unimportant. The importance of density-dependent dispersal in population dynamics was judged from studies (1) examining the link between density-dependent population suppression and density-dependent dispersal (e.g., Denno *et al.*, 1994), (2) employing an experimental analysis of population change following density-dependent dispersal (e.g., Edson, 1985; Lamb and MacKay, 1987; Fujisaki, 1994), and (3) assessing whether density-dependent dispersal was a major "mortality source" through life table or key factor analysis (e.g., Wright and Cone, 1988). Authors' conclusions in this regard were often explicit. For example, McClure (1991) concluded that "the density-dependent production of winged sexuparae played a major role in the decline of [the adelgid] *Adelges tsugae* populations." In contrast, Kidd (1988) maintained that "density-dependent emigration by flight has little effect in reducing population numbers [of the aphid, *Cinara pinea* because] . . . alatae are not produced for long enough or in adequate quantities to have any regulative significance." After assigning species to the appropriate category, we then compared the highest incidence of migratory forms occurring in field populations between the two groups of species. We hypothesized that when density-dependent dispersal contributes substantially to local population suppression, the incidence of migratory forms should be greater than when density-dependent dispersal has minor impact on population change.

The maximum incidence of migratory forms was much higher (81%) for species in which dispersal was shown to be an important restraining process (Table 3). In contrast, for species in which dispersal was deemed inconsequential, the percentage of migratory forms averaged only 11%. This difference was highly significant (*t*-test on arcsine-transformed proportions, $t = 9.06$, $P < 0.001$). Thus, density-dependent dispersal appeared to contribute to local population decline only when levels of dispersal were high.

An examination of the population biology of three species of wing-dimorphic planthoppers provides insight into the relationship between density-dependent dispersal and the density-dependent suppression of populations. All three of these species exhibit a density-dependent production of migratory forms in field populations (Figs. 2A–2C). However, *Nilaparvata lugens* and *Prokelisia marginata* are very migratory, macropters are triggered at very low densities (Figs. 2A and 2B), and average levels of macroptery in the field are high, ranging from 70 to 90%. In contrast, *Prokelisia dolus* is quite sedentary, macropters are triggered at a much higher density (Fig. 2C), and levels of macroptery in the field average <15%.

For mobile species such as *N. lugens* and *P. marginata,* we hypothesized that dispersal may be the most important factor contributing to density-dependent change in planthopper populations. If this were so, then density dependence should be more evident in populations of migratory species compared to sedentary ones.

Using Varley and Gradwell's (1963) technique, we examined population data from the three species of planthoppers for density dependence. With this technique, density-dependent suppression is indicated when a plot of population size in the subsequent generation (N_{t+1}) against population size in the preceding generation (N_t) results in a slope significantly <1.

Specifically for these three trivoltine species, we asked if population growth from the first to third generation was adversely affected by density. For both migratory species (Figs. 2D and 2E), the regression slope was significantly and substantially <1, suggesting strong density dependence. In contrast for the sedentary *P. dolus,* there was no evidence for density-dependent suppression; the slope of the regression was not significantly different from 1 (Fig. 2F). These

TABLE 3

Incidence of Migratory Forms (%) at High Population Densities in the Field and the Importance of Dispersal in Local Population Suppression

Taxon	Dispersal important	Dispersal unimportant	Reference
HOMOPTERA			
Adeligidae			
Adelges tsugae	100		McClure (1991)
Aphididae			
Acyrthosiphon pisum	80		Lamb and MacKay (1987)
Acyrthosiphon spartii	70		Waloff (1968)
Aphis fabae	87		Way (1968)
Aphis gossypii		5	Tamaki and Allen (1969); Vehrs *et al.* (1992)
Brachycorynella asparagi	70		Wright and Cone (1988)
Brevicoryne brassicae	50		Way (1968)
Cinara pinea		20	Kidd and Tozier (1984); Kidd (1988)
Macrosiphoniella sanborni		5	Tamaki and Allen (1969)
Uroleucon caligatum	90		Cappuccino (1987)
Uroleucon gravicorne	60		Moran (1986)
Uroleucon nigrotuberculatum	90		Edson (1985); Cappuccino (1987)
Delphacidae			
Javesella pellucida	90		Raatikainen (1967)
Nilaparvata lugens	95		Kisimoto (1965); Kuno and Hokyo (1970)
Prokelisia dolus		15	Denno and Roderick (1992); Denno *et al.* (1994)
Prokelisia marginata	85		Denno and Roderick (1992); Denno *et al.* (1994)
HETEROPTERA			
Lygaeidae			
Cavelerius saccharivorus	90		Fujisaki (1985, 1994)
Mean ± SE	81 ± 4	11 ± 4	

data for planthoppers display a link between density-dependent dispersal and density-dependent population suppression, but only in those species that readily produce dispersers.

That dispersal is an important factor contributing to population decline can be seen in *Nilaparvata lugens,* for which there is a significant negative relationship between population growth rate and the percentage of macropterous adults in the population (Fig. 3A). Similarly, for the wing-polymorphic lygaeid *Cave-*

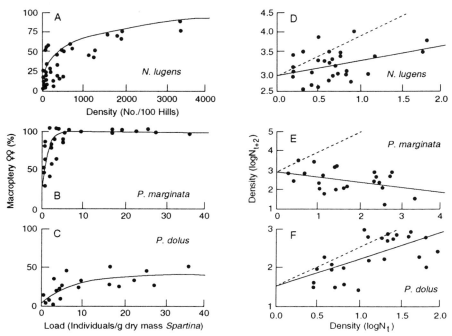

Figure 2. The association between the intensity of density-dependent dispersal and density-dependent population suppression in three planthopper species. The relationship between macroptery (%) and density in field populations of (A) the brown planthopper *Nilaparvata lugens* (adapted from Denno *et al.*, 1994) and the salt marsh planthoppers (B) *Prokelisia marginata* and (C) *Prokelisia dolus* (adapted from Denno and Roderick, 1992). Evidence for density-dependent population suppression in (D) *N. lugens* and (E) *P. marginata* [density dependence is indicated by slopes significantly <1 following Varley and Gradwell (1963)]. (F) Lack of density-dependent suppression in *P. dolus* (slope not significantly different from 1). D–F adapted from Denno *et al.* (1994).

lerius saccharivorus, populations decline more rapidly as the dispersal ability of adults (indexed as relative wing length) increases in the population (Fig. 3B). Also, for the cabbage aphid *Brevicoryne brassicae,* a balance between additions attributable to birth and losses due to emigrating alates contributes to local population equilibrium (Fig. 3C). When populations are small, births outweigh emigration and populations grow; in contrast, when populations are large, emigration exceeds birth and population growth is suppressed.

At high densities, populations increase less both because fewer adults remain in the local habitat to reproduce and because the macropters or alates that do remain are inherently less fecund than are flightless brachypters (reviewed in Denno, 1994). For example, a phenotypic trade-off between flight capability and

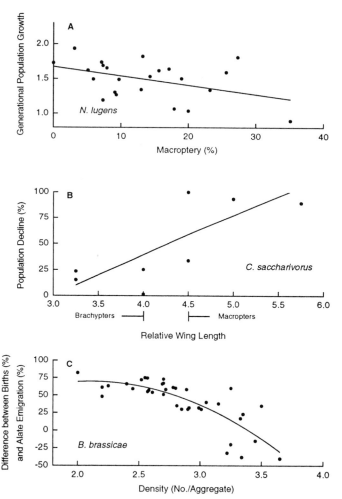

Figure 3. Relationship between (A) generational population growth and macroptery (%) in field populations of *Nilaparvata lugens* (adapted from Denno *et al.*, 1994) and (B) population decline (%) and relative wing length (index of macroptery) of *Cavelerius saccharivorus* (adapted from Fujisaki, 1985). (C) Relationship between population density and the difference between additions by birth (%) and losses due to alate emigration (%) in the cabbage aphid *Brevicoryne brassicae* (adapted from Way, 1968).

reproductive success imposes constraints on the fecundity of the dispersal morph in many wing-dimorphic sap-feeders, including planthoppers (Denno *et al.*, 1989), aphids (Dixon, 1985), and some heteropterans (Fujisaki, 1986). Reproductive penalties associated with dispersal are also very evident for monomorphic sap-feeders such as leafhoppers and lygaeid bugs (Solbreck, 1978, 1985; Waloff, 1980).

Without a doubt, factors other than dispersal also contribute to the density-dependent suppression of sap-feeder populations. For example, reduced survival and extended development often accompany the production of migratory forms under crowded conditions, and all of these factors act in concert to suppress population growth (Dixon, 1985; Denno *et al.*, 1994). Nevertheless, in many of the studies we reviewed, density-dependent dispersal was targeted as the intraspecific factor most responsible for the suppression of sap-feeder populations (Table 3).

IV. Conclusions and Prospectus

In summary, we found widespread evidence of density-dependent dispersal in sap-feeding insects, as measured by both increased emigration of individuals and heightened production of migratory forms. Among the taxa we reviewed, density-dependent dispersal occurred with similar frequency at the intraplant, interplant, and interhabitat spatial scales. The most important factor mediating density-dependent emigration or the production of migratory forms was declining host-plant quality, a bottom-up influence. Finally, we found strong indication that density-dependent dispersal can suppress the growth of populations, but only in those species that readily produce dispersers and emigrate.

We recognize that dispersal can occur independent of density in sap-feeders, both in species that exhibit obligate seasonal migration (Solbreck, 1985) and in species for which dispersal is largely under genetic control (Solbreck, 1986). In addition, density-independent dispersal has been reported for other phytophagous insects (Farrow, 1982; Brower, 1985).

There are several reasons to suspect that density-dependent dispersal may be a more frequent phenomenon in sap-feeders than in other groups of phytophagous insects. Because sap-feeding insects are typically aggregative and exhibit rapid population growth (reviewed in Denno and Roderick, 1992), they may reach densities at which emigration is favored more readily than do other phytophagous insects. Furthermore, it has been documented that populations of sap-feeding insects are overrepresented in the literature on interspecific competition (Denno *et al.*, 1995), suggesting that sap-feeders, compared to other phytophages, may be more often limited by density-dependent processes in general.

Despite the reasons for thinking that density-dependent dispersal may be more widespread in sap-feeders, this phenomenon has been documented for a variety of other phytophagous insects, including forest-defoliating lepidopterans (reviewed in Lance and Barbosa, 1979), dipteran and coleopteran crop pests (e.g., Utida, 1972; Kowalski and Benson, 1978), and grasshoppers (reviewed in

McAnelly, 1985; Joern and Gaines, 1990). Thus, the emerging pattern is that density-dependent dispersal may be an important suppressing process in the population dynamics of a wide range of phytophagous insects.

Acknowledgments

We thank Ann Herzig and Christer Solbreck for their comments on an earlier draft of this chapter. This research was supported in part by National Science Foundation Grants BSR-8206603 and BSR-8614561 to R. F. D. and Maryland Agricultural Experiment Station Competitive Grant ENTO-95-10 to M. A. P. and R. F. D. This is Contribution Number 8827 of the Maryland Agricultural Experiment Station, Department of Entomology.

References

Antolin, M. F., and Addicott, J. F. (1988). Habitat selection and colony survival of *Macrosiphum valeriani* Clarke (Homoptera: Aphididae). *Ann. Entomol. Soc. Am.* **81**, 245–251.

Aryawan, I. G. N., Widiarta, I. N., Suzuki, Y., and Nakasuji, F. (1993). Life table analysis of the green rice leafhopper, *Nephotettix virescens* (Distant) (Hemiptera: Cicadellidae), an efficient vector of rice tungro disease in asynchronous rice fields in Indonesia. *Res. Popul. Ecol.* **35**, 31–43.

Baumgärtner, J., and Yano, E. (1990). Whitefly population dynamics and modelling. *In* "Whiteflies: Their Bionomics, Pest Status and Management" (D. Gerling, ed.), pp. 123–146. Intercept, Andover, UK.

Brower, L. P. (1985). New perspectives on the migration biology of the monarch butterfly, *Danaus plexippus* L. *Contrib. Mar. Sci.* **27**, 748–785.

Bull, R. M. (1981). Population studies on the sugar cane leafhopper (*Perkinsiella saccharicida* Kirk.) in the Bundaberg District. *Proc. Aust. Soc. Sugar Cane Technol.* **10**, 293–303.

Cappuccino, N. (1987). Comparative population dynamics of two goldenrod aphids: Spatial patterns and temporal constancy. *Ecology* **68**, 1634–1646.

Crespi, B. J. (1988). Adaptation, compromise, and constraint: The development, morphometrics, and behavioral basis of a fighter-flier polymorphism in male *Hoplothrips karnyi* (Insecta: Thysanoptera). *Behav. Ecol. Sociobiol.* **23**, 93–104.

Delisle, J., Cloutier, C., and McNeil, J. N. (1983). Precocene II-induced alate production in isolated and crowded alate and apterous virginoparae of the aphid, *Macrosiphum euphorbiae*. *J. Insect Physiol.* **6**, 477–484.

Dempster, J. P. (1983). The natural control of populations of butterflies and moths. *Biol. Rev. Cambridge Philos. Soc.* **58**, 461–481.

Dempster, J. P., and Pollard, E. (1986). Spatial heterogeneity, stochasticity and the detection of density dependence in animal populations. *Oikos* **46**, 413–416.

Denno, R. F. (1976). Ecological significance of wing-polymorphism in Fulgoroidea which inhabit salt marshes. *Ecol. Entomol.* **1**, 257–266.

Denno, R. F. (1994). Life history variation in planthoppers. *In* "Planthoppers: Their Ecology and Management" (R. F. Denno and T. J. Perfect, eds.), pp. 163–215. Chapman & Hall, New York.

Denno, R. F., and Roderick, G. K. (1992). Density-related dispersal in planthoppers: Effects of interspecific crowding. *Ecology* **73**, 1323–1334.

Denno, R. F., Douglass, L. W., and Jacobs, D. (1985). Crowding and host plant nutrition: Environmental determinants of wing-form in *Prokelisia marginata. Ecology* **66**, 1588–1596.

Denno, R. F., Olmstead, K. L., and McCloud, E. S. (1989). Reproductive cost of flight capability: A comparison of life history traits in wing dimorphic planthoppers. *Ecol. Entomol.* **14**, 31–44.

Denno, R. F., Roderick, G. K., Olmstead, K. L., and Döbel, H. G. (1991). Density-related migration in planthoppers (Homoptera: Delphacidae): The role of habitat persistence. *Am. Nat.* **138**, 1513–1541.

Denno, R. F., Cheng, J., Roderick, G. K., and Perfect, T. J. (1994). Density-related effects on the components of fitness and population dynamics of planthoppers. *In* "Planthoppers: Their Ecology and Management" (R. F. Denno and T. J. Perfect, eds.), pp. 257–281. Chapman & Hall, New York.

Denno, R. F., McClure, M. S., and Ott, J. R. (1995). Interspecific interactions in phytophagous insects: Competition reexamined and resurrected. *Annu. Rev. Entomol.* **40**, 297–331.

Dingle, H., and Arora, G. (1973). Experimental studies of migration in bugs of the genus *Dysdercus. Oecologia* **12**, 119–140.

Dixon, A. F. G. (1969). Population dynamics of the sycamore aphid *Drepanosiphum platanoides* (Schr.) (Hemiptera: Aphididae): Migratory and trivial flight activity. *J. Anim. Ecol.* **38**, 585–606.

Dixon, A. F. G. (1972). Crowding and nutrition in the induction of macropterous alatae in *Drepanosiphum dixoni. J. Insect Physiol.* **18**, 459–464.

Dixon, A. F. G. (1985). "Aphid Ecology." Blackie, London.

Dixon, A. F. G., and Barlow, N. D. (1979). Population regulation in the lime aphid. *Zool. J. Linn. Soc.* **67**, 225–237.

Dixon, A. F. G., and Glenn, D. M. (1971). Morph determination in the bird cherry-oat aphid, *Rhopalosiphum padi* L. *Ann. Appl. Biol.* **68**, 11–21.

Drosopoulos, S. (1977). Biosystematic studies on the *Muellerianella* Complex (Delphacidae, Homoptera, Auchenorrhyncha). *Meded. Landbouwhogesch., Wageningen* **284**, 77–114.

Edson, J. L. (1985). The influences of predation and resource subdivision on the coexistence of goldenrod aphids. *Ecology* **66**, 1736–1743.

Eichhorn, O. (1969). Problems of the determination of morphs in the genus *Dreyfusia* CB (Hopoptera: Adelgidae). *Z. Angew. Entomol.* **64**, 437–446.

Everett, T. R. (1969). Vectors of hoja blanca virus. *In* "The Virus Diseases of the Rice Plant," pp. 111–121. Johns Hopkins Univ. Press, Baltimore.

Farrow, R. A. (1982). Population dynamics of the Australian plague locust, *Chortoicetes terminifera* (Walker), in central western New South Wales. III. Analysis of population processes. *Aust. J. Zool.* **30**, 199–222.

Fisk, J., Bernays, E. A., Chapman, R. F., and Woodhead, S. (1981). "Report of Studies on the Feeding Biology of *Peregrinus maidis*," COPR Core Programme Project 27. Centre for Overseas Pest Research/International Crops Research Institute for Semi-Arid Tropics, London.

Fujisaki, K. (1985). Ecological significance of the wing polymorphism of the oriental chinch bug, *Cavelerius saccharivorus* Okajima (Heteroptera: Lygaeidae). *Res. Popul. Ecol.* **27**, 125–136.

Fujisaki, K. (1986). Reproductive properties of the oriental chinch bug, *Cavelerius saccharivorus* Okajima (Heteroptera: Lygaeidae) in relation to its wing polymorphism. *Res. Popul. Ecol.* **28**, 43–52.

Fujisaki, K. (1994). Wing polymorphism in the oriental chinch bug, *Cavelerius saccharivorus:* Its adaptive significance and ecological consequences. *Proc. Saturday Sem. Entomol. Okayama University, 100th,* Okayama, Japan, pp. 109–125.

Gadgil, M. (1971). Dispersal: Population consequences and evolution. *Ecology* **52**, 253–261.

Gibson, C. W. D. (1980). Niche use patterns among some Stenodemini (Heteroptera: Miridae) of limestone grassland, and an investigation of the possibility of inter-specific competition between *Notostira elongata* Geoffroy and *Megaloceraea recticornis* Geoffroy. *Oecologia* **47**, 352–364.

Gibson, C. W. D., and Visser, M. (1982). Interspecific competition between two field populations of grass-feeding bugs. *Ecol. Entomol.* **7**, 61–67.

Hanski, I. (1991). Single-species metapopulation dynamics: Concepts, models and observations. *Biol. J. Linn. Soc.* **42**, 17–38.

Hassell, M. P. (1985). Insect natural enemies as regulating factors. *J. Anim. Ecol.* **54**, 323–334.

Itô, Y. (1960). Ecological studies on population increase and habitat segregation among barley aphids. *Bull. Natl. Inst. Agric. Sci., Ser. C* **11**, 45–130.

Iwanaga, K., Nakasuji, F., and Tojo, S. (1987). Wing polymorphism in Japanese and foreign strains of the brown planthopper, *Nilaparvata lugens. Entomol. Exp. Appl.* **43**, 3–10.

Joern, A., and Gaines, S. B. (1990). Population dynamics and regulation in grasshoppers. *In* "Biology of Grasshoppers" (R. F. Chapman and A. Joern, eds.), pp. 415–482. Wiley, New York.

Johnson, B. (1965). Wing polymorphism in aphids. II. Interaction between aphids. *Entomol. Exp. Appl.* **8**, 49–64.

Judge, F. D., and Schaefers, G. A. (1971). Effects of crowding on alary polymorphism in the aphid, *Chaetosiphon fragaefolii. J. Insect Physiol.* **17**, 143–148.

Kamm, J. A. (1972). Environmental influence on reproduction, diapause and morph determination of *Anaphothrips obscurus* (Thysanoptera: Thripidae). *Environ. Entomol.* **1**, 16–19.

Kehat, M., and Wyndham, M. (1973). The relation between food, age, and flight in the rutherglen bug, *Nysius vinitor* (Hemiptera: Lygaeidae). *Aust. J. Zool.* **21**, 427–434.

Kidd, N. A. C. (1988). The large pine aphid on Scots pine in Britain. *In* "Dynamics of Forest Insect Populations: Patterns, Causes, and Implications" (A. Berryman, ed.), pp. 111–128. Plenum, New York.

Kidd, N. A. C., and Cleaver, A. M. (1986). The control of migratory urge in *Aphis fabae* Scopoli (Hemiptera: Aphididae). *Bull. Entomol. Res.* **76**, 77–87.

Kidd, N. A. C., and Tozier, D. J. (1984). Host plant and crowding effects in the induction of alatae in the large pine aphid, *Cinara pinea*. *Entomol. Exp. Appl.* **35**, 37–42.

Kisimoto, R. (1965). Studies on the polymorphism and its role playing in the population growth of the brown planthopper, *Nilaparvata lugens* Stål. *Bull. Shikoku Agric. Exp. Stn.* **13**, 1–106.

Kowalski, R., and Benson, J. F. (1978). A population dynamics approach to the wheat bulb fly *Delia coarctata* problem. *J. Appl. Ecol.* **15**, 89–104.

Kuno, E. (1984). Pest status, dynamics and control of rice planthopper and leafhopper populations in Japan. *Prot. Ecol.* **7**, 129–145.

Kuno, E., and Hokyo, N. (1970). Comparative analysis of the population dynamics of rice leafhoppers, *Nephotettix virescens* Uhler and *Nilaparvata lugens* Stål, with special reference to natural regulation of their numbers. *Res. Popul. Ecol.* **12**, 154–184.

Lamb, R. J., and MacKay, P. A. (1987). *Acythosiphon kondoi* influences alata production by the pea aphid, *A. pisum*. *Entomol. Exp. Appl.* **45**, 195–198.

Lance, D., and Barbosa, P. (1979). Dispersal of larval Lepidoptera with special reference to forest defoliators. *Biologist* **61**, 90–110.

Matsumura, M. (1994). Genetic basis of wing dimorphism and responses to wing-form selection on life history traits in the whitebacked planthopper, *Sogatella furcifera*. *Proc. Saturday Semin. Entomol. Okayama Univ.*, *100th*, Okayama, Japan, pp. 47–57.

May, Y. Y. (1975). Study of two forms of the adult *Stenocranus minutus*. *Trans. Ry. Entomol. Soc. London* **127**, 241–254.

McAnelly, M. L. (1985). The significance and control of migratory behavior in the grasshopper *Melanoplus sanguinipes*. *Contrib. Mar. Sci.* **27**, 687–703.

McClure, M. S. (1977). Population dynamics of the red pine scale, *Matsucoccus resinosae* (Homoptera: Margarodidae): The influence of resinosis. *Environ. Entomol.* **6**, 789–795.

McClure, M. S. (1979). Self-regulation in populations of the elongate hemlock scale, *Fiorinia externa* (Homoptera: Diaspididae). *Oecologia* **39**, 25–36.

McClure, M. S. (1980). Competition between exotic species: Scale insects on hemlock. *Ecology* **61**, 1391–1401.

McClure, M. S. (1984). Influence of cohabitation and resinosis on site selection and survival of *Pineus boerneri* Annand and *P. coloradensis* (Gillette) (Homoptera: Adelgidae) on red pine. *Environ. Entomol.* **13**, 657–663.

McClure, M. S. (1989). Biology, population trends, and damage of *Pineus boerneri* and *P. coloradensis* (Homoptera: Adelgidae) on red pine. *Environ. Entomol.* **18**, 1066–1073.

McClure, M. S. (1990). Cohabitation and host species effects on the population growth of *Matsucoccus resinosae* (Homoptera: Margarodidae) and *Pineus boerneri* (Homoptera: Adelgidae) on red pine. *Environ. Entomol.* **19**, 672–676.

McClure, M. S. (1991). Adelgid and scale insect guilds on hemlock and pine. *USDA For. Serv. Northeast For. Exp. Stn.*, *Gen. Tech. Rep.* NE-153, 256–270.

McLain, D. K., and Shure, D. J. (1990). Spatial and temporal density dependence of host plant patch use by the seed bug, *Neacoryphus bicrucis* (Hemiptera: Lygaeidae). *Oikos* **58**, 306–312.

Metcalfe, J. R. (1972). An analysis of the population dynamics of the Jamaican sugar-cane pest *Saccharosydne saccharivora* (Westw.) (Hom., Delphacidae). *Bull. Entomol. Res.* **62**, 73–85.

Mochida, O. (1973). The characters of the two wing-forms of *Javesella pellucida* (F.) (Homoptera: Delphacidae), with special reference to reproduction. *Trans. R. Entomol. Soc. London* **125**, 177–225.

Moran, N. (1986). Benefits of host plant specificity in *Uroleucon* (Homoptera: Aphididae). *Ecology* **67**, 108–115.

Mori, K., and Nakasuji, F. (1991). Effects of day length and density on development and wing form of small brown planthopper, *Laodelphax striatellus* (Hemiptera: Delphacidae). *Appl. Entomol. Zool.* **26**, 557–561.

Murai, M. (1977). Population studies of *Cavelerius saccharivorus* Okajima (Heteroptera: Lygaeidae): Adult dispersal in relation to the density. *Res. Popul. Ecol.* **18**, 147–159.

Murdoch, W. W., and Reeve, J. D. (1987). Aggregation of parasitoids and the detection of density dependence in field populations. *Oikos* **50**, 137–140.

Raatikainen, M. (1967). Bionomics, enemies and population dynamics of *Javesella pellucida* (F.) (Homoptera, Delphacidae). *Ann. Agric. Fenn.* **6**, 1–149.

Roderick, G. K. (1987). Ecology and evolution of dispersal in Californian populations of a salt marsh insect, *Prokelisia marginata*. Ph.D. Thesis, University of California, Berkeley.

Shaw, M. J. P. (1973). Effects of population density on alienicolae of *Aphis fabae* Scop. IV. The expression of the migratory urge among alatae in the field. *Ann. Appl. Biol.* **74**, 1–7.

Solbreck, C. (1978). Migration, diapause, and direct development as alternative life histories in a seed bug, *Neacoryphus bicrucis*. *In* "The Evolution of Insect Migration and Diapause" (H. Dingle, ed.), pp. 195–217. Springer-Verlag, New York.

Solbreck, C. (1985). Insect migration strategies and population dynamics. *Contrib. Mar. Sci.* **27**, 648–652.

Solbreck, C. (1986). Wing and flight muscle polymorphism in a lygaeid bug, *Horvathiolus gibbicollis:* Determinants and life history consequences. *Ecol. Entomol.* **11**, 435–444.

Solbreck, C., and Sillén-Tullberg, B. (1990). Population dynamics of a seed-feeding bug, *Lygaeus equestris*. 1. Habitat patch structure and spatial dynamics. *Oikos* **58**, 199–209.

Stiling, P. D. (1987). The frequency of density dependence in insect host–parasitoid systems. *Ecology* **68**, 844–856.

Stiling, P. D. (1988). Density-dependent processes and key factors in insect populations. *J. Anim. Ecol.* **57**, 581–593.

Tamaki, G., and Allen, W. W. (1969). Competition and other factors influencing the population dynamics of *Aphis gossypii* and *Macrosiphoniella sanborni* on greenhouse chrysanthemums. *Hilgardia* **39**, 447–505.

Toba, H. H., Paschke, J. D., and Friedman, S. (1967). Crowding as the primary factor of the agamic alate form of *Therioaphis maculata* (Homoptera: Aphididae). *J. Insect Physiol.* **13**, 381–396.

Utida, S. (1972). Density dependent polymorphism in the adult of *Callosobruchus maculatus* (Coleoptera: Bruchidae). *J. Stored Prod. Res.* **8**, 111–126.

Varley, G. C., and Gradwell, G. R. (1963). Predatory insects as density dependent mortality factors. *Proc. Int. Congr. Zool., 16th,* Washington, DC, 1963, Vol. 1, p. 240.

Varty, I. W. (1956). *Adelges* insects of silver firs. *For. Comm. Bull. (Edinburgh, Scotland)* **60.**

Vehrs, S. L. C., Walker, G. P., and Parrella, M. P. (1992). Comparison of population growth rate and within-plant distribution between *Aphis gossypii* and *Myzus persicae* (Homoptera: Aphididae) reared on potted chrysanthemums. *J. Econ. Entomol.* **85,** 799–807.

Walde, S. J., and Murdoch, W. W. (1988). Spatial density dependence in parasitoids. *Annu. Rev. Entomol.* **33,** 441–466.

Waloff, N. (1968). Studies on the insect fauna on Scotch broom *Sarothamnus scoparius* (L.) Wimmer. *Adv. Ecol. Res.* **5,** 87–209.

Waloff, N. (1980). Studies on grassland leafhoppers (Auchenorrhyncha, Homoptera) and their natural enemies. *Adv. Ecol. Res.* **11,** 81–215.

Watt, A. D., and Dixon, A. F. G. (1981). The role of cereal growth stages and crowding in the induction of slatae in *Sitobium avenae* and its consequences for population growth. *Ecol. Entomol.* **6,** 441–447.

Way, M. J. (1968). Intra-specific mechanisms with special reference to aphid populations. *In* "Insect Abundance" (T. R. E. Southwood, ed.), pp. 18–36. Blackwell, Oxford.

Whitham, T. G. (1986). Costs and benefits of territoriality: Behavioral and reproductive release by competing aphids. *Ecology* **67,** 139–147.

Widiarta, I. N., Susuki, Y., Fujisaki, K., and Nakasuji, F. (1992). Comparative population dynamics of green leafhoppers in paddy fields of the tropics and temperate regions. *JARQ* **26,** 115–123.

Willard, J. R. (1973). Wandering time of the crawlers of California red scale, *Aonidiella aurantii* (Mask.) (Homoptera: Diaspididae), on citrus. *Aust. J. Zool.* **21,** 217–229.

Williams, A. G., and Whitham, T. G. (1986). Premature leaf abscission: An induced plant defense against gall aphids. *Ecology* **67,** 1619–1627.

Wright, L. C., and Cone, W. W. (1988). Population dynamics of *Brachycorynella asparagi* (Homoptera: Aphididae) on undisturbed asparagus in Washington State. *Environ. Entomol.* **17,** 878–886.

Chapter 7

Using Density-Manipulation Experiments to Study Population Regulation

Susan Harrison and Naomi Cappuccino

I. Introduction

The nature of population stability has long been a subject of debate in ecology (see review in Murdoch, 1994). At issue has been whether or not populations are regulated by density-dependent processes, as is assumed in most ecological theory of the past 50 years (Egerton, 1973). Many ecologists view such regulation as self-evidently necessary, since unregulated populations will tend to become extinct or infinitely large. But dissenters have long emphasized that little evidence exists for density dependence in natural populations (Andrewartha and Birch, 1954; Ehrlich and Birch, 1967; Dempster, 1983; Strong, 1986; den Boer, 1991). Alternative ideas involving nonequilibrium (Hubbell and Foster, 1990; Pickett et al., 1992) or metapopulation dynamics (Hanski and Gilpin, 1991; Hastings and Harrison, 1994) have in fact come to dominate North American ecological thinking over the past decade or two (Koetsier et al., 1990).

Empirical evidence on density dependence has mostly come from census and life table analyses, which have been the subject of much controversy over statistics and interpretation (e.g., Gaston and Lawton, 1987; Stiling, 1988; Hassell et al., 1989; Turchin, 1990; Woiwod and Hanski, 1992; Dennis and Taper, 1994). An alternative approach, in keeping with the present experimental emphasis in ecology, is to alter the densities of natural populations and test for density-dependent changes in rates of birth, death, and change in population size. Such density-perturbation or "convergence" experiments were first proposed by Nicholson (1957), and later advocated by Murdoch (1970) and others (Gaston and Lawton, 1987; Strong, 1987; Sinclair, 1989). Yet surprisingly few such studies have been done. Perhaps this is because the regulation debate arose in a less experimental era and is now viewed as fully resolved by many ecologists— both those who embrace omnipresent regulation and those who reject it!

We view the either–or attitude toward density dependence as outdated, and the "resolution" of the regulation question as premature. Current theory suggests that populations may be regulated by very infrequent (e.g., Nisbet and Gurney, 1982) or spatially localized (e.g., Pulliam, 1988) density dependence, or by complex interactions between local and metapopulation processes (e.g., Reeve, 1988; Murdoch, 1994). This raises a plethora of unanswered empirical questions about when, where, how often, and by what mechanisms real populations are regulated. Density-perturbation experiments, we argue, are a promising tool for addressing these questions. We see testing for density dependence not as an end in itself, but as a first step in analyzing why a real population behaves as it does.

In this chapter, we review the handful of experimental studies of density dependence. We search for generalities about its prevalence, and about the relative importance of resources, enemies, and other potentially regulating forces. We examine certain issues of design and interpretation, and the limitations of the experimental approach. Finally, we discuss alternatives to traditional density-dependent regulation, which can be addressed by experiments as well.

We pay particular attention to the patterns shown by herbivorous insects, since these have been the subjects of much of the debate over regulation. Andrewartha and Birth (1954) argued that insect herbivores often appear to fluctuate independently of density, whereas Hairston et al. (1960) claimed that most are regulated by enemies. Newer work has stressed the importance of resource quality for herbivorous insects (Hunter and Price, 1992), but it remains unclear whether this influence is often regulatory or just has a large density-independent effect.

We will sometimes use the terms "density dependence" and "regulation" interchangeably, but we note that this is imprecise. To regulate, density dependence must be direct (lead to lower birth or higher death rates as densities increase), not inverse (the opposite). Moreover, direct density dependence does not always regulate; it may be undercompensating and fail to prevent random fluctuations, or overcompensating and lead to deterministic instability. Thus, although empirical studies usually focus on testing the statistical significance of density dependence, it is equally important to measure its form and strength. Finally, regulation may take place at the level of metapopulations rather than, or in addition to, that of local populations; we discuss this issue later.

II. Review of Experimental Studies of the Past Twenty-Four Years

We surveyed the literature in two ways for density-perturbation experiments. First, starting with Murdoch's (1970) call for such experiments and working up to the present, we skimmed *Ecology* and the *Journal of Animal Ecology* for papers with the words population regulation, population stability, density dependence, density perturbation, or intraspecific competition in the titles, key words, or abstracts. Second, we used Science Citation Abstracts to find all papers citing Murdoch (1970). We retained all those reporting the results of density-perturbation experiments, regardless of their spatial scale and whether they were performed in the laboratory or field. Although this search was not exhaustive, it should be representative.

We found 60 studies on a wide range of animal taxa, including 11 on herbivorous insects. Most authors (75%) cited population regulation as one of the subjects of their study. Study systems included laboratory microcosms and artificial ponds (18), field enclosures (11), and open-field systems (31). Initial densities were set by the experimenters (58) or natural catastrophes (2).

Typical response variables were survivorship, weight or size, development time, fecundity, and dispersal. Others included strength of mutualistic interaction (Breton and Addicott, 1992), cannibalism (Orr *et al.*, 1990), and wing form (Denno and Roderick, 1992). We classified a response as potentially regulatory if it produced, or was likely to produce, direct density dependence in the rate of population growth. Since many studies examined more than one kind of regulatory force, Table 1 reports a total of 84 attempts to detect density dependence.

We classified potential regulating agents as either bottom-up (prey, hosts, or host plants), top-down (predators, pathogens, or parasitoids), or "lateral" forces. So-called lateral forces are those that do not directly involve species at other trophic levels, such as interference competition, cannibalism, territoriality, and dispersal. Of course, these processes sometimes indirectly reflect food limitation, but in other cases they act at densities well below those where food is limiting.

Our results were analyzed as a series of contingency tables. Because of the low cell frequencies in many tables, we analyzed them using Monte Carlo simulations (Monte Carlo R × C contingency table test, Version 1.0, developed by Bill Engels, University of Wisconsin, Madison, Wisc., U.S.A.). This procedure creates 1000 random tables having the same marginal sums as the observed ones, and calculates the proportion that deviate as much as or more than the observations.

The first striking result was a strong preponderance of direct density dependence (Table 1). Of the 84 effects examined in the 60 studies, fully 79% showed direct density dependence. Inverse density dependence was found twice. Of course, the biased reporting of results is a perennial problem in literature analyses such as this one, but we think the recent trend toward nonequilibrial theory should have encouraged the publication of negative results on regulation.

Another clear result was that bottom-up forces were more likely to produce density dependence than top-down ones (Table 1). Fully 89% of studies looking for resource-based density dependence found it, compared with only 38.5% of those examining enemy-based effects. Bottom-up effects were prevalent even

TABLE 1
Number of Density-Perturbation Studies[a] **(N) That Investigated Bottom-Up, Top-Down,**
and Lateral Regulatory Forces, and the Percentage of These
Studies Documenting Direct Density Dependence

	Bottom-up		Top-down		Lateral		Total	
	N	%	N	%	N	%	N	%
Invertebrates	25	88.0	10	30.0	9	77.7	44	72.7
Herbivorous insects	9	66.7	6	33.3	5	80.0	20	60.0
Fish, amphibians, reptiles	18	94.4	1	100.0	9	77.8	28	89.3
Birds, mammals	4	75.0	2	50.0	6	83.3	12	75.0
Total	47	89.4	13	38.5	24	79.2	84	78.6

[a]Data were taken from the following sources: Alatalo and Lundberg (1984), Baars and van Dijk (1984), Boonstra and Rodd (1983), Breton and Addicott (1992), Brunsting and Heessen (1984), Butler (1976), Cappuccino (1992), Carpenter (1983), Choquenot (1991), Cronin and Strong (1990), Denno and Roderick (1992), Doherty (1983), Ekman *et al.* (1981), Evans and Pienkowski (1982), Faeth and Simberloff (1981), Fletcher (1988), Forrester (1990), Freeland and Choquenot (1990), Gill (1979), Gould *et al.* (1990), Harrison (1994), Hassal and Dangerfield (1990), Hurd and Eisenberg (1984), Jaffee *et al.* (1992), Jones (1990), Klenner (1991), Levitan (1989), Livdahl (1982), Luckinbill and Fenton (1978), Mills *et al.* (1979), Morin (1986), Murdoch and Scott (1984), Newton and Marquis (1991), Ohgushi and Sawada (1985), Ólafsson (1986), Orr *et al.* (1990), Petranka (1989), Petranka and Sih (1986), Scott (1990, 1994), Searcy (1988), Semltsh (1987), Slade and Balph (1974), Smith (1983), Smith and Arcese (1989), Török and Tóth (1988), Wesołowski (1981), Wilbur (1977a,b, 1982), and Wilson (1983).

among insect herbivores, albeit at a lower frequency than in other taxa (66.7% versus 94.7%; $P = 0.04$). For example, Cappuccino (1992) found density-dependent survival in immature goldenrod gallers (*Eurosta solidaginis,* Tephritidae), evidently produced by a defensive reaction in the host plant. Harrison (1994) found density-dependent larval survival and adult fecundity in an outbreak population of the western tussock moth (*Orgyia vetusta,* Lymantriidae), caused by an absolute shortage of food.

Top-down effects were studied only 28% as frequently as bottom-up ones (Table 1). In the seven field studies in which losses could be ascribed to enemies, four found direct density dependence. Two of the four involved herbivorous insects; Cappuccino (1987) found density-dependent rates of disease in aphids and Gould *et al.* (1990) reported density-dependent attack by parasitoids on pupal gypsy moths. Two studies on herbivorous insects (Cappuccino, 1987; Cronin and Strong, 1985) found inverse density dependence in predation and parasitism. Density-perturbation studies involving predators and parasitoids are often difficult because of these species' mobility, which imposes problems of inadequate spatial scale (see following discussion). Pathogens may generally be more tractable, as shown in the studies by Dwyer (1991), Jaffee *et al.* (1992), and Grosholz (1993).

Density dependence imposed by "lateral" factors was found in 79.2% of the studies that searched for it. In a number of studies of territorial birds or mammals, the experimenter removed territory-holding individuals and observed their replacement by territoryless "floaters." If floaters would otherwise die or fail to reproduce, as found by Smith and Arcese (1989) for sparrows, Ekman *et al.* (1981) for willow tits, and Kluver (1991) for red squirrels, then territory limitation can be said to regulate the population.

Other studies examined lateral regulation by density-dependent dispersal; for example, Ohgushi and Sawada (1985) found that the lady beetle *Henosepilachna niponica* disperses density dependently before resources become limiting, and attributed this insect's remarkable population stability to this mechanism. Lateral regulation through interference competition and cannibalism have been demonstrated in water bugs (Orr *et al.*, 1990) and larval *Ambystoma* salamanders (Petranka, 1989; Scott, 1990, 1994), using manipulations of population density in whole ponds.

Several striking conclusions emerge from this survey of experimental studies. First, the results appear to refute the idea that density dependence is rare in natural populations. Second, they indicate that resources are a much more common regulating force than natural enemies. Third, the fact that these conclusions hold for herbivorous insects as well as other taxa would seem to overthrow both of the traditional opposing views: that populations of herbivorous insects are wholly unregulated, and that they are usually regulated by enemies!

However, the strength of these conclusions is limited not only by the small number of studies, and the potential biases in selection of study systems and reporting of results, but also by the incompleteness of most studies with respect to drawing solid conclusions about regulation. Since many of the experiments were designed for other purposes, this is not meant as criticism. However, in the next section we propose guidelines for tackling the regulation question effectively with experiments.

III. Guidelines for Experimental Studies of Regulation

In this section we discuss general methodological issues such as the choice of appropriate time scales, spatial scales, and treatment levels. We then address specific problems associated with studies of bottom-up, top-down, and lateral regulation.

A. How Long Do Experiments Need to Run?

Experiments lasting many generations are ideal, because these can show not only whether density-dependent forces operate, but whether their action in fact leads to convergence and stability. However, experiments lasting one generation can suffice to test for regulation. One approach is to use a wide range of initial densities, and fit the results to a model such as the logistic or Ricker to determine whether the parameters lie in the stabilizing realm. In a species with a short generation time, the resulting predictions can be tested against observed population behavior (e.g., Prout and McChesney, 1985; Dennis *et al.*, 1995). This approach has mainly been used in the laboratory, but it could be applied in the field as well.

Half the studies we surveyed looked for density dependence during only a fraction of the species' life cycle. Such studies are not fully conclusive, in that the regulatory effects of density dependence at one life stage may be exacerbated or canceled by processes at other stages. At a minimum, experiments must last as long as a given regulating agent takes to act. For example, though rates of cannibalism may respond quickly to changing population density, the numerical response of natural enemies has an intrinsic time delay, so that experiments concerning enemies must last at least one enemy (as well as prey) generation.

Most experiments on herbivorous insects have focused on competition or predation affecting the larval stages. However, by manipulating densities of adult leaf-mining moths (*Cameraria* sp.) in sleeve cages on oak branches, Bultman and Faeth (1988) detected interference between ovipositing females, a potentially regulating effect that experiments on larvae alone would have overlooked.

Species with many generations per year are attractive subjects for population experiments. However, some of them (e.g., aphids) show strong seasonal trends and great variation in yearly peak values. In this case, if the species' long-term dynamics are to be understood, regulation from one yearly peak to the next is of greater interest than regulation from one within-season generation to the next.

B. How Large Should Experimental Units Be and Should They Be Caged?

The spatial scale of density manipulations is a key and difficult issue. Ideally, densities are manipulated within replicated areas that correspond to closed populations of the target species, and of any natural enemies that are also subjects of the experiment. When experimental units are smaller than this, any density dependence that is found may be a spatial effect caused by immigration or emigration, which does not necessarily lead to temporal density dependence (as we discuss in the following). Enclosures such as fences, cages, or laboratory microcosms are one way to alleviate this problem; if density dependence occurs at realistic densities within enclosures, it must be due to altered rates of birth or death rather than movement. Although cages may alter the microclimate, this is not a fatal problem unless cage effects interact with and confound the effects of population density.

Cages have often been used successfully in studies of insect populations; for example, Faeth and Simberloff (1981) enclosed entire oak trees in mesh cages to study the dynamics of a leaf-mining moth (*Cameraria* spp. nov.), and Brunsting and Heesen (1984) used field enclosures to isolate populations of carabid beetles (*Pterostichus oblongipunctatus*).

Most studies focus on the most sedentary life stages and choose a spatial scale appropriate to the needs and behaviors of that stage, for example, cattle tanks for tadpoles (Morin, 1986), ponds for larval salamanders (Petranka, 1989), or host-plant stems for homopteran nymphs (Denno and Roderick, 1992). When the authors mention the more mobile life stages at all, it is usually to claim that nothing potentially regulating goes on at that stage! However, caution must be used in the interpretation of such experiments. A 50% increase in the number of larvae per plant or pond may not be a good surrogate for a 50% temporal increase in the density of the whole population, since the mobile adults can distribute their offspring in such a way as to minimize crowding.

C. What Treatment Levels Should Be Used?

Since a realistic choice of initial densities is crucial to meaningful results, it is essential to present data on the range of densities encountered in the field. Most authors (83.3%) did this, although fewer (38.3%) indicated whether their species appeared stable at these levels.

Which life stage is the best basis for choosing "realistic" densities? Returning to the example of herbivorous insects, it seems apparent that (alas) the most mobile life stage is the best. For example, a doubling in the population density of a butterfly is best mimicked by placing twice as many adult butterflies in a field cage, rather than twice as many eggs or larvae on a single plant, because of the adults' ability to minimize the crowding of their offspring through "normal" local movement.

Often there is significant spatial variation in natural densities owing to variation in habitat or resource quality. When this is the case, tests for regulation may be made more powerful by using relative rather than absolute treatments, that is, by raising or lowering natural densities by fixed percentages.

D. Statistical Analyses

Analytical techniques must of course be tailored to each study. Here we discuss one issue that has plagued many studies of density dependence, that of false or biased correlations. Biologically, density dependence is a causal inverse relationship between population density and per capita birth rate, death rate, or rate of net change in density. Statistically, false correlations are possible, since population density is both the independent variable and the denominator of the dependent variable(s). Also, the nonindependence of population sizes in successive years leads to a bias in the estimate of the slope of the density-dependent relationship (Dennis and Taper, 1994).

For census and life table analyses, various alternative statistical techniques have been proposed (e.g., Pollard *et al.*, 1987; Dennis and Taper, 1994), some of which may be used in experimental studies as well. However, a major advantage of the experimental approach is that initial density is fixed, or often can be measured with little error. This greatly reduces the risk of false or biased correlations (Prairie and Bird, 1989; Reddingius and den Boer, 1989; Cappuccino, 1992).

When considerable mortality occurs during the life cycle, it may be well to take repeated censuses, and to use the density at the beginning of each life stage (e.g., each larval instar), rather than the original densities imposed by the experimenter (e.g., the number of eggs) when testing for density dependence at each stage; this will increase the power of tests for density dependence (e.g., Brunsting and Heesen, 1984).

E. Bottom-Up Regulation

Interactions between a species and its food resources are a relatively straightforward experimental subject. Unlike predator abundance, resource abundance changes immediately in response to the density of a population, and thus may provoke immediate change in rates of survival and fecundity. However, delayed responses are possible too, such as when plants mobilize chemical defenses in response to herbivory (Karban and Myers, 1989).

When density dependence is detected, one way to identify the limiting resource is through additional experiments in which the resource supply is altered. This has often been done with granivores (e.g., Brown *et al.*, 1986), for example. However, it is more difficult in the case of herbivorous insects, which usually live on the plants they consume, so that it borders on trivial to show that adding more plants increases the number of insects.

F. Top-Down Regulation and the Problem of Predator Mobility

Compared with resources, predation may be a more difficult regulating force to detect with experiments. In principle, the design is the same: expose varying initial densities of prey to a fixed initial number of predators, and test for increased per capita predation with increasing prey densities. However, in practice the issue is more complicated.

Predation may be density dependent for three reasons: because predators reproduce in response to abundant prey (numerical response), because they switch to the most abundant prey species (one type of functional response), or because they immigrate to areas of high prey density (spatial response). Under the first and second mechanisms, finding density dependence in a manipulation experiment indicates the potential for regulation. That is, whether prey density varies from one year to the next, or from one treatment unit to the next, per capita rates of predation should respond similarly.

Under the third (spatial) mechanism, however, the link between experimentally detected density dependence and temporal population regulation is unclear. In its simplest form, spatially density-dependent predation merely redistributes mortality without changing its total amount. It does not lead to temporal density dependence because if prey abundance increased everywhere at once, there would be no area of low prey density from which predators could immigrate. [We note that in a well-studied class of models, spatial density dependence can stabilize predator–prey dynamics; however, the actual regulation in these models is produced not by spatial effects, but by the numerical responses of prey and predator to one another. See Taylor (1993) for a lucid review.]

To detect the potential for regulation by natural enemies, experiments must be done at the spatial scale of entire, closed enemy populations, as well as the time scale of at least one enemy generation. Unfortunately, these scales may be very large. In one of the most ambitious experiments ever done on regulation, Gould et al. (1990) manipulated densities of pupal gypsy moths in 100 × 100-m plots and found that parasitism by tachinid flies was strongly density dependent. However, this was probably just a spatial response caused by immigration by the foraging tachinids. When constant densities of neonate larvae were placed out in the plots the next season, to test for temporal density dependence, there was no significant relationship between densities in the first season and parasitism in the second season (Elkinton et al., 1990).

The need to enclose populations in order to study top-down regulation seems almost inescapable. Notable examples include work by Jaffee et al. (1992) on pathogens attacking nematodes and Mitchell et al. (1992) on parasitoids of *Drosophila*.

G. Lateral Regulation

Cannibalism and interference competition can readily be studied with density manipulations, although a problem mentioned earlier may arise: these processes often occur among siblings, during the aggregated immature stages of a species that has mobile adults. If so, they are more relevant to the evolution of clutch size than to the regulation of population size. Again, this problem is avoided by manipulating the densities of the mobile (usually adult) life stage over a realistic range.

Dispersal is a trickier issue. Rates of emigration are often density dependent at small spatial scales, but whether this is indeed regulatory depends on the fate of dispersers. Only when dispersal leads to substantially increased mortality or lost fecundity on the part of dispersers does it regulate a population's size, rather than simply redistributing the population. One way to avoid this problem is to manipulate densities either at a large scale relative to the species' dispersal ability or within cages. Another option is to follow the demographic fates of dispersers.

IV. Beyond Local Regulation

Clearly, some experiments will turn up little evidence for density dependence, even if they are well designed and have adequate statistical power. In other studies, the degree of density dependence found may be inadequate to regulate the population. How should such results be interpreted?

Alternatives to local population regulation have become the subjects of much theory; Caswell (1978) provides a useful classification. A single, regulated population is a "closed, equilibrium" system; a metapopulation that persists because local extinctions are balanced by recolonization is an "open, equilibrium" system; a single unregulated population is "closed, nonequilibrium"; a set of connected populations that takes a very long time to reach global extinction is an "open, nonequilibrium" system.

Testing the open, equilibrium model requires measuring rates of extinction and colonization in a set of interconnected local populations (e.g., Schoener and Spiller, 1987; Harrison et al., 1988; Thomas and Harrison, 1992; Sjögren, 1994), and/or manipulating such key variables as the number of populations and the rate of dispersal (e.g., Walde, 1991, and Chapter 9, this volume). Simply failing to show that local regulation occurs is not enough (e.g., Menges, 1990; Stacey and Taper, 1992). At present, only a few experimental studies have found evidence for the existence of classic metapopulations (e.g., Hanski et al., 1994; Hanski and Kuussaari, Chapter 8, this volume; Sjögren, 1994; reviews in Harrison, 1991; Hastings and Harrison, 1994).

Closed, nonequilibrium models propose that single populations may exist for many generations without regulation (Nisbet and Gurney, 1982). Strong (1986) argues that many populations of herbivorous insects experience density dependence very infrequently, at extremely high or low densities. Open, nonequilibrium models add the spatial dimension to explain how such unregulated or weakly regulated populations can persist. Hubbell and Foster (1990) suggest that tropical forest trees can survive in a "random walk" of population densities, perhaps for the entire lifetimes of species, because of trees' individual longevity and the broad geographic distributions of species. But nonequilibrium hypotheses are very difficult to test, given the long time scales that are typically involved.

We wish to stress several points. First, classic regulation through locally acting density dependence is not a logical necessity; certainly, it need not operate constantly or homogeneously for a population to remain extant. But second, the absence of evidence for local regulation is not evidence of absence; alternatives such as nonequilibrium or metapopulation persistence must be tested as well. Third, regulation through local and spatial (metapopulation) processes is not mutually exclusive. Indeed, we believe much of the challenge for future studies of population dynamics lies in linking its local and spatial aspects in experimental studies (e.g., Fahrig and Paloheimo, 1988; Walde, 1991, and Chapter 9, this volume; Hanski *et al.*, 1994; Hanski and Kuussaari, Chapter 8, this volume; Harrison, 1994; Harrison *et al.*, 1995). A promising general approach is to compare stability among local populations that vary, either naturally or experimentally, in the amount of exchange they experience with other populations.

V. How Far Can Experiments Get Us?

Ideally, the kinds of studies we have described will someday be done on enough species, some eruptive and others more stable, that we can begin to generalize about the prevalence of various modes of population regulation in nature. Unfortunately, satisfactory experimental studies of population and metapopulation dynamics will probably be confined largely to abundant, short-lived, and not very mobile species. For many other species, and for cases of regulation by mobile predators or parasitoids, we may have to be content with nonexperimental answers. Particularly compelling are careful life table studies that take advantage of natural perturbations of population density (e.g., Ohgushi and Sawada, 1985; Ohgushi, Chapter 15, this volume; Solbreck and Sillén-Tullberg, 1990; Solbreck, 1991, and Chapter 14, this volume).

Despite these limitations, much progress remains to be made through experiments on when, where, how often, and by what mechanisms real populations are regulated. We conclude with the plea that population regulation be treated neither as a self-evident fact nor as a quaint anachronism in ecological theory, but as an exciting and unresolved question at the heart of population and community ecology, conservation biology, and evolution.

References

Alatalo, R. V., and Lundberg, A. (1984). Density-dependence in breeding success of the pied flycatcher (*Ficedula hypoleuca*). *J. Anim. Ecol.* **53**, 969–977.

Andrewartha, H. G., and Birch, L. C. (1954). "The Distribution and Abundance of Animals." Univ. of Chicago Press, Chicago.

Baars, M. A., and van Dijk, T. S. (1984). Population dynamics of two carabid beetles at a Dutch heathland. II. Egg production and survival in relation to density. *J. Anim. Ecol.* **53**, 389–400.

Boonstra, R., and Rodd, F. H. (1983). Regulation of breeding density in *Microtus pennsylvanicus. J. Anim. Ecol.* **52**, 757–780.

Breton, L. M., and Addicott, J. F. (1992). Density-dependent mutualism in an aphid–ant interaction. *Ecology* **73**, 2175–2180.

Brown, J. H., Davidson, D. W., Munger, J. C., and Inouye, R. S. (1986). Experimental community ecology: The desert granivore system. *In* "Community Ecology" (J. Diamond and T. J. Case, eds.), pp. 41–62. Harper & Row, New York.

Brunsting, A. M. H., and Heessen, H. J. L. (1984). Density regulation in the carabid beetle *Pterostichus oblongopunctatus. J. Anim Ecol.* **53**, 751–760.

Bultman, T. L., and Faeth, S. H. (1988). Abundance and mortality of leaf miners on artificially shaded Emory oak. *Ecol. Entomol.* **13**, 131–142.

Butler, A. J. (1976). A shortage of food for the terrestrial snail *Helicella virgata* in South Australia. *Oecologia* **25**, 349–371.

Cappuccino, N. (1987). Comparative population dynamics of two goldenrod aphids: Spatial patterns and temporal constancy. *Ecology* **68**, 1634–1646.

Cappuccino, N. (1992). The nature of population stability in *Eurosta solidaginis*, a nonoutbreaking herbivore of goldenrod. *Ecology* **73**, 1792–1801.

Carpenter, S. R. (1983). Resource limitation of larval treehole mosquitoes subsisting on beech detritus. *Ecology* **64**, 219–223.

Caswell, H. (1978). Predator-mediated coexistence: A nonequilibrium model. *Am. Nat.* **112**, 127–154.

Choquenot, D. (1991). Density-dependent growth, body condition, and demography in feral donkeys: Testing the food hypothesis. *Ecology* **72**, 805–813.

Cronin, J. T., and Strong, D. R. (1990). Density-independent parasitism among host patches by *Anagrus delicatus* (Hymenoptera: Mymaridae); Experimental manipulation of hosts. *J. Anim. Ecol.* **59**, 1019–1026.

Dempster, J. P. (1983). The natural control of populations of butterflies and moths. *Biol. Rev. Cambridge Philos. Soc.* **58**, 461–481.

den Boer, P. J. (1991). Seeing the trees for the wood: Random walks or bounded fluctuations in population size? *Oecologia* **86**, 484–491.

Dennis, B., and Taper, M. L. (1994). Density dependence in time series observations of natural populations: Estimation and testing. *Ecol. Monogr.* **64**, 205–224.

Dennis, B., Desharnais, R. A., Cushing, J. M., and Constantino, R. F. (1995). Nonlinear demographic dynamics: Mathematical models, statistical methods, and biological experiments. *Ecology* (in press).

Denno, R. F., and Roderick, G. K. (1992). Density-related dispersal in planthoppers: Effects of interspecific crowding. *Ecology* **73**, 1323–1334.

Doherty, P. J. (1983). Tropical territorial damselfishes: Is density limited by aggression or recruitment? *Ecology* **64**, 176–190.

Dwyer, G. (1991). The roles of density, stage and patchiness in the transmission of an insect virus. *Ecology* **72**, 559–574.

Egerton, F. N. (1973). Changing concepts of the balance of nature. *Q. Rev. Biol.* **48**, 322–350.

Ehrlich, P. R., and Birch, L. C. (1967). The "balance of nature" and "population control." *Am. Nat.* **101**, 97–107.

Ekman, J., Cederholm, G., and Asken, C. (1981). Spacing and survival in winter groups of willow tit *Parus montanus* and crested tit *P. cristatus*—A removal study. *J. Anim. Ecol.* **50**, 1–9.

Elkinton, J. S., Gould, J. R., Ferguson, C. S., Liebhold, A. M., and Wallner, W. E. (1990). Experimental manipulations of gypsy moth density to assess impact of natural enemies. *In* "Population Dynamics of Forest Insects" (A. D. Watt, S. R. Leather, M. D. Hunter, and N. A. C. Kidd, eds.), pp. 275–287. Intercept, Andover, UK.

Evans, P. R., and Pienkowski, M. W. (1982). Behaviour of shelducks *Tadorna tadorna* in a winter flock: Does regulation occur? *J. Anim. Ecol.* **51**, 241–262.

Faeth, F. S., and Simberloff, D. (1981). Population regulation of a leaf-mining insect, *Cameraria* sp. nov., at increased field densities. *Ecology* **62**, 620–624.

Fahrig, L., and Paloheimo, J. (1988). Effect of spatial arrangement of habitat patches on local population size. *Ecology* **69**, 468–475.

Fletcher, W. J. (1988). Intraspecific interactions between adults and juveniles of the subtidal limpet, *Patelloida mufria*. *Oecologia* **75**, 272–277.

Forrester, G. E. (1990). Factors influencing the juvenile demography of a coral reef fish. *Ecology* **71**, 1666–1681.

Freeland, W. J., and Choquenot, D. (1990). Determinants of herbivore carrying capacity: Plants, nutrients, and *Equus asinus* in northern Australia. *Ecology* **71**, 589–597.

Gaston, K. J., and Lawton, J. H. (1987). A test of statistical techniques for detecting density dependence in sequential censuses of animal populations. *Oecologia* **74**, 404–410.

Gill, D. E. (1979). Density dependence and homing behavior in adult red-spotted newts *Notophthalmus viridescens* (Rafinesque). *Ecology* **60**, 800–813.

Gould, J. R., Elkinton, J. S., and Wallner, W. E. (1990). Density-dependent suppression of experimentally created gypsy moth *Lymantria dispar* (Lepidoptera: Lymantriidae) populations by natural enemies. *J. Anim. Ecol.* **59**, 213–233.

Grosholz, E. D. (1993). The influence of habitat heterogeneity on host–pathogen population dynamics. *Oecologia* **96**, 347–353.

Hairston, N. G., Smith, F. E., and Slobodkin, L. B. (1960). Community structure, population control, and competition. *Am. Nat.* **94**, 421–425.

Hanski, I., and Gilpin, M. E. (1991). Metapopulation dynamics: Brief history and conceptual domain. *In* "Metapopulation Dynamics: Empirical and Theoretical Investigations" (M. E. Gilpin and I. Hanski, eds.), pp. 3–16. Academic Press, London.

Hanski, I., Kuussaari, M., and Nieminen, M. (1994). Metapopulation structure and migration in the butterfly *Melitaea cinxia*. *Ecology* **75**, 747–762.

Harrison, S. (1991). Local extinction and metapopulation persistence: An empirical evaluation. *In* "Metapopulation Dynamics: Empirical and Theoretical Investigations" (M. E. Gilpin and I. Hanski, eds.), pp. 73–88. Academic Press, London.

Harrison, S. (1994). Resources and dispersal as factors limiting a population of the tussock moth (*Orgyia vetusta*), a flightless defoliator. *Oecologia* **99**, 27–34.

Harrison, S., Murphy, D. D., and Ehrlich, P. R. (1988). Distribution of *Euphydryas editha bayensis*: Evidence for a metapopulation model. *Am. Nat.* **132,** 360–382.

Harrison, S., Thomas, C. D., and Lewinsohn, T. M. (1995). Testing a metapopulation model of coexistence in the insect community on ragwort (*Senecio jacobaea*). *Am. Nat.* **145,** 546–562..

Hassal, M., and Dangerfield, J. M. (1990). Density-dependent processes in the population dynamics of *Armadillidium vulgare* (Isopoda: Oniscidae). *J. Anim. Ecol.* **59,** 941–958.

Hassell, M. P., Latto, J., and May, R. M. (1989). Seeing the wood for the trees: Detecting density dependence from existing life-table studies. *J. Anim. Ecol.* **58,** 883–892.

Hastings, A., and Harrison, S. (1994). Metapopulation dynamics and genetics. *Annu. Rev. Ecol. Syst.* **25,** 167–188.

Hubbell, S. P., and Foster, R. B. (1990). Structure, dynamics, and equilibrium status of old-growth forest on Barro Colorado Island. *In* "Four Neotropical Rainforests" (A. H. Gentry, ed.), pp. 522–541. Yale Univ. Press, New Haven, CT.

Hunter, M. D., and Price, P. W. (1992). Playing chutes and ladders: Heterogeneity and the relative roles of bottom-up and top-down forces in natural communities. *Ecology* **73,** 724–732.

Hurd, L. E., and Eisenberg, R. M. (1984). Experimental density manipulations of the predator *Tenodera sinensis* (Orthoptera: Mantidae) in an old-field community. I. Mortality, development and dispersal of juvenile mantids. *J. Anim. Ecol.* **53,** 269–281.

Jaffee, B., Phillips, R., Muldoon, A., and Mangel, M. (1992). DensityZ-dependent host–pathogen dynamics in soil microcosms. *Ecology* **73,** 495–506.

Jones, G. P. (1990). The importance of recruitment to the dynamics of a coral reef fish population. *Ecology* **71,** 1691–1698.

Karban, R., and Myers, J. H. (1989). Induced plant responses to herbivory. *Annu. Rev. Ecol. Syst.* **20,** 331–348.

Klenner, W. (1991). Red squirrel population dynamics. II. Settlement patterns and the response to removals. *J. Anim. Ecol.* **60,** 979–993.

Koetsier, P., Dey, P., Mladenka, G., and Check, J. (1990). Rejecting equilibrium theory—A cautionary note. *Bull. Ecol. Soc. Am.* **71,** 229–230.

Levitan, D. R. (1989). Density-dependent size regulation in *Diadema antillarum*: Effects on fecundity and survivorship. *Ecology* **70,** 1414–1424.

Livdahl, T. P. (1982). Competition within and between hatching cohorts of a treehole mosquito. *Ecology* **63,** 1751–1760.

Luckinbill, L. S., and Fenton, M. M. (1978). Regulation and environmental variability in experimental populations of protozoa. *Ecology* **59,** 1271–1276.

Massot, M., Clobert, J., Pilorge, T., Lecomte, J., and Barbault, R. (1992). Density dependence in the common lizard: Demographic consequences of a density manipulation. *Ecology* **73,** 1742–1756.

Menges, E. S. (1990). Population viability analysis for an endangered plant. *Conserv. Biol.* **4,** 52–62.

Mills, C. A., Anderson, R. M., and Whitfield, P. J. (1979). Density-dependent survival and reproduction within populations of the ectoparasitic diagenean *Transversotrema patialense* on its fish host. *J. Anim. Ecol.* **48,** 383–399.

Mitchell, P., Wallace, A., and Farrow, M. (1992). An investigation of population limitation using factorial experiments. *J. Anim. Ecol.* **61**, 591–598.

Morin, P. J. (1986). Interactions between intraspecific competion and predation in an amphibian predator prey system. *Ecology* **67**, 713–720.

Murdoch, W. W. (1970). Population regulation and population inertia. *Ecology* **51**, 497–502.

Murdoch, W. W., and Scott, M. A. (1984). Stability and extinction of a laboratory populations of zooplankton preyed on by backswimmer *Notonecta*. *Ecology* **65**, 1231–1248.

Newton, I., and Marquis, M. (1991). Removal experiments and the limitation of breeding density in sparrowhawks. *J. Anim. Ecol.* **60**, 535–544.

Nicholson, A. J. (1957). The self adjustment of populations to change. *Cold Spring Harbor Symp. Quant. Biol.* **22**, 153–172.

Nisbet, R. M., and Gurney, W. S. C. (1982). "Modeling Fluctuating Populations." Wiley, Chichester.

Ohgushi, T., and Sawada, H. (1985). Population equilibrium with respect to available food resource and its behavioural basis in an herbivorous lady beetle, *Henosepilachna niponica*. *J. Anim. Ecol.* **54**, 781–796.

Ólafsson, E. B. (1986). Density dependence in suspension-feeding and deposit-feeding populations of the bivalve *Macoma balthica*: A field experiment. *J. Anim. Ecol.* **55**, 517–526.

Orr, B. K., Murdoch, W. W., and Bence, J. R. (1990). Population regulation, convergence, and cannibalism in *Notonecta* (Hemiptera). *Ecology* **71**, 68–82.

Petranka, J. W. (1989). Density-dependent growth and survival of larval *Ambystoma*: Evidence from whole-pond manipulations. *Ecology* **70**, 1752–1767.

Petranka, J. W., and Sih, A. (1986). Environmental instability, competition, and density-dependent growth and survivorship of a stream-dwelling salamander. *Ecology* **67**, 729–736.

Pickett, S. T. A., Parker, T., and Fiedler, P. L. (1992). The new paradigm in ecology: Implications for conservation biology above the species level. *In* "Conservation Biology: The Theory and Practice of Nature Conservation, Preservation and Management" (P. L. Fiedler and S. K. Jain, eds.), pp. 65–90. Chapman & Hall, New York.

Pollard, E., Lakhani, K. H., and Rothery, P. (1987). The detection of density-dependence from a series of annual censuses. *Ecology* **68**, 2046–2055.

Prairie, Y. T., and Bird, D. F. (1989). Some misconceptions about the spurious correlation problem in the ecological literature. *Oecologia* **81**, 285–288.

Price, P. W. (1987). The role of natural enemies in insect populations. *In* "Insect Outbreaks" (P. Barbosa and J. C. Schultz, eds.), pp. 287–312. Academic Press, New York.

Prout, T., and McChesney, F. (1985). Competition among immatures affects their adult fertility: Population dynamics. *Am. Nat.* **126**, 521–558.

Pulliam, R. (1988). Sources, sinks and population regulation. *Am. Nat.* **132**, 652–661.

Reddingius, J., and den Boer, P. J. (1989). On the stabilization of animal numbers. Problems of testing. 1. Power estimates and estimation errors. *Oecologia* **78**, 1–8.

Reeve, J. D. (1988). Environmental variability, migration and persistence in host–parasitoid systems. *Am. Nat.* **132**, 810–836.

Schoener, T. W., and Spiller, D. A. (1987). High population persistence in a system with high turnover. *Nature (London)* **330**, 474–477.

Scott, D. E. (1990). Effects of larval density in *Ambystoma opacum:* An experiment in large-scale field enclosures. *Ecology* **71**, 296–306.

Scott, D. E. (1994). The effect of larval density on adult demographic traits in *Ambystoma opacum*. *Ecology* **75**, 1383–1396.

Searcy, W. A. (1988). Do female red-winged blackbirds limit their own breeding densities? *Ecology* **69**, 85–95.

Semlitsh, R. (1987). Paedomorphosis in *Ambystoma talpoideum:* Effects of density, food, and pond drying. *Ecology* **68**, 994–1002.

Sinclair, A. R. E. (1989). Population regulation in animals. *In* "Ecological Concepts" (J. M. Cherrett, ed.), pp. 197–241. Blackwell, Oxford.

Sjögren, P. (1994). Distribution and extinction patterns within a northern metapopulation case of the pool frog, *Rana lessonae*. *Ecology* **75**, 1357–1367.

Slade, N. A., and Balph, D. F. (1974). Population ecology of Uinta ground squirrels. *Ecology* **55**, 989–1003.

Smith, D. C. (1983). Factors controlling tadpole populations of the chorus frog (*Pseudoacris triseriata*) on Isle Royale, Michigan. *Ecology* **64**, 501–510.

Smith, J. N. M., and Arcese, P. (1989). How fit are floaters? Consequences of alternative territorial behavior in a non-migratory sparrow. *Am. Nat.* **133**, 830–845.

Solbreck, C. (1991). Unusual weather and insect population dynamics: *Lygaeus equestris* during an extinction and recovery period. *Oikos* **60**, 343–350.

Solbreck, C., and Sillén-Tullberg, B. (1990). Population dynamics of a seed-feeding bug, *Lygaeus equestris*. I. Habitat patch structure and spatial dynamics. *Oikos* **58**, 199–209.

Stacey, P. B., and Taper, M. (1992). Environmental variation and the persistence of small populations. *Ecol. Appl.* **2**, 18–29.

Stiling, P. (1988). Density-dependent processes and key factors in insect populations. *J. Anim. Ecol.* **57**, 581–593.

Strong, D. R. (1986). Density-vague population change. *Trends Ecol. Evol.* **1**, 39–42.

Strong, D. R. (1987). Population theory and understanding pest outbreaks. *In* "Ecological Theory and Integrated Pest Management" (M. Kogan, ed.), pp. 37–58. Wiley, New York.

Taylor, A. D. (1993). Heterogeneity in host–parasitoid interactions: "Aggregation of risk" and the "$CV^2 < 1$ rule." *Trends Ecol. Evol.* **8**, 400–405.

Thomas, C. D., and Harrison, S. (1992). Spatial dynamics of a patchily-distributed butterfly species. *J. Anim. Ecol.* **61**, 437–446.

Török, J., and Tóth, L. (1988). Density dependence in reproduction of the collared flycatcher (*Ficedula albicollis*) at high population levels. *J. Anim. Ecol.* **57**, 251–258.

Walde, S. J. (1991). Patch dynamics of a phytophagous mite population: Effect of number of subpopulations. *Ecology* **72**, 1591–1598.

Wesołowski, T. (1981). Population restoration after removal of wrens (*Troglodytes troglodytes*) breeding in primaeval forest. *J. Anim. Ecol.* **50**, 809–814.

Wilbur, H. M. (1977a). Density-dependent aspects of growth and metamorphosis in *Bufo americanus*. *Ecology* **58**, 196–200.

Wilbur, H. M. (1977b). Interactions of food level and population density in *Rana syl-vatica*. *Ecology* **58**, 206–209.

Wilbur, H. M. (1982). Competition between tapdoes of *Hyla femoralis* and *Hyla gratiosa* in laboratory experiments. *Ecology* **63**, 278–282.

Wilson, W. H., Jr. (1983). The role of density dependence in a marine infaunal community. *Ecology* **64**, 295–306.

Woiwod, I. P., and Hanski, I. (1992). Patterns of density dependence in moths and aphids. *J. Anim. Ecol.* **61**, 619–629.

Chapter 8

Butterfly Metapopulation Dynamics

Ilkka Hanski and Mikko Kuussaari

I. Introduction

The metapopulation concept is swiftly spreading into common usage in population (Gilpin and Hanski, 1991; McCauley, 1993) and conservation biology (Western and Pearl, 1989; Falk and Holsinger, 1991; Fiedler and Jain, 1992). The term "metapopulation" itself is not new, as it was coined by Richard Levins (1970) 25 years ago; but for reasons discussed elsewhere (Hanski, 1995), it took 20 years before biologists at large began to employ it. In this chapter, we restrict the term metapopulation to its original meaning (Levins, 1969, 1970): a set of local populations that persists in a balance between stochastic local extinctions and establishment of new local populations (Hanski and Gilpin, 1991). A critical element in the metapopulation concept thus defined is the notion that the long-term persistence of a metapopulation cannot be explained by the regulation and consequent persistence of local populations. This is the novel aspect in comparison with most of the past and present research on population dynamics, ecology, and biology, which is traditionally focused on single populations.

The purpose of this chapter is to review the use of the metapopulation approach in butterfly population ecology. We could have as well selected some other group of arthropods (Schoener and Spiller, 1987; McCauley, 1993), or even mammals (Smith, 1974; Hanski, 1992, 1993; Verboom et al., 1991) or birds (van Dorp and Opdam, 1987), for this chapter, or we could have written a more general review without emphasizing a particular taxon (Harrison, 1991, 1994; Hastings and Harrison, 1995; Hanski, 1995). There are several reasons for our choice. Butterfly biologists have been particularly quick to turn to metapopulation studies, following the pioneering investigations by Shapiro (1979) and Ehrlich (1984), and there are currently more relevant empirical results for butterflies than perhaps for any other taxon. Equally importantly, influential current critics of excessive enthusiasm for metapopulation dynamics (for which one of us could be blamed) have emerged from butterfly biology (Harrison, 1991, 1994;

C. D. Thomas, 1994a,b). We find it simply challenging to apply the general concepts to butterflies, many of which appear to fit rather well the assumptions of metapopulation models. The metapopulation approach may be useful in conservation, and here again butterflies represent an important case example. Our challenge to our readers not working on butterflies is to try these ideas on other kinds of organisms.

We commence by enumerating four necessary conditions for metapopulation-level persistence of species (following Hanski *et al.*, 1995a). Second, we outline a realistic and practical modeling approach to metapopulation dynamics, and we use the model parameters to draw a list of the key processes of which information is needed in metapopulation studies. We then move on to empirical work. We describe in some detail a case study from our own research on the Glanville fritillary *Melitaea cinxia,* followed by a review of other metapopulation studies on butterflies. Finally, we summarize the strengths and the limitations of the metapopulation approach to population dynamics and conservation of butterflies and other taxa.

II. Four Necessary Conditions for Metapopulation-Level Regulation and Persistence of Species

To demonstrate that the long-term persistence of a species living in a fragmented landscape is dependent on metapopulation-level processes, rather than on local population regulation, one should demonstrate that the following four conditions are met (Hanski *et al.,* 1995a):

Condition 1: The species has local breeding populations in relatively discrete habitat patches. —This condition does not exclude even substantial migration among local populations (and empty habitat patches), but it stresses that the metapopulation is spatially structured: most individuals interact with others in the natal habitat patch. One counterexample is a population using a patchy resource, with individuals routinely and repeatedly moving among the resource patches (Harrison, 1991, 1994) but not forming "local populations" in these patches. It is futile to argue how much migration may occur among local populations for the assemblage still to be called a metapopulation. One should recognize that there is a continuum of spatial population structures, and the real concern is which kind of conceptual and modeling framework is most useful in the study of particular questions with particular species in particular places. The metapopulation concept may be useful even when a large fraction of individuals migrates in each generation, provided that migration distances are restricted. The metapopulation concept may not be helpful in situations in which the majority of individuals migrate long distances in comparison with the spatial scale of interest.

Condition 2: No single local population is large enough to have a long expected lifetime in comparison with the expected lifetime of the metapopulation. —If the metapopulation includes one or more local populations that are likely to persist for a very long time (often called "mainland" populations), metapopulation persistence can be ascribed to the persistence of these populations (Harrison, 1991, 1994). Such population structures are common, and they are not uninteresting, but this chapter is primarily concerned with metapopulations without mainlands.

Condition 3: The habitat patches are not too isolated to prevent recolonization. —This is an obvious necessary condition for metapopulation persistence in the face of unstable local dynamics. There are situations in which this condition is not met, called nonequilibrium metapopulations by Harrison (1991, 1994). There is, however, no important conceptual distinction between equilibrium and nonequilibrium metapopulations, even if only the former are expected to persist for much longer than the lifetime of local populations. Increasing habitat fragmentation may increase isolation beyond a point where recolonization rate is not sufficient to allow metapopulation survival, at which point an "equilibrium" metapopulation turns to a "nonequilibrium" one.

Condition 4: Local dynamics are sufficiently asynchronous to make simultaneous extinction of all local populations unlikely. —If there is complete synchrony in local dynamics, the metapopulation will persist only as long as the local population with the smallest risk of extinction. Generally, we would expect that metapopulation persistence is increased by increasing asynchrony in local dynamics (Harrison and Quinn, 1989; Gilpin, 1990).

III. The Incidence Function Model

In this section we review a recent modeling approach to metapopulation dynamics called the incidence function model (Hanski, 1994a,b). As its name implies, the model is based on a generalized incidence function (Diamond, 1975), and it provides a simple yet realistic and practical means of modeling real metapopulations. The small number of model parameters can be used to draw up a list of critical metapopulation processes for empirical studies. As the incidence function model has been described in detail elsewhere (Hanski, 1994a,b; see also Hanski, 1992, 1993), it suffices here to summarize the rationale behind the model and its main assumptions.

The incidence function model belongs to the family of so-called "patch models," which focus, for mathematical simplicity, on the presence or absence of species in habitat patches and ignore local population dynamics. This assumption can best be defended when the habitat patches are relatively small, in which case local populations may quickly reach the local "carrying capacity" following establishment. Technically, the incidence function model is a first-order linear

Markov chain, which means that changes in the state of a habitat patch (occupied or not) depend on the current state only. In particular, we assume that if patch i is presently empty (unoccupied), there is a constant probability C_i of recolonization in unit time (typically one year in practical applications to butterflies). Similarly, if patch i is presently occupied, there is a constant probability E_i of stochastic extinction in unit time. With these assumptions, the stationary probability of patch i being occupied is given by

$$J_i = \frac{C_i}{C_i + E_i}. \tag{1}$$

From here we proceed in three steps:

(1) Specific assumptions are made about the effects of landscape structure on the colonization and extinction probabilities. To keep the model simple, Hanski (1994a,b) assumed that the extinction probability depends on patch area only (because the extinction probability depends on population size, which depends on patch area):

$$E_i = \min\left[\frac{e}{A^x}, 1\right], \tag{2}$$

where A_i is the area of patch i and e and x are two parameters. Hanski (1994a,b) assumed that the colonization probability depends on the spatial locations and areas of the presently occupied patches:

$$C_i = \frac{M_i^2}{M_i^2 + y^2}, \tag{3a}$$

$$M_i = \beta \sum_{j=1}^{n} p_j e^{-\alpha d_{ij}} A_j$$

$$= \beta\, S_i, \tag{3b}$$

where M_i is the number of migrants arriving at patch i in unit time, p_j equals 1 for occupied and 0 for empty patches, d_{ij} is the distance between patches i and j, and y, β, and α are three model parameters.

One could make some other assumptions about the effects of landscape structure on C_i and E_i. The purpose of these assumptions is to transform Eq. (1) into a parameterized model, which can be fitted to empirical patch occupancy data to estimate model parameters (Hanski, 1994a,b).

(2) Plugging assumptions (2) and (3) into Eq. (1), and assuming that immi-

gration decreases the risk of extinction (the rescue effect), we obtain the model (Hanski, 1994a)

$$J_i = \frac{1}{1 + \dfrac{e'}{S_i^2 A_i^x}}, \tag{4}$$

where $e'=ey'$ and $y'= y/\beta$. The model parameters can be estimated with nonlinear regression and using data on the pattern of patch occupancy (for details, see Hanski, 1994a). In parameter estimation, the unknown incidence J_i is replaced by the observed state of patch i, occupied ($p_i=1$) or not (0). A great advantage of the incidence function approach is that parameter values can be estimated from "snapshot" presence/absence data, which are much easier to obtain than information on the rates of extinction and colonization.

The model parameters summarize essential information about metapopulation dynamics. Thus, the value of e gives the probability of extinction per unit time in a patch of unit size; x gives the rate of change in extinction probability with increasing patch area; α describes the effect of distance on migration rate; y describes colonization efficiency; and β is a compound parameter including the density of individuals in habitat patches and the rate of emigration. Note that the values of y and β cannot be estimated independently [Eq. (3)], hence no straightforward interpretation of their product (y') is possible without some extra information. Furthermore, when the rescue effect is included in the model (which is generally preferable, unless the turnover rate is low), the values of e and y' cannot be estimated independently either [Eq. (4)]. This is unfortunate, because knowledge of e and y' is needed for numerical iteration of the Markov chain (see the following). To tease apart the values of e and y', one may either use information about population turnover between two or more years (Hanski, 1994a) or one may, for instance, make an extra assumption about the minimum patch area A_0 for which the extinction probability E_i equals 1 (Hanski *et al.*, 1995b; then $e=A_i^x$).

(3) When the model parameters α, x, e, and y' have been estimated, one may numerically iterate metapopulation dynamics in the original or in some other patch network to generate quantitative predictions about transient dynamics and the stochastic steady state (Hanski, 1994a,b). This is an exciting prospect, because it means that we may draw inferences about the processes of extinction and colonization from information on patch occupancy only.

It is worth emphasizing the key assumption underlying parameter estimation. The metapopulation from which model parameters are estimated by fitting the incidence function to patch occupancy data is assumed to be at a stochastic steady state. Unfortunately, this may not apply to many rare and declining species.

We shall now turn to empirical butterfly studies employing the conceptual framework developed in the previous section and the modeling approach outlined in this section. We shall first summarize our own results on one species, the Glanville fritillary *Melitaea cinxia,* which we have studied in Finland since 1991 (Kuussaari *et al.,* 1993; Hanski *et al.,* 1994, 1995a,b). In the subsequent section we review other pertinent data from a selection of studies on other butterfly species.

IV. The Glanville Fritillary *Melitaea cinxia*

Melitaea cinxia is one of the many European butterflies that have declined during the past decades (Heath *et al.,* 1984; van Swaay, 1990; Warren, 1993). Figure 1 summarizes its current patchy geographical distribution in Europe. In Finland, *M. cinxia* went extinct on the mainland in the late 1970s (Marttila *et al.,* 1990), but it has survived on the Åland islands. A comprehensive survey of the Åland revealed a total of 1502 habitat patches (dry meadows) suitable for the species, of which 536 were occupied in late summer 1993 (Fig. 2; Hanski *et al.,* 1995a). With this practically complete knowledge of the population structure of the species in Finland, supplemented with the results of more intensive studies conducted since 1991, we are ready to examine to what extent the four necessary conditions for metapopulation-level persistence are met by *Melitaea cinxia* in Finland.

A. The Necessary Conditions for Metapopulation-Level Persistence

Condition 1: On the Åland islands, *M. cinxia* has two larval host plants, *Plantago lanceolata* and *Veronica spicata,* which both grow on dry meadows. The meadows are generally well delimited from the surrounding environment and they are small in size (mean, median, and maximum areas of 0.13, 0.03, and 6.80 ha, $n = 1502$; Hanski *et al.,* 1995a). We have estimated that about 80% of the butterflies spend their entire lifetime in the natal patch (Hanski *et al.,* 1994). We conclude that *M. cinxia* in Finland is structured into local breeding populations living in discrete habitat patches.

Condition 2: In 1994, the largest local population had about 500 butterflies. We have observed a population of about 650 butterflies in 1991 to go extinct by 1993 (Hanski *et al.,* 1994, 1995a), strongly suggesting that even the largest local populations on the Åland are so small that they have a substantial risk of extinction.

Condition 3: In a mark–recapture study in a 50-patch network in 1991 (shaded region in Fig. 2), a total of 1731 butterflies were marked and released

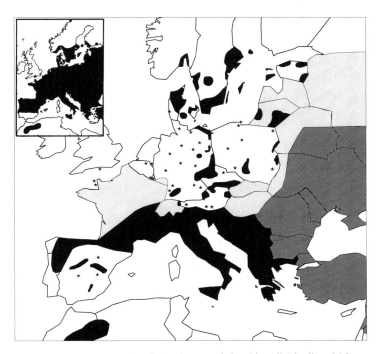

Figure 1. The fragmental geographical distribution of the Glanville fritillary *Melitaea cinxia* in Europe. The smaller map (insert) gives the distribution in the late 1960s as presented by Higgins and Riley (1970). Solid shading indicates areas of continuous distribution and the asterisks show the locations of isolated populations or metapopulations. Light and darker shadings indicate scarce and presumed continuous distribution, respectively, but in both areas the knowledge of the current occurrence of M. *cinxia* is inadequate. We thank the following individuals for information: Kaare Aagaard, Karl-Olof Bergman, Jaroslaw Buszko, John G. Coutsis, Henri Descimon, Claude Dutreix, Claes Eliasson, Andreas Erhardt, Reinart Feldmann, Philippe Goffart, Povilas Ivinskis, Predrag Jaksic, Diego Jordano, Ib Kreutzer, Zdravko Lorkovic, Marc Meyer, Edward Palik, Regina Pauler, E. R. Reichl, Rolf Reinhardt, Toomas Tammaru, Karin Verspui, and Hikmet Özbek. We have also used literature data in the construction of this map.

(Hanski *et al.*, 1994). We obtained 741 recaptures, of which 9% were from a new patch. The mean, median, and maximum distances moved by the migrants among patches were 590, 330, and 3050 m, with 20% of the migrants having moved >1 km but only 3% having moved >2 km. As the mean nearest-neighbor distance is only 240 m among the 1502 meadows (median 128 m, maximum 3870 m), we may conclude that the meadows are generally not too isolated to prevent recolonization, even if some meadows are so isolated that the recolonization probability is small.

Condition 4: The intensive study of the metapopulation in the 50-patch

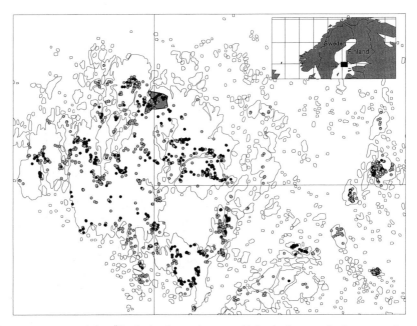

Figure 2. A map of the Åland islands, southwestern Finland, showing the locations of empty (shaded dots) and occupied meadows (solid dots) in late summer 1993 (from Hanski *et al.*, 1995a). The shaded region includes a network of 50 patches that has been studied since 1991. The vertical and horizontal lines delimit the western, northeastern, and southeastern parts of Åland, discussed in the text.

network revealed largely asynchronous local dynamics in 1991–1993 (Hanski *et al.*, 1995a), in spite of a general decline in density during this period, due to two successive unfavorable summers. Asynchrony appears to be maintained by two factors, interaction between the effects of weather and habitat quality, and the action of two specialist parasitoids, which themselves show distinct metapopulation structures (Hanski *et al.*, 1995a; Lei and Hanski, 1996).

In summary, *M. cinxia* in Finland satisfies all four necessary conditions for metapopulation-level persistence.

B. Metapopulation Patterns and Processes

We have analyzed the survey results (Fig. 2) by dividing the entire study area (50 by 70 km²) into 2×2-km² squares, which roughly correspond in size to the movement range of the species (Hanski *et al.*, 1994; see the foregoing). The results demonstrate that the fraction of occupied patches, denoted by P, increases

TABLE 1
**Effects of Patch Area and Density on Extinction Rate between 1991 and 1993
and on Patch Occupancy in 1993**[a]

	Patch area				Patch density			
	Extinctions		Occupancy			Extinctions		Occupancy	
Area (ha)	n_1	%	n_2	P	Number of patches (per 4 km²)	n_1	%	n_2	P
<0.01	4	50	23	0.24	1	1	100	61	0.21
0.01–0.1	15	73	138	0.24	2–3	1	100	70	0.32
0.10–1.0	20	30	88	0.40	4–7	7	71	58	0.25
>1.0	3	0	6	0.56	>7	33	36	66	0.41

[a]From Hanski *et al.* (1995a). n_1 is the number of occupied patches in the 50-patch network (shaded in Fig. 2) in 1991, of which the percentage (%) went extinct by 1993. Occupancy was calcuated for 2 × 2-km² squares, into which the entire study area was divided. n_2 is the number of squares and P is the mean fraction of occupied patches in a square. In the case of occupancy, patch area is the average patch area in the square. In the case of extinctions, patch density was measured by the number of patches in a 2 × 2-km² square centered around the focal patch. The effects of patch area and density on the probability of extinction were tested with the logistic regression model, logit (extinction) = constant + area + density. Both effects were significant (P = 0.016 and 0.024; no significant interaction; deviance = 44.76, df = 39, P = 0.24). The effects of average patch area and patch density on occupancy (P) were tested with ANOVAs on ranks, using the four patch area and density classes shown in the table. Both effects were highly significant (area: $F_{3,251}$ = 5.69, P = 0.001; density: $F_{3,251}$ = 4.21, P = 0.006; no significant interaction).

with increasing regional density of suitable patches and with increasing average area of the patches (Table 1), in agreement with model predictions (Levins, 1969, 1970; Hanski, 1991). Our results also support the presumed reasons for these patterns: increasing patch density facilitates recolonization and lowers extinction rate, and hence increases *P;* and increasing average patch area decreases extinction rate and hence increases *P* (Table 1; Hanski *et al.*, 1995b).

We have estimated the parameter values of the incidence function model using material collected from the 50-patch network in 1991 (Table 2). Using these parameter values, we predicted patch occupancy in the rest of the Åland islands, and compared model predictions with observations made in 1993. Figure 3 shows that for most of the main Åland the match between the predicted and observed *P* values is good, but in the southeastern part of the study area (Fig. 2) the predictions largely fail (Hanski *et al.*, 1995b). We do not yet know the reasons for this failure. In the Western part, where observations agreed with model predictions (Fig. 3), no parameters of habitat quality explained variation in patch occupancy, but in the eastern part, patch occupancy depended on several aspects of patch quality (Hanski *et al.*, 1995b). This suggests that regional

TABLE 2
Parameters of the Incidence Function Model for Three Species
of Butterflies[a]

Parameter[b]	Melitaea cinxia[c]	Hesperia comma	Scolitantides orion
α	-1	-2	-2
x	0.952 (0.271)	1.103 (0.234)	1.339 (0.135)
e'	0.158 (0.166)	0.084 (0.048)	0.001 (0.001)
e	0.010	0.008	0.021
y'	4.04	3.19	0.259

[a]From Hanski (1994a) and Hanski et al. (1995b). Asymptotic standard errors in parentheses.
[b]For the interpretation of these parameters see Section III.
[c]These values from Hanski et al. (1995b) are slightly different from those reported in Hanski (1994a) because of revised patch areas and some changes in parameter estimation.

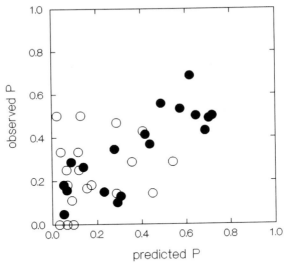

Figure 3. Comparison between the predicted and observed fractions of occupied patches P in western Åland (to the left of the vertical line in Fig. 2). For the purposes of this analysis, the Åland islands was divided into 4 × 4-km² squares, for which the predicted P values were generated with the incidence function model (Hanski et al., 1995b). Open circles indicate squares with 4 to 15 meadows, and solid circles are for squares with >15 meadows (squares with <4 meadows were omitted). Weighted linear regression of observed P against predicted P (weighted with the number of meadows per square): $F = 33.82$, $P < 0.0001$, $n = 43$ squares, $R^2 = 0.44$ ($R^2 = 0.61$ for squares with >15 meadows).

variation in environmental conditions affects extinction and/or colonization rates and hence the pattern of patch occupancy. We also suspect that some metapopulations in the physically fragmented southeastern Åland (Fig. 2) are not at a steady state as assumed by the model. These issues are analyzed and discussed more fully in Hanski *et al.* (1995b).

We conclude that the *Melitaea cinxia* metapopulations on the Åland islands exhibit patterns that are qualitatively (Table 1) and, with some reservations, quantitatively (Fig. 3) consistent with model predictions.

V. Other Butterfly Metapopulations

A. Three Necessary Conditions for Metapopulation-Level Persistence in Finnish Butterflies

To answer the question of how widely the results on *Melitaea cinxia* are applicable to other butterflies, we have evaluated to what extent the necessary conditions for metapopulation-level persistence are satisfied in the Finnish butterfly fauna, which includes 94 species with permanently breeding populations in Finland. Condition 4, the among-population synchrony in temporal dynamics, could not be evaluated.

According to the results in Table 3, most Finnish butterfly species (88%) are

TABLE 3
Numbers of Finnish Butterfly Species That Satisfy (Y)/Do Not Satisfy (N) the Three Conditions for Metapopulation-Level Persistence

Conditions[a]			No. of species	No. of declining species[b]	No. of threatened species	Representative examples
1	2	3				
Y	Y	Y	61	32	18	*Melitaea cinxia, Scolitantides orion*
Y	Y	N	10	9	5	*Maculinea arion, Pseudophilotes baton*
Y	N	Y	12	0	0	*Callophrys rubi, Plebejus argus*
Y	N	N	0	0	0	
N	Y	Y	4	0	0	*Papilio machaon, Anthocharis cardamines*
N	N	Y	7	0	0	*Pieris napi, Gonepteryx rhamni*
N	Y	N	0	0	0	
N	N	N	0	0	0	
Total			94	41	23	

[a]Condition 1: Local breeding populations in relatively discrete habitat patches. Condition 2: No large "mainland" populations. Condition 3: Habitat patches not too isolated to prevent recolonization.
[b]Number of declining species includes all the threatened species in Finland (according to Marttila *et al.* 1990; Rassi *et al.* 1992). Two representative species are given for each group of species.

structured into local populations. Our classification of species with and without local populations in more or less discrete habitat patches is very similar to the classification of species with "closed" versus "open" population structure (e.g., J. A. Thomas, 1984; Warren, 1992a,b; see also Section V, C and Hanski *et al.*, 1994). The proportion of species considered to have local or "closed" populations was very similar in Finland and England (88 and 84% of the permanently breeding species, respectively), and only one species occurring in both countries was classified differently (*Papilio machaon*). The species that did not satisfy our first condition are often very mobile (e.g., *Anthocharis cardamines, Gonepteryx rhamni*) and breed in extensive areas where the larval host plants occur sparsely without being conspicuously aggregated (e.g., *Papilio machaon, Polygonia c-album*).

More than half of the species not forming local populations failed also to satisfy the second condition, because they have large "mainland" populations. These species are the most common Finnish butterflies (*Pieris napi, Gonepteryx rhamni*, and *Aglais urticae*). They lack any strict environmental requirements for breeding and form continuous populations in large areas of heterogeneous habitat. Another group of species with very large ("mainland") populations consists of habitat specialists that still have large continuous areas of suitable habitat in Finland. Some of these species are quite sedentary (e.g., *Callophrys rubi, Plebejus argus, Vacciniina optilete, Brenthis ino, Aphantopus hyperantus*). This group includes species living in pine forests (*Plebejus argus, Callophrys rubi*), forest clearings (*Brenthis ino*), bogs (*Vacciniina optilete, Clossinia euphrosyne*), and meadows and abandoned fields (*Thymelicus lineola, Heodes virgaureae*). The habitats of these species can be distinguished from unsuitable habitat, which makes delimitation of local populations possible.

A clear majority of Finnish butterfly species (76%) satisfied the first two conditions for metapopulation-level persistence. All the declining species and the species classified as threatened in Finland belong to this group (Table 3). Sixty-one of these 71 species were considered to satisfy also the third condition: the habitat patches are not generally too isolated to prevent recolonization. Without having detailed knowledge of the spatial distribution of suitable habitat for all Finnish butterflies, the fulfillment of this condition was difficult to evaluate. In our classification the species was considered to satisfy the third condition if there are (thought to be) several suitable habitat patches in a reachable distance from the existing populations in the core areas of the butterfly's range in Finland.

All 10 species that did not satisfy the third condition, because their habitat patches are too isolated to make (re)colonization possible, are rare species in Finland. Nine of the 10 species are declining and 5 of them are considered threatened. All of these species have strict habitat requirements, including species of dry sandy meadows (*Maculinea arion, Pseudophilotes baton*) and undisturbed bogs (*Pyrgus centaureae, Clossinia freija, C. frigga, Erebia embla*). (The status of the bog specialists was evaluated only for southern Finland, where only small isolated fragments of untouched bogs are left. In Lapland these species still have a lot of suitable habitat.) All the species in this group satisfied the first two conditions, and they used to be more common before the increase in fragmentation of their habitats during this century. They appear to represent species occurring as nonequilibrium metapopulations (Harrison, 1994) on their way to extinction.

Although we found a number of butterfly species that did not satisfy the conditions for metapopulation-level persistence, in more than half of the Finnish species (65%) metapopulation-level regulation may be important. This result encourages the use of the metapopulation approach in conservation studies of butterflies more generally, especially in the case of threatened species with strict environmental requirements.

It is necessary to emphasize that our classification was made without detailed knowledge of every species' habitat requirements and spatial distribution in Finland. Therefore the results should be interpreted with caution. It is also clear that it would not be as simple to apply the metapopulation approach to all the species considered to satisfy the three conditions as it has been in the case of *Melitaea cinxia*. Even if all of these species can be considered to breed only in a particular kind of habitat, the distinction between habitat and nonhabitat would be more difficult for most species other than *M. cinxia*.

B. Butterfly Metapopulation Studies

During the past 10 years the metapopulation approach has been applied in several butterfly studies in both North America and Europe. These studies have documented frequent turnover events (local extinctions and recolonizations) as well as significant effects of isolation and patch area on patch occupancy.

The long-term population studies of checkerspot butterflies in California, particularly with *Euphydryas editha,* have led to the conclusion that local butterfly populations do not generally persist irrespective of each other (Singer and Ehrlich, 1979; Ehrlich *et al.,* 1980; Ehrlich, 1984; Ehrlich and Murphy, 1987). Relatively frequent extinctions especially in small populations are compensated for with effective recolonization in years of exceptional weather (Ehrlich *et al.,* 1980; Murphy and White, 1984). However, when the dynamics of *Euphydryas editha* populations in one network of habitat patches were studied in detail, the persistence of the metapopulation turned out to be largely due to one very large "mainland" population (Harrison *et al.,* 1988).

In Europe, butterfly metapopulation studies have been made with Hesperiidae (C. D. Thomas and Jones, 1993; C. D. Thomas *et al.,* 1992), Lycaenidae (C. D. Thomas *et al.,* 1992; C. D. Thomas and Harrison, 1992; Saarinen, 1993; Verspui, 1993, 1994), Nymphalidae (Baguette and Néve, 1994; Hanski *et al.,* 1994,

TABLE 4
List of Butterfly Metapopulation Studies Known to Us with Some Comparative Statistics[a,b]

Species	No. of patches	Turnover Ext	Turnover Col	Study period	Country
Hesperiidae					
Thymelicus acteon[1,2]	74 (1)	—[c]	—	1978	England
Hesperia comma[1,3]	197 (4)	10	29	1982–91	England
Lycaenidae					
Callophrys rubi[4]	57 (1)	—	—	1991–92	Netherlands
Scolitantides orion[5]	71 (1)	5	17	1991–92	Finland
Plebejus argus[6,1]	157 (9)	15	27	1983–90	Wales
Nymphalidae					
Proclossiana eunomia[7,8]	36 (3)	4	26	1970–93	Belgium and France
Melitaea cinxia[9,10]	1502 (>10)	22[d]	6[d]	1991–93	Finland
Mellicta athalia[1,11]	59 (3)	28	11	1980–89	England
Eurodryas aurinia[12]	432[e]	Many	Some	1983–90	Great Britain
Euphydryas editha[13,14]	60 (1)	0	2	1986–88	United States

[a]Number of metapopulations is indicated in brackets following the total number of patches studied.
Ext = number of local extinctions, Col = number of local colonizations.
[b]References: (1) C. D. Thomas *et al.* (1992); (2) J. A. Thomas (1983a); (3) C. D. Thomas and Jones
(1993); (4) Verspui (1993); (5) Saarinen (1993); (6) C. D. Thomas and Harrison (1992); (7) Baguette
and Néve (1994); (8) G. Néve *et al.* (1995); (9) Hanski *et al.* (1995a); (10) Hanski *et al.* (1994); (11)
Warren (1991); (12) Warren (1994); (13) Harrison *et al.* (1988); (14) Harrison (1989).
[c]No data available.
[d]Turnover of *M. cinxia* was surveyed in one metapopulation of 50 suitable patches.
[e]Number of local populations instead of suitable patches.

1995a,b; Warren, 1994), and Satyridae (Weidner, 1994; Verspui, 1994). These
studies have typically considered sets of local populations of threatened butterflies without large "mainland" populations (Table 4). Increasing probability of
patch occupancy with increasing patch area and with decreasing isolation from
the nearest population or populations have been reported for several species
(C. D. Thomas *et al.,* 1992; C. D. Thomas and Harrison, 1992; C. D. Thomas
and Jones, 1993; Hanski *et al.,* 1994, 1995a; Verspui, 1994). Turnover events
have been mostly observed in the smallest habitat patches (C. D. Thomas and
Harrison, 1992; C. D. Thomas and Jones, 1993; Hanski *et al.,* 1994, 1995a).
Many European studies have emphasized the importance of maintaining connectivity between suitable habitat patches for butterfly conservation.

Table 5 demonstrates that local extinctions of butterfly populations are not
rare events since they have commonly occurred even on nature reserves that have
been established especially for butterfly conservation (J. A. Thomas, 1984,

TABLE 5
Extinctions of Rare Butterflies on Nature Reserves in Britain between 1960 and 1982[a]

Species	Number of colonies on nature reserves	Extinctions on nature reserves in 1960–1982	
		n	*%*
Lycaena dispar	1	1	100
Maculinea arion	4	4	100
Carterocephalus palaemon	4	4	100
Melitaea cinxia	7	0	0
Mellicta athalia	7	7	100
Satyrium pruni	11	0	0
Hesperia comma	33	9	24
Lysandra bellargus	19	4	27
Total	86	29	34

[a]Modified from Warren (1992a).

1991; Warren, 1992b, 1993). These extinctions probably consist of two kinds: deterministic extinctions due to inappropriate management of early-successional habitats, and extinctions due to stochasticity in population dynamics. An extreme suggestion is that the extinctions have occurred mostly or entirely because of natural succession and lack of appropriate management before the subtle ecological requirements of the caterpillars became better known in the 1980s (J. A. Thomas, 1991; Warren, 1992b; C. D. Thomas, 1994b,c).

The first attempts to model the dynamics of butterfly metapopulations have produced encouraging results for *Euphydryas editha* in California (Harrison *et al.*, 1988), *Melitaea cinxia* and *Scolitantides orion* in Finland (Hanski, 1994a; Hanski *et al.*, 1994; 1995a,b), and *Hesperia comma* and *Plebejus argus* in England (Hanski, 1994a; Hanski and Thomas, 1994). With the exception of *E. editha,* the long-term persistence of all of these species appears to hinge on a balance between local extinctions and recolonizations of empty patches.

Table 2 gives the parameter values of the incidence function model for three species of butterflies (for more details, see Hanski, 1994a). We draw attention to the similarities in these values. For instance, the estimated probability of local extinction in a habitat patch of 1 ha was 1–2% per year in all three species. The relative uniformity of these results suggests that, as a first approximation, one could iterate the metapopulation dynamics of many specialist butterflies in fragmented landscapes using just one set of parameter values. Hanski (1994a) suggested the following values: $\alpha = 2$, $y' = 1$, $e = 0.01$, and $x = 1$ (for a more detailed analysis of *M. cinxia* with general implications for the practical application of the incidence function model see Hanski *et al.*, 1995b).

C. Migration Rate in Butterflies

Migration between habitat patches is a key process in metapopulation dynamics. We will therefore make a few comments on the existing data on butterfly movements from the metapopulation perspective.

Our studies on *M. cinxia* (Hanski *et al.*, 1994, 1995a) and other studies (e.g., Dempster, 1989, 1991; Shreeve, 1992) suggest that the mobility of many butterflies forming local populations may have been underestimated, probably because of the difficulty of observing small numbers of butterflies moving relatively long distances in mark–release–recapture (MRR) studies, and the lack of laborous MRR studies conducted simultaneously in many habitat patches in large areas. Table 6 summarizes results on migration distances observed in MRR studies that have been conducted simultaneously in more than one local population. Several species that were previously considered to be very sedentary have nonetheless been observed to disperse over one or several kilometers.

More reliable data on the longest possible movements of butterflies with well-defined local populations can be gathered from surveys of colonizations of previously empty patches, especially following (re)introduction of a species into an area where it was missing from an entire patch network. The scarce data that exist typically indicate longer migration distances than have been observed in MRR studies (Table 6).

It is instructive to compare the occurrence of an extremely sedentary blue-wing butterfly, *Plebejus argus,* in Britain and in Finland. Careful MRR studies and surveys of the colonization of empty patches have revealed that in Britain, where the butterfly is rare, individuals typically do not move more than some tens of meters during their lifetime and that the rare migrants are capable of colonization over distances of only about one kilometer from the source population (C. D. Thomas, 1985; C. D. Thomas and Harrison, 1992). In Finland, *P. argus* is a very common and abundant species of dry pine forests and rocky areas with *Calluna vulgaris,* the larval host plant (Marttila *et al.,* 1990). There is plenty of suitable habitat for the butterfly throughout the country, including thousands of small rocky islands in an archipelago in southwestern Finland. The butterfly occurs even on some of the most isolated small islands in the outer archipelago, where the colonization distances must have been several kilometers or more (M. Nieminen, unpublished). It remains an open question whether the butterfly is more mobile in Finland than in England, as the distributional data appear to suggest. Alternatively, the extremely large "mainland" populations on the south coast of Finland may produce so large a number of migrants that even the most distant islands are eventually colonized.

There is surprisingly little empirical knowledge of the factors affecting emigration and immigration behavior in butterflies (Shreeve, 1992). To study these questions we conducted a large-scale experiment with *Melitaea cinxia* by releasing

TABLE 6
Selected Mark–Release–Recapture Studies (MRR) in Which Several Local
Populations Were Studied Simultaneously[a,b]

Species	Number of local populations (or patches) in the MRR study	Longest observed migration distance (km)	Longest observed colonization distance (km)
Hesperiidae			
Hesperia comma[1,2]	2	0.3	8.6
Pieridae			
Colias alexandra[3]	11	8.0	—
Colias meadii[3]	3	1.3	—
Lycaenidae			
Lysandra bellargus[4]	3	0.3	—
Scolitantides orion[5]	58	1.5	—
Pseudophilotes baton[6]	2	1.5	—
Maculinea arion[7]	2	5.7	—
Plebejus argus[8,9]	4	0.05	ca. 1
Nymphalidae			
Proclossiana eunomia[10,11]	13	4.7	>5
Melitaea cinxia[12]	50	3.0	—
Mellicta athalia[13]	3	1.5	2.5
Eurodryas aurinia[14]	—	—	15–20
Euphydryas editha[15,16]	9	5.6	4.4
Euphydryas anicia[17]	4	3.0	—
Satyridae			
Melanargia galathea[18]	44	7.3	—

[a]The table gives the observed longest migration distance (MRR) and the observed longest colonization distance (not necessarily from the same study).
[b]References: (1) J. A. Thomas *et al.* (1986); (2) C. D. Thomas and Jones (1993); (3) Watt *et al.* (1977); (4) J. A. Thomas (1983b); (5) Saarinen (1993); (6) Väisänen *et al.* (1994); (7) Pajari (1992); (8) C. D. Thomas (1985); (9) C. D. Thomas and Harrison (1992); (10) Baguette and Néve (1994); (11) G. Néve *et al.* (1995); (12) Hanski *et al.* (1994); (13) Warren (1987); (14) Warren (1994); (15) Harrison *et al.* (1988); (16) Harrison (1989); (17) White (1980); (18) Weidner (1994).

nearly 1000 marked butterflies into a network of discrete, empty habitat patches (Kuussaari *et al.*, 1993). Several patch attributes had a significant effect on migration behavior: high density of butterflies and flowers as well as increasing patch area tended to keep the butterflies longer in the patch, whereas open landscape around the patch increased emigration. Numbers of immigrants increased with increasing patch area and butterfly density, as well as with decreasing isolation in relation to the release points (Kuussaari *et al.*, 1993). More data of this type are needed.

We conclude by suggesting that most of the butterfly species with well-defined local populations are not as sedentary as often thought but that typically some 10–20% of individuals from (small) local populations move distances up to 0.5–2 km. Furthermore, most species are capable of rare colonization events of several kilometers or more from the source population.

VI. Discussion

It is not generally agreed that metapopulation dynamics in the sense described in this chapter are often critical for long-term persistence of butterflies and other taxa. Harrison (1991, 1994) and C. D. Thomas (1994a,b) have expressed two opposing viewpoints. Harrison believes that only seldom does the spatial structure of natural populations agree with conditions 1 to 3 in Section II. According to C. D. Thomas (1994b, p. 374), "most extinctions of any but the smallest populations are determined by persistent changes in the local environment." C. D. Thomas thus argues that stochastic extinctions are exceptional, though he recognizes that species' reproduction and dispersal capacities may critically affect colonization, and hence that long-term persistence may depend on regional patch configuration.

It is difficult to disagree about rampant environmental changes, and it is equally clear that many populations have been exterminated because the habitat became unsuitable. However, we do not agree with C. D. Thomas's assumptions that only very small populations (say <100 individuals) have a substantial risk of extinction in stationary environments. We have presented results suggesting that "stochastic" extinctions also occur in larger populations of *M. cinxia*. *Melitaea cinxia* may be exceptional, but until a score of other equally comprehensive studies have been completed, there is no reason to rush to a general conclusion. Here, we wish to make a number of points to clarify the issues.

1. To some extent what constitutes a "stochastic" extinction is a semantic issue. No environment remains entirely unchanged, and many "stochastic" extinctions may be considered to involve some element of temporally changing habitat quality. For some ecologists, this is part of the stochastic process, for others this is "deterministic." What matters is whether the habitat remains, or does not remain, suitable for the species following local extinction (if not, then the cause of extinction was "deterministic").

2. Even if the cause of local extinctions were often deterministic in the preceding sense, this would not necessarily make any material difference to the applicability of metapopulation models, provided that the environment is stationary at the landscape level: the loss of habitat is balanced by the appearance of suitable habitat elsewhere. However, if extinctions and colonizations are essentially driven by environmental changes, one would expect metapopulation dynamics to be erratic, because habitat recovery is generally unlikely to even

roughly match habitat loss within relatively small areas and during relatively short periods of time.

3. Therefore, if suitable but empty habitat patches due to stochastic extinctions were not common, it would be surprising that patch occupancy should so generally increase with increasing patch area and especially with decreasing isolation in *Melitaea cinxia* (Table 1; Hanski *et al.*, 1994) and in many other butterflies (C. D. Thomas and Harrison, 1992; C. D. Thomas *et al.*, 1992; C. D. Thomas and Jones, 1993; Hanski and Thomas, 1994; Hanski, 1994a). Furthermore, metapopulation models assuming stochastic extinctions and recolonizations have been reasonably successful in predicting patterns of patch occupancy (Table 1 and Fig. 3; Hanski *et al.*, 1995a,b) and even transient dynamics (Hanski and Thomas, 1994) in some species of butterflies. The models may generate right predictions for wrong reasons, but more likely the assumed mechanisms reflect, even if only imperfectly, the reality.

4. We agree, though, that many landscapes now change so fast that environmental changes may dominate the dynamics of many metapopulations. In other words, habitat patches turn unsuitable before populations have time to go extinct for other reasons, and metapopulations have no time to reach a stochastic steady state. But not all metapopulations are like this. The reason why models appear to predict quite successfully patterns of patch occupancy in *Melitaea cinxia* in Finland (Table 1 and Fig. 3) is, most likely, that this system exemplifies a metapopulation in which stochastic extinction–colonization dynamics are fast in comparison with the rate of environmental change.

5. In summary, a major limitation of the current metapopulation approach is the difficulty of analyzing the modeling metapopulation dynamics in landscapes with fast and idiosyncratic changes in their structure. On the other hand, a strength of the metapopulation approach lies in parameterized models that can be used to make quantitative predictions. In conservation biology, what matters is our ability to predict the phenomena of interest, for if we can do that, we can determine the environmental consequences of the various actions that we humans may ponder about. In population biology, there is additional emphasis on understanding the causes of phenomena, but even here one could take the view that a useful test of our understanding of the causes is the accuracy of our predictions. In this perspective, the metapopulation approach to population dynamics makes a valuable contribution not because nature is patchy but if, and only if, this approach improves our ability to understand and predict patterns of distribution and abundance of species.

Acknowledgments

We thank Lauri Kaila, Jaakko Kullberg, Marko Nieminen, and Juha Pöyry for their help in the classification of Finnish butterflies, and N. Cappuccino, S. Harrison, S. Walde, and especially C. D. Thomas for helpful comments on the manuscript.

References

Baguette, M., and Néve, G. (1994). Adult movements between populations in the specialist butterfly *Proclossiana eunomia* (Lepidoptera, Nymphalidae). *Ecol. Entomol.* **19**, 1–5.

Dempster, J. P. (1989). Insect introductions: Natural dispersal and population persistence in insects. *Entomologist* **108**, 5–13.

Dempster, J. P. (1991). Fragmentation, isolation and mobility of insect populations. *In* "The Conservation of Insects and Their Habitats" (N. M. Collins and J. A. Thomas, eds.), pp. 143–153. Academic Press, San Diego, CA.

Diamond, J. M. (1975). Assembly of species communities. *In* "Ecology and Evolution of Communities" (M. L. Cody and J. M. Diamond, eds.), pp. 342–444. Harvard Univ. Press, Cambridge, MA.

Ehrlich, P. R. (1984). The structure and dynamics of butterfly populations. *In* "The Biology of Butterflies" (R. I. Vane-Wright and P. R. Ackery, eds.), pp. 25–40. Academic Press, London.

Ehrlich, P. R., and Murphy, D. D. (1987). Conservation lessons from long-term studies of checkerspot butterflies. *Conserv. Biol.* **1**, 122–131.

Ehrlich, P. R., Murphy, D. D., Singer, M. C., Sherwood, C. B., White, R. R., and Brown, I. L. (1980). Extinction, reduction, stability and increase: The responses of checkerspot butterfly (*Euphydryas*) populations to California drought. *Oecologia* **46**, 101–105.

Falk, D. A., and Holsinger, K. E. (1991). "Genetics and Conservation of Rare Plants." Oxford Univ. Press, Oxford.

Fiedler, P. L., and Jain, S. K., eds. (1992). "Conservation Biology: The Theory and Practice of Nature Conservation, Preservation and Management." Chapman & Hall, London.

Gilpin, M. (1990). Extinction of finite metapopulations in correlated environments. *In* "Living in a Patchy Environment" (B. Shorrocks and I. R. Swingland, eds.), pp. 177–186. Oxford Univ. Press, Oxford.

Gilpin, M., and Hanski, I., eds. (1991). "Metapopulation Dynamics: Empirical and Theoretical Investigations." Academic Press, London.

Hanski, I. (1991). Single-species metapopulation dynamics: Concepts, models and observations. *In* "Metapopulation Dynamics: Empirical and Theoretical Investigations" (M. E. Gilpin and I. Hanski, eds.), pp. 17–38. Academic Press, London.

Hanski, I. (1992). Inferences from ecological incidence functions. *Am. Nat.* **139**, 657–662.

Hanski, I. (1993). Dynamics of small mammals on islands. *Ecography* **16**, 372–375.

Hanski, I. (1994a). A practical model of metapopulation dynamics. *J. Anim. Ecol.* **63**, 151–162.

Hanski, I. (1994b). Patch-occupancy dynamics in fragmented landscapes. *Trends Ecol. Evol.* **9**, 131–135.

Hanski, I. (1995). Metapopulation ecology. *In* "Spatial and Temporal Aspects of Population Processes" (O. E. Rhodes, Jr., R. K. Chesser, and M. H. Smith, eds.). Univ. of Chicago Press, Chicago. (In press.)

Hanski, I., and Gilpin, M. (1991). Metapopulation dynamics: Brief history and conceptual domain. *In* "Metapopulation Dynamics: Empirical and Theoretical Investigations" (M. Gilpin and I. Hanski, eds.), pp. 3–16. Academic Press, London.

Hanski, I., and Thomas, C. D. (1994). Metapopulation dynamics and conservation: A spatially explicit model applied to butterflies. *Biol. Conserv.* **68**, 167–180.

Hanski, I., Kuussaari, M., and Nieminen, M. (1994). Metapopulation structure and migration in the butterfly *Melitaea cinxia*. *Ecology* **75**, 747–762.

Hanski, I., Pakkala, T., Kuussaari, M., and Lei, G. (1995a). Metapopulation persistence of an endangered butterfly in a fragmented landscape. *Oikos* **72**, 21–28.

Hanski, I., Moilanen, A., Pakkala, T., and Kuussaari, M. (1995b). Metapopulation persistence of an endangered butterfly: A test of the quantitative incidence function model. *Conserv. Biol.* (in press).

Harrison, S. (1989). Long-distance dispersal and colonization in the bay checkerspot butterfly, *Euphydryas editha bayensis*. *Ecology* **70**, 1236–1243.

Harrison, S. (1991). Local extinction in metapopulation context: An empirical evaluation. *In* "Metapopulation Dynamics: Empirical and Theoretical Investigations" (M. Gilpin and I. Hanski, eds.), pp. 73–88. Academic Press, London.

Harrison, S. (1994). Metapopulations and conservation. *In* "Large-Scale Ecology and Conservation Biology" (P. J. Edwards, R. M. May, and N. R. Webb, eds.), pp. 111–128. Blackwell, Oxford.

Harrison, S., and Quinn, J. F. (1989). Correlated environments and the persistence of metapopulations. *Oikos* **56**, 293–298.

Harrison, S., Murphy, D. D., and Ehrlich, P. R. (1988). Distribution of the bay checkerspot butterfly, *Euphydryas editha bayensis:* Evidence for a metapopulation model. *Am. Nat.* **132**, 360–382.

Hastings, A., and Harrison, S. (1994). Metapopulation dynamics and genetics. *Annu. Rev. Ecol. Syst.* **25**, 167–188.

Heath, J., Pollard, E., and Thomas, J. (1984). "Atlas of Butterflies in Britain and Ireland." Viking Penguin Books Ltd., Middlesex, England.

Higgins, L. G., and Riley, N. D. (1970). "A Field Guide to the Butterflies of Britain and Europe." Collins, London.

Kuussaari, M., Nieminen, M., and Hanski, I. (1993). An experimental study of migration in the Glanville fritillary. *Proc. Conf. Ecol. Conserv. Butterflies.* Keele University, Abstr. p. 24.

Lei, G., and Hanski, I. (1995). Metapopulation structure of *Cotesia melitaearum*, a parasitoid of the butterfly *Melitaea cinxia*. (In preparation.)

Levins, R. (1969). Some demographic and genetic consequences of environmental heterogeneity for biological control. *Bull. Entomol. Soc. Am.* **15**, 237–240.

Levins, R. (1970). Extinction. *In* "Some Mathematical Problems in Biology" (M. Gesternhaber, ed.), pp. 77–107. Am. Math. Soc., Providence, RI.

Marttila, O., Haahtela, T., Aarnio, H., and Ojalainen, P. (1990). "Suomen Päiväperhoset (Finnish Butterflies)." Kirjayhtymä, Helsinki.

McCauley, D. E. (1993). Genetic consequences of extinction and recolonization in fragmented habitats. *In* "Biotic Interactions and Global Change" (P. M. Kareiva, J. G. Kingsolver, and R. B. Huey, eds.), pp. 217–233. Sinauer Assoc., Sunderland, MA.

Murphy, D. D., and White, R. R. (1984). Rainfall, resources, and dispersal in southern populations of *Euphydryas editha* (Lepidoptera: Nymphalidae). *Pan-Pac. Entomol.* **60**, 350–354.

Néve, G., Barascud, B., Hughes, R., Baguette, M., Descimon, H., and Lebrun, P. (1995). Dispersal, colonisation power and metapopulation structure in the vulnerable butterfly *Proclossiana eunomia* (Lepidoptera, Nymphalidae). *J. Appl. Ecol.* (in press).

Pajari, M. (1992). Muurahaissininisiiven (*Maculinea arion* (L.)) populaatiokoon arviointi ja habitaattivaatimusten tutkiminen kesällä 1990 Pohjois-Karjalan Liperissä. Unpublished M.Sc. Thesis, University of Joensuu, Finland.

Rassi, P., Kaipiainen, H., Mannerkoski, I., and Ståhls, G., eds. (1992). "Report on the monitoring of Threatened Animals and Plants in Finland (in Finnish with English summary)," Komiteanmietintö 1991, 30. Ympäristöministeriö, Helsinki.

Saarinen, P. (1993). Kalliosinisiiven (*Scolitantides orion*) ekologia ja esiintyminen Lohjalla. Unpublished M.Sc. Thesis, University of Helsinki, Finland.

Schoener, T. W., and Spiller, D. A. (1987). High population persistence in a system with high turnover. *Nature (London)* **330**, 474–477.

Shapiro, A. M. (1979). Weather and the lability of breeding populations of the checkered white butterfly, *Pieris protodice* Boisduval and LeConte. *J. Res. Lepid.* **17**, 1–23.

Shreeve, T. G. (1992). Monitoring butterfly movements. *In* "The Ecology of Butterflies in Britain" (R. L. H. Dennis, ed.), pp. 120–128. Oxford Univ. Press, Oxford.

Singer, M. C., and Ehrlich, P. R. (1979). Population dynamics of the checkerspot butterfly *Euphydryas editha*. *Fortschr. Zool.* **25**, 53–60.

Smith, A. T. (1974). Temporal changes in insular populations of the pika *Ochotona princeps*. *Ecology* **61**, 8–13.

Thomas, C. D. (1985). Specializations and polyphagy of *Plebejus argus* (Lepidoptera: Lycaenidae) in North Wales. *Ecol. Entomol.* **10**, 325–340.

Thomas, C. D. (1994a). Local extinctions, colonizations and distributions: Habitat tracking by British butterflies. *In* "Individuals, Populations and Patterns in Ecology" (S. R. Leather, A. D. Watt, N. J. Mills, and K. F. A. Walters, eds.), pp. 319–336. Intercept, Andover, UK.

Thomas, C. D. (1994b). Extinction, colonization, and metapopulations: Environmental tracking by rare species. *Conserv. Biol.* **8**, 373–378.

Thomas, C. D. (1994c). Difficulties in deducing dynamics from static distributions. *Trends Ecol. Evol.* **9**, 300.

Thomas, C. D., and Harrison, S. (1992). Spatial dynamics of a patchily-distributed butterfly species. *J. Anim. Ecol.* **61**, 437–446.

Thomas, C. D., and Jones, T. M. (1993). Partial recovery of a skipper butterfly (*Hesperia comma*) from population refuges. *J. Anim. Ecol.* **62**, 472–481.

Thomas, C. D., Thomas, J. A., and Warren, M. S. (1992). Distributions of occupied and vacant butterfly habitats in fragmented landscapes. *Oecologia* **92**, 563–567.

Thomas, J. A. (1983a). The ecology and status of *Thymelicus acteon* (Lepidoptera: Hesperiidae) in Britain. *Ecol. Entomol.* **8**, 427–435.

Thomas, J. A. (1983b). The ecology and conservation of *Lysandra bellargus* (Lepidoptera: Lycaenidae) in Britain. *J. Appl. Ecol.* **20**, 59–83.

Thomas, J. A. (1984). The conservation of butterflies in temperate countries: Past efforts and lessons for the future. *In* "The Biology of Butterflies" (R. I. Vane-Wright and P. R. Ackery, eds.), pp. 333–353. Academic Press, London.

Thomas, J. A. (1991). Rare species conservation: Case studies of European butterflies. *In*

"The Scientific Management of Temperate Communities for Conservation" (I. F. Spellerberg, F. B. Goldsmith, and M. G. Morris, eds.), pp. 149–197. Blackwell, Oxford.

Thomas, J. A., Thomas, C. D., Simcox, D. J., and Clarke, R. T. (1986). The ecology and declining status of the silver-spotted skipper butterfly (*Hesperia comma*) in Britain. *J. Appl. Ecol.* **23**, 365–380.

Väisänen, R., Kuussaari, M., Nieminen, M., and Somerma, P. (1994). Biology and conservation of *Pseudophilotes baton* in Finland. *Ann. Zool. Fenn.* **31**, 145–156.

van Dorp, D., and Opdam, P. F. M. (1987). Effects of patch size, isolation and regional abundance on forest bird communities. *Landscape Ecol.* **1**, 59–73.

van Swaay, C. A. M. (1990). An assessment of the changes in butterfly abundance in the Netherlands during the 20th century. *Biol. Conserv.* **52**, 287–302.

Verboom, J., Lankester, K., and Metz, J. A. J. (1991). Linking local and regional dynamics in stochastic metapopulation models. *In* "Metapopulation Dynamics: Empirical and Theoretical Investigations" (M. Gilpin and I. Hanski, eds.), pp. 39–55. Academic Press, London.

Verspui, K. (1993). The green hairstreak in a fragmented landscape. *Proc. Conf. Ecol. Conserv. Butterflies,* Keele University, Abstr., p. 26.

Verspui, K. (1994). Butterfly metapopulations in a landscape with heathland patches. *Proc. Conf. Butterfly Ecol. Evol.,* Stockholm University, Abstr., Poster 29.

Warren, M. S. (1987). The ecology and conservation of the heath fritillary, *Mellicta athalia*. II. Adult population structure and mobility. *J. Appl. Ecol.* **24**, 483–498.

Warren, M. S. (1991). The successful conservation of an endangered species, the heath fritillary butterfly *Mellicta athalia,* in Britain. *Biol. Conserv.* **55**, 37–56.

Warren, M. S. (1992a). Butterfly populations. *In* "The Ecology of Butterflies in Britain" (R. L. H. Dennis, ed.), pp. 73–92. Oxford Univ. Press, Oxford.

Warren, M. S. (1992b). The conservation of British butterflies. *In* "The Ecology of Butterflies in Britain" (R. L. H. Dennis, ed.), pp. 246–274. Oxford Univ. Press, Oxford.

Warren, M. S. (1993). A review of butterfly conservation in central southern Britain: I. Protection, evaluation and extinction on prime sites. *Biol. Conserv.* **64**, 25–35.

Warren, M. S. (1994). The UK status and suspected metapopulation structure of a threatened European butterfly, the marsh fritillary *Eurodryas aurinia. Biol. Conserv.* **67**, 239–249.

Watt, W. B., Chew, F. D., Snyder, L. R. G., Watt, A. G., and Rothschild, D. E. (1977). Population structure of *Pierid* butterflies. I. Numbers and movements of some montane *Colias* species. *Oecologia* **27**, 1–22.

Weidner, A. (1994). The ecology and population biology of *Melanargia galathea. Proc. Conf. Butterfly Ecol. Evol.,* Stockholm University, Abstr., poster 33.

Western, D., and Pearl, M. (1989). "Conservation Biology in the 21st Century." Oxford Univ. Press, Oxford.

White, R. R. (1980). Inter-peak dispersal in alpine checkerspot butterflies (Nymphalidae). *J. Lepid. Soc.* **34**, 353–362.

Internal Dynamics and Metapopulations: Experimental Tests with Predator–Prey Systems

Sandra J. Walde

I. Introduction

Although the idea of a metapopulation as a collection of local populations linked by dispersal dates back at least to the theoretical explorations of Levins (1969), there has recently been a renewed effort to determine the role and importance of interactions among as well as within populations (Gilpin and Hanski, 1991). This is due partly to the fact that metapopulation theory deals with extinctions in fragmented ecosystems and thus is of relevance to conservation biology (e.g., Quinn and Hastings, 1987; Harrison, 1994), and partly to the discontent of some ecologists with the universal application of traditional equilibrium theory to predator–prey systems within biological control (e.g., Murdoch *et al.*, 1985).

Considerable theory has been developed that explores the dynamics of metapopulations of individual species (e.g., Gurney and Nisbet, 1978a; Hanski, 1989, 1991; Hastings, 1991) and of predator–prey interactions within a metapopulation structure (Vandermeer, 1973; Roff, 1974; Hastings, 1977; Gurney and Nisbet, 1978b; Crowley, 1981; Nachman, 1987a,b; Sabelis and Diekmann, 1988; Reeve, 1988). The broad conclusion from these studies has been that stability is possible at the metapopulation level despite unstable local population dynamics, provided that there is some density dependence at the local population level, that migration rates are appropriate, and that there is sufficient stochasticity to keep local population dynamics unsynchronized. Spatially structured models to date have only scratched the surface of a complex array of possible dynamic outcomes, as illustrated by Hastings (1993), who showed that observed dynamics may depend not only on dispersal but also on initial conditions. Predator–prey modeling within a metapopulation structure is also being expanded to include the plant as a third, interactive trophic level (Sabelis *et al.*, 1991; Jansen and Sabelis, 1992).

POPULATION DYNAMICS
Copyright © 1995 by Academic Press, Inc. All rights of reproduction in any form reserved.

Metapopulation models have been of two general types, those that model the internal dynamics of the local populations only in terms of presence/absence of predator and prey, and those that include a more detailed description of local predator–prey interactions (Taylor, 1988). The latter have demonstrated that metapopulation dynamics can be affected by internal dynamics. In models stabilized by aggregation within subpopulations, persistence at the metapopulation level was influenced by the degree of stability at the local level, by prey growth rates, and by variability in prey growth rate and predator attack rate (Reeve, 1988). On the other hand, in fairly complex simulation studies, Sabelis and Laane (1986) found that although factors normally important to the stability of local populations such as density-dependent mortality did not affect regional stability, other factors such as delay in predator numerical response and variation in the length of the predator–prey interaction period could be very important. Other simulation studies have shown that predator consumption rate, reproductive capacities, and prey preferences could also affect dynamics (Fujita et al., 1979; Takafuji et al., 1983).

As might be expected, empirical testing of metapopulation theory has lagged behind the development of the theory. There have been a number of published studies that are suggestive of metapopulation dynamics, but unambiguous examples of the persistence of predator–prey interactions due to a metapopulation structure seem to be lacking (Taylor, 1991). In this chapter I do not try to evaluate whether studies or systems are good examples of metapopulations dynamics. Rather, my objective is to describe the type and range of dynamics shown by predator–prey interactions that occur within a metapopulation structure. I will limit my discussion of the dynamic effects of metapopulation structure to one group of organisms—spider mites and their predators. This is a predator–prey interaction that frequently has a metapopulation structure and has received considerable attention from this perspective. So far, only a few species have been studied in detail, but these species differ in traits that could have major impacts on population dynamics. Differences in biological characteristics can thus be associated with differences in population dynamics and these patterns compared with those predicted by theoretical investigations. I will begin by presenting in some detail my own experimental work with one spider mite–predaceous mite pair. I will then compare results from a suite of experimental studies on mites to theoretical predictions.

II. Experimental Manipulations of
Panonychus ulmi on Apple

A metapopulation structure has several defining characteristics: (1) the presence of spatially separated local populations, (2) partially independent dynamics for the local populations, and (3) an impact of migration on local dynamics.

(These correspond, roughly, to the first, third, and fourth criteria for metapopulation dynamics as set out by Hanski and Kuussaari, Chapter 8, this volume). To determine the extent to which a metapopulation structure has an impact on population dynamics of a plant-feeding mite and its predators, I manipulated the number of connected patches and monitored effects on densities of prey and predator. The first set of experiments was conducted at a relatively small spatial scale, using closely spaced apple saplings in a newly planted orchard. Details of the methodology and results of these experiments are described elsewhere (Walde, 1991, 1994; Walde *et al.*, 1992). Here I summarize only the most significant findings, and with the benefit of hindsight, present some new analyses that are relevant to the issue of metapopulation structure. I then present a new set of experiments conducted at a much larger spatial scale, and point out where the results confirm, contradict, or extend the conclusions drawn from the smaller-scale experiments.

A. Biology of *Panonychus ulmi* and *Typhlodromus pyri*

Panonychus ulmi is an economically important, plant-feeding mite found on apple and other related deciduous trees throughout the apple-growing regions of the world. In Nova Scotia (Canada), *P. ulmi* completes three to five generations per season (Herbert, 1970). It moves within trees by crawling and among trees by aerial dispersal. *Panonychus* species spin silk threads and descend in response to deterioration of the food supply, and under conditions of gentle wind (Wanibuchi and Saito, 1983; Johnson and Wellington, 1984). Most dispersal, both within and among trees, is by mated adult females. Although *P. ulmi* does not form colonies, its tendency to move short distances under uncrowded conditions produces a patchy distribution within trees and within orchards (Nyrop, 1988).

The pest status of *P. ulmi* arises from current orchard practices; the use of organophosphates and pyrethroids to control pests interferes with the natural control of the pest by predators. In northeastern North America the predaceous mite of greatest importance is *Typhlodromus pyri* (Acari: Phytoseiidae). *Typhlodromus pyri* is a generalist predator, but *P. ulmi* is its preferred prey (Dicke, 1988; Dicke and DeJong, 1988; Nyrop, 1988). *Typhlodromus pyri* is fairly mobile within a tree; average residence time on a leaf under field conditions is 6 h (Lawson and Walde, 1993). It disperses less among trees than some other common phytoseiids (Dunley and Croft, 1990).

B. Experimental Manipulations in a Young Orchard

I varied migration rates among local populations by planting trees in groups containing one (G-1), four (G-4), or sixteen (G-16) trees. I obtained an index of immigration by suspending sticky cards on four sides of the trees. Immigration rate of *P. ulmi* increased with group size, indicating that the experimental design was successful in manipulating prey dispersal (Walde, 1991, 1994). Both *P. ulmi* and *T. pyri* were already present on the trees in low numbers, and I augmented *P. ulmi* densities early in the season. I monitored densities over two seasons, and then looked at the effect of group size on density and stability (temporal variability, persistence, and tendency to converge to equilibrium) of the local and metapopulations.

The presence of a metapopulation structure significantly affected prey but not predator densities. Cumulative densities of *P. ulmi* were highest on G-16 trees, whereas densities of *T. pyri* did not vary consistently with group size (Walde, 1991, 1994).

To measure temporal variability I used an estimator based on the standard deviation of the logarithm of the sequential densities (Stewart-Oaten *et al.*, 1995). This estimator eliminates biases due to differences in spatial heterogeneity or sampling intensity, but does not necessarily eliminate dependence on the mean. I therefore regressed the estimate of variability against mean density, and then compared the residuals across group sizes. I found that local populations of *P. ulmi* within a metapopulation structure were not less variable than isolated populations; temporal variability of individual populations of *P. ulmi* did not vary significantly with group size (mean residual $\pm$ SE, G-1: -0.019 ± 0.034, G-4: -0.013 ± 0.036, G-16: 0.036 ± 0.051) ($P = 0.582$). Similarly, metapopulations were not less variable than isolated populations (G-1: 0.010 ± 0.034, G-4: -0.013 ± 0.036, G-16: -0.010 ± 0.034) ($P = 0.882$). Thus immigration at the levels generated by this experiment did not tend to damp fluctuations in density for local populations nor did it affect temporal variability of the metapopulation.

I then examined differences in persistence between isolated and linked populations. I defined persistence for the first year as the presence of at least one mite in a sample taken after mid-September, and for the second year as the presence of at least one mite in any sample taken during the season. I found that a metapopulation structure did tend to increase persistence of individual populations. In the first year, 68% of the populations of *P. ulmi* on G-1 trees were persistent as compared with 81% of the G-16 populations ($P = 0.12$). In the second year, 30% of the G-1 populations were persistent as compared with 67% of the G-16 populations ($P = 0.042$). *Typhlodromus pyri* populations were much more persistent than those of their prey (Walde *et al.*, 1992).

Finally I looked at whether a metapopulation structure affected the tendency of individual populations to converge to an equilibrium after a perturbation. Early in the second year I augmented *P. ulmi* densities in half of the blocks, and then compared densities in augmented populations late in the season with those of natural density populations. I found significantly faster convergence in isolated populations, but the equilibrium density toward which the populations were converging appeared to be zero. Faster convergence therefore suggests less stability in isolated populations (Walde, 1994).

In summary, I found that a metapopulation structure had no effect on temporal variability, that it increased persistence, and that it slowed the tendency to converge to an equilibrium density of zero. Thus it can be concluded that immigration within a metapopulation did contribute to stabilizing populations, but at the spatial scale of the experiment (maximum 16 populations per metapopulation) the populations were not stable.

There is considerable indirect evidence that the instability of *P. ulmi* populations is caused by its predators. Most importantly, populations of *P. ulmi* do not go extinct in the absence of predators, as shown by predator-removal experiments (Oatman, 1973; Aliniazee and Cranham, 1980; Walde *et al.*, 1992). Several aspects of the biology of *T. pyri* may contribute to an unstable predator–prey interaction. Predator aggregation can stabilize predator–prey interactions under at least some circumstances (Hassell, 1978; Murdoch and Stewart-Oaten, 1989; Godfray and Pacala, 1992; Murdoch *et al.*, 1992), and *T. pyri* does not aggregate in prey patches very effectively (Nyrop, 1988; Lawson and Walde, 1993). Second, *T. pyri* shows a classic Type II functional response to changes in *P. ulmi* density (Walde *et al.*, 1992), also suggesting an unstable interaction. It should be noted, however, that the functional response in the presence of alternate prey is not known. Third, although *T. pyri* does respond numerically to increases in prey density, its reproductive rate lags behind that of its prey, and a destabilizing, delayed density-dependent response is usually seen (Walde, 1995). Finally, *T. pyri* is a generalist predator that supplements its diet of *P. ulmi* with other mites (Dicke and DeJong, 1988) and can survive and reproduce on pollen (Duso and Camporese, 1991). In untreated orchards it is common to find *T. pyri* at higher densities than *P. ulmi* (Strickler *et al.*, 1987). A generalist habit coupled with a preference for *P. ulmi* will also contribute to the instability of the interaction. Once established, *T. pyri* is always present and can quickly quash any incipient establishment of *P. ulmi* via immigration.

In sum, the instability of local populations of *P. ulmi* can be attributed in part, if not in whole, to the presence of the predator *T. pyri*. The fact that connected populations converged more slowly suggests that higher levels of immigration lead to higher effective population growth rates and as a result *T. pyri* requires more time to drive *P. ulmi* extinct.

C. Experimental Manipulations in a Mature Orchard

The preceding set of experiments was conducted in a very young orchard, on trees less than 1.5 m in height and spaced 2 m apart within blocks. To determine if similar patterns would be seen at a larger spatial scale, I conducted a second experiment with a similar design in an orchard more than 50 years old, where trees were spaced at 3.5 m within rows and rows were 6 m apart. I cut down the trees, leaving 12 blocks of 1 tree (G-1), 5 blocks of 4 trees (4 adjacent trees within a row: G-4), and 3 blocks of 12 trees (3 rows of 4 trees: G-12). Mite

populations were left unmanipulated, and both *P. ulmi* and *T. pyri* were present
in the orchard.

Populations were monitored over two seasons. All G-1 and G-4 trees, and
6 trees from each G-12 block (the central row and two from a side row) were
sampled, a total of 50 trees. Eight clusters of leaves (average of six leaves per
cluster) were taken from each tree at biweekly intervals (two clusters from each
cardinal direction) and examined for mites in the laboratory. Immigration was
monitored from early July to late August in the first season only. Sticky cards
(100 cm²) were hung on four sides of all G-1 trees, and the central two trees in
each G-4 and G-12 group. Cards were put out a total of six times, each time for
one week.

There were several potentially significant differences between this experiment and the previous one. (1) The spatial scale was much larger. A single
branch of a mature tree was larger than a whole tree in the young orchard. (2) Because of their size, the branches of trees touched within rows, and mites could
thus move among trees by crawling along rows, as well as by aerial dispersal as
in the young orchard. (3) The experiment was begun with somewhat reduced
numbers of *T. pyri* due to the application of an organophosphate (Imidan) twice
early in the season of the first year. (4) There was a second predaceous mite,
Zetzellia mali (Acari: Stigmaeidae), present in considerable numbers. It is a
generalist predator that feeds on *P. ulmi*, but is much less voracious and more
sedentary than *T. pyri*. *Zetzellia mali* prefers the egg state of *P. ulmi*, whereas
T. pyri prefers motile stages, and in addition *P. ulmi* is not the preferred food of
Z. mali (Clements and Harmsen, 1990; Walde *et al.*, 1995).

The experimental manipulation was once again successful in that per capita
immigration of *P. ulmi* was significantly lower for isolated (G-1) populations
than for G-4 or G-12 populations ($P = 0.013$). Aerial dispersal of both predators
was much lower than that of *P. ulmi;* a seasonal total of only four *T. pyri* and
three *Z. mali* were caught immigrating. In the following sections I examine the
effect of metapopulation structure on change in density of prey and predators,
and on two aspects of stability, temporal variability and persistence.

1. Density

The initial distribution of *P. ulmi* within the orchard was very patchy, and
despite random (as far as possible) allocation of treatment blocks, initial densities
were quite different among treatments. Thus rather than analyzing density per se,
I looked at change in density from the first to the second year (difference of log
cumulative density for each season). In addition, the patterns seen depended on
initial prey density, and thus I analyzed separately populations that had low
(<175 mite-days) versus high cumulative densities (>175 mite-days) of *P. ulmi*
in the first year. I examined the effect of increased connectedness on density of
populations of *P. ulmi* and predators at the spatial scale of the tree, and at the
scale of patches within trees.

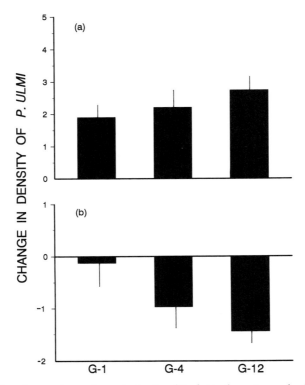

Figure 1. Effect of group size on change in density of *P. ulmi* in the mature orchard for populations with (a) low and (b) high initial densities. Change in density was calculated as (log N_{1992} − log N_{1991}), where log N_{1992} is log of cumulative densities in 1992.

a. Low-Density Populations. Group size did not significantly affect change in density for populations that were initially small ($P = 0.447$). Migration of mites among trees in the mature orchard thus appeared to have less effect on density of *P. ulmi* at the spatial scale of a tree than in the young orchard, although the trend was in the same direction (Fig. 1a).

I then looked at the effect of connectedness at a smaller spatial scale, one that more closely approximated that of the younger orchard. Because the trees were so large, different areas of the tree might be expected to have somewhat independent dynamics, and thus I analyzed the four sides of the trees as separate units or "patches." Some of these patches were more connected to populations on other trees, since trees touched within rows. I compared change in density in patches that touched other trees and thus had the potential for crawling as well as aerial dispersal, with patches that had no intermingling leaves. *Panonychus ulmi*

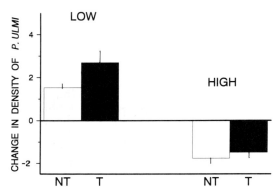

Figure 2. Effect of connectedness among patches on change in density of *P. ulmi* in the mature orchard. Patches either did not touch (NT) or did touch (T) patches in adjacent trees. Results for low and high initial densities are presented separately.

did increase significantly more in patches that touched other trees ($P = 0.018$, Fig. 2). Thus a higher level of connectedness did lead to higher densities of *P. ulmi* in the mature orchard as well as in the young orchard, but the effects were seen at the spatial scale of patches within trees rather than the entire tree.

The presence of a metapopulation structure did not affect changes in density of the predators at the spatial scale of whole trees or of patches within trees. Predators did not increase more in connected versus more isolated patches (*T. pyri*: $P = 0.683$; *Z. mali*: $P = 0.913$), and at the spatial scale of the tree, increased least in G-12 populations and most in G-4 populations (*T. pyri*: $P = 0.054$; *Z. mali*: $P < 0.001$). Predator dynamics may not have been directly affected by immigration because migration rates were so low.

Changes in density of *P. ulmi*, however, were linked to predator densities, but again these effects were seen only at the spatial scale of patches within trees. The largest declines in *P. ulmi* density occurred in patches with high densities of *T. pyri*, but only for those patches that did not touch adjacent trees (Fig. 3a). Thus predators had a much greater impact on prey densities in the more isolated patches, that is, increased levels of prey immigration appeared to swamp predator effects.

In conclusion, better connected patches of *P. ulmi* can attain higher densities because (1) higher levels of immigration lead to a higher net rate of population growth and (2) these higher rates of population growth reduce the effectiveness of the predators.

b. High-Density Populations. The dynamics of populations with high initial numbers of prey were not affected by connectedness with other populations at either the spatial scale of the tree or the scale of patches within trees. Change

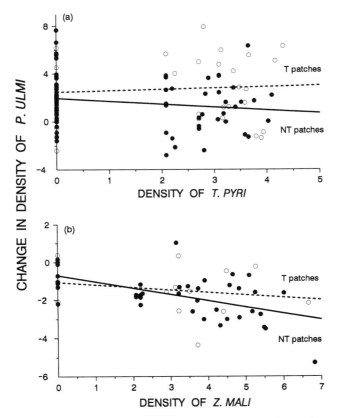

Figure 3. Relationship between change in density of *P. ulmi* and predator density for touching (T = open circles) and nontouching (NT = solid circles) patches in the mature orchard. (a) Change in density of *P. ulmi* versus density of *T. pyri* for all patches in low-density populations (NT patches: slope = 0.376, $P = 0.016$; T patches: slope = 0.100, $P = 0.738$). (b) Change in density of *P. ulmi* versus density of *Z. mali* for high-density patches (NT patches: slope = 0.331, $P < 0.001$; T patches: slope = 0.133, $P = 0.482$).

in density of *P. ulmi* again did not vary significantly among group sizes ($P = 0.196$), although the G-12 populations tended to have larger declines, opposite to the trend seen for low-density populations (Fig. 1). At the spatial scale of patches within trees, increased connectedness did not affect change in density of *P. ulmi* ($P = 0.466$, Fig. 2). Immigration thus appeared to have little impact when prey densities were initially high, and processes internal to the populations were likely responsible for the variation in dynamics observed.

Declines in *P. ulmi* abundance in high-density populations were unrelated to density of *T. pyri* ($P = 0.567$), but were significantly associated with high densities of *Z. mali* ($P < 0.001$). Again, predator impact was strongest in the more isolated patches (Fig. 3b). For unknown historical reasons, initial predator densities were higher on G-12 trees ($P < 0.001$). The tendency for a greater decline of *P. ulmi* in these populations can probably be explained by a higher potential for predation unrelated to my manipulation of immigration. Predator roles also differed among groups when *P. ulmi* densities were high. Over all patches with high initial densities of *P. ulmi*, *Z. mali* was significantly associated with declines in density of *P. ulmi* ($P < 0.001$) and *T. pyri* was not ($P = 0.402$). However, in G-12 populations, both *T. pyri* ($P - 0.004$) and *Z. mali* ($P = 0.003$) played significant roles. Thus higher initial densities probably allowed *T. pyri* to play a more significant role in G-12 populations. The higher predator abundances likely also affected the results of low-density populations, but the immigration effect was still strong enough to be detectable at the patch level.

In conclusion, then, impacts of immigration on change in prey density could only be detected for populations with initially low population densities, and effects were strongest at the spatial scale of patches within trees. In high-density populations, internal dynamics dominated to the extent that effects of immigration on prey density could not be detected. However, in all populations, higher levels of immigration reduced the impact of predation.

2. Stability

Temporal variability was assessed for local and metapopulations of *P. ulmi* in the second season of the experiment. The patterns in the mature orchard coincide with those seen for the younger orchard. Local populations were not less variable as a result of being within a metapopulation structure (mean residual $\pm$ SE) (G-1: -0.050 ± 0.042, G-4: 0.010 ± 0.041, G-12: 0.024 ± 0.028) ($P = 0.422$), and metapopulations were not less variable than isolated populations, although the trend was in this direction (G-1: 0.041 ± 0.041, G-4: -0.046 ± 0.075, G-12: -0.086 ± 0.066) ($P = 0.316$). Therefore it is clear that presence of a metapopulation structure does not damp fluctuations in density of local populations. The slight tendency toward a decrease in variability for metapopulations as a whole suggests some asynchrony among local populations within a group. However, this effect, if it exists, is small compared to the seasonal dynamics common to all populations and the latter obviously dominate.

Because of an untimely end to the experiment (the trees were removed in July of the third season), I obtained only a very inadequate estimate of persistence for these populations. Initial *T. pyri* densities were low owing to the organophosphate treatment, and so *P. ulmi* increased from the first to the second season. The predators began to show a major impact on the prey populations by the second season, and the samples obtained in late June of the third season indicated that the *P. ulmi* population had declined dramatically. The proportion of trees that had no *P. ulmi* in any sample was much higher than it had been at the same time the previous year (96 versus 29%). These data are clearly inconclusive, but are consistent with the hypothesis that the prey populations were headed toward elimination. Previous work has suggested that *P. ulmi* can go extinct at the spatial scale of mature trees (Walde *et al.*, 1992; Walde, 1994).

D. Conclusions

What can we conclude about the effect of metapopulation structure on the dynamics of *P. ulmi* and its predators? The relevant interactions in these populations, immigration and predation, occur at spatial scales smaller than that of a mature tree, and thus population dynamics must also be considered at these smaller spatial scales. Local populations should probably be defined at the spatial scale of a sapling or of patches within trees, as has been suggested for a related species, *P. citri* (Wanibuchi and Saito, 1983).

Dispersal rate of the prey is higher than that of the predator, and thus manipulation of interpopulation distances affected prey migration rate more than that of the predators. An increase in prey migration relative to predator migration is expected to lead to higher prey densities (Reeve, 1988), as was seen in these studies. An increase in prey migration is also expected to damp fluctuations in local population density (Reeve, 1988), but this was not observed, possibly because predators were less effective under these conditions. The predator–prey interaction appears to be unstable and thus in the presence of *T. pyri,* local populations of *P. ulmi* tend to go extinct. Similar conclusions were reached by Johnson and Wellington (1984) in field observations of *P. ulmi* and two other species of *Typhlodromus*. Linking local populations via immigration does contribute to stability in that it slows the decline to extinction, but in these experiments, the immigration was not of the magnitude or form to actually stabilize the populations.

III. Metapopulation, Structure, Theoretical Predictions, and Spider Mite Populations

Models of predator–prey interactions within a metapopulation structure range from relatively simple cell-occupancy models, to fairly general models that include some explicit internal dynamics, to complex simulation models aimed at describing particular systems. Taylor (1988) summarizes the predictions that these models have in common, and where they diverge. Some of the variation is related to the inclusion or not of internal dynamics, and some to the nature of the internal dynamics. Here I will consider how two predictions concerning internal dynamics correspond to patterns seen in spider mite metapopulation studies:

TABLE 1
Summary of Experimental Manipulations of Spider Mites and Their Predators within a Metapopulation Context[a]

P. persimilis		
Local population		Prey extinction, followed by predator
Metapopulations		
Small	32 (bean)	Prey extinction, followed by predator[1]
	8 groups of 4	Imminent prey and predator extinction[1]
	Variable	Prey extinction, followed by predator[2]
Med	72 (strawberry)	Prey extinction, followed by predator[3]
	72 (rose)	Unstable prey oscillations, predator decline[4]
Large	720 (cucumber)	Persisted 6 months, large oscillations[5]
T. occidentalis		
Local population		Prey extinction, followed by predator
Metapopulations		
Small	120 (oranges)	Extinction after three oscillations[6]
Med	72 (strawberry)	Prey and predator persistence?[3]
Large	6 groups of 48	Prey and predator persistence[7]
	(apple seedlings)	
T. pyri		
Local population		Prey extinction, predator persistence
Metapopulations		
Small–med	4–16 saplings	Prey extinction, predator persistence[8]
Large	12 (trees)	Prey extinction?, predator persistence[8]

[a]References: (1) Takafuji (1977); (2) Takafuji *et al.* (1983); (3) Laing and Huffaker (1969); (4) Burnett (1979); (5) Nachman (1981, 1991); (6) Huffaker (1958); (7) van de Klashorst *et al.* (1992); (8) this study.

(1) increasing the stability of the local predator–prey interaction leads to greater persistence of the metapopulation, and to lower temporal variability of the prey (Reeve, 1988); and (2) models that include explicit internal dynamics predict that low levels of migration lead to greatest persistence, whereas persistence in cell-occupancy models is generally favored by high migration rates (Taylor, 1988). I will draw on the data presented in the previous sections as well that presented in other acarine metapopulation studies, summarized in Table 1.

Spider mites and their predators have been more extensively studied from the point of view of metapopulation dynamics than have other predator–prey systems, owing to their economic importance in agroecosystems and their tendency to form spatially structured populations. In natural ecosystems, densities of spider mite populations are typically low, probably due both to predation and to a heterogeneous distribution of food (Kennedy and Smitley, 1985). Traits that have helped them to cope with heterogeneity in food supply, including arrhenotoky, general host plant requirements, and high intrinsic rates of increase due to high reproductive rates and female-biased sex ratios, have allowed very successful exploitation of agroecosystems. In addition, the spatial structure of spider mite populations has led to well-developed dispersal capabilities. Despite their small size and lack of wings, dispersal is nonrandom; spider mites tend to move in response to changes in host plant quality (Bernstein, 1984). Specialized dispersal behaviors to increase the probability of wind transport are often present, including descent on silk threads and posturing (raising the forelegs) at the tips of leaves (Kennedy and Smitley, 1985).

Predaceous mites of the family Phytoseiidae often play an important role in the population dynamics of spider mites, and a number of species are widely used as biological control agents. Some species have also specialized dispersal behaviors (Johnson and Croft, 1976). Both predators and prey show a wide range of life-history characteristics that are likely to influence population dynamics (Sabelis, 1985a,b,c), and predators also vary widely in prey preference, generality of diet, and prey search behaviors.

Most of the work on spider mite–phytoseiid predators within a metapopulation context, other than my own with *T. pyri,* has concentrated on two species of predaceous mites, *Phytoseiulus persimilis* and *Typhlodromus occidentalis.* The dynamics of the interaction between the two-spotted spider mite (*Tetranychus urticae*) and *P. persimilis* is the most extensively studied of any spider mite–phytoseiid pair, and factors contributing to dispersal are also known for both predator and prey (Bernstein, 1984; Sabelis and van der Meer, 1986). At a small spatial scale (a colony or patch containing several colonies), the predator–prey interaction is clearly unstable. Predators, upon entering a colony of *T. urticae,* proceed to decimate it. Predator dispersal begins before all prey are eaten, but some remain until the prey have been eliminated (Takafuji *et al.,* 1983; Sabelis and van der Meer, 1986). Prey colonies or patches are also unstable in the absence of predators, as *T. urticae* typically destroys its host plant (Sabelis and Laane, 1986).

The first explicit test of the effect of dispersal on *P. persimilis—Tetranychus* dynamics was conducted with *T. kanzawai* (Takafuji, 1977, Table 1). Predator and prey were introduced onto two arrangements of plants, producing either a single population or a metapopulation structure. In the former, prey were eliminated in 30 days and the predators died out soon after. In the metapopulation, prey and predators persisted for 80 days, but because of the deterioration of the plants, it was clear that both prey and predator would soon be eliminated. Takafuji *et al.* (1983) then conducted a similar experiment for *P. persimilis* and *T. urticae,* where they varied the number of plants and the rate of dispersal of the predator. The predator again always drove the prey extinct, and increasing the number of plants or decreasing dispersal rate of the predator increased persistence only slightly. Mechanisms that reduced overexploitation of the host plant by the prey were more successful in increasing persistence.

Investigations of the *P. persimilis–T. urticae* interaction had been conducted earlier at a somewhat larger spatial scale (Table 1; Laing and Huffaker, 1969; Burnett, 1979). The predator–prey interaction was still clearly unstable; *P. persimilis* either eliminated *T. urticae* from all plants and then went extinct itself or the prey underwent oscillations of increasing amplitude (Table 1). In the largest spatial scale study of these species to date (Table 1, Nachman, 1981, 1991), conspicuous oscillations in population density were seen, but neither species went extinct over a 6-month period. The importance of the metapopulation structure was indicated by the fact that the oscillations in different parts of the arena were somewhat out of phase and were greater than that for the entire "metapopulation."

The first study with *T. occidentalis* is also the best-known study on mite population dynamics (Table 1; Huffaker, 1958). These were relatively small-scale manipulations, with different numbers and arrangements of oranges. The most stable system was the largest and most complex one—an arena of 120 oranges with enhanced prey (*Eotetranychus sexmaculatus*) dispersal and reduced predator dispersal. However, the interaction did not appear to be stable; the oscillations were of increasing amplitude, and persistence of the predator through three oscillations was due to the chance survival of a single female during the crashes.

A larger-scale experiment aimed at getting persistence between *T. occidentalis* and *T. urticae* was more successful (Table 1; Laing and Huffaker, 1969). In contrast to the *P. persimilis–T. urticae* study conducted at the same spatial scale, *T. occidentalis* did not go extinct over a 2-year period. The details of the experimental design (additions of prey part way through the experiment) make it difficult to evaluate the stability of the interaction, but the authors did not think that *T. urticae* had been driven extinct.

A much more explicit test of the effects of metapopulation structure on *T. occidentalis–T. urticae* dynamics has been conducted by van de Klashorst *et al.* (1992) (Table 1). Predator dispersal rate was manipulated within blocks of potted apple seedlings. Increased connectedness within blocks produced lower prey densities, due to more rapid discovery by the predators, but there was no evidence that stability was affected. Both predator and prey persisted over a period encompassing 50 generations of prey. Dispersal among the blocks may also have contributed to persistence. A weakness with this study was that in order to keep apple foliage present year-round, alternate trees had to be substantially cut back every 6 weeks. This disturbance removed both predators and prey from the system at regular, nonnatural intervals, and could have aided prey persistence if there was a disproportionate removal of older, predator-dominated patches prior to predator dispersal.

Thus, as has been predicted by all metapopulation models to date (Taylor, 1988), increasing spatial scale tended to increase the persistence of the predator–prey interaction for all three predator–prey systems. For the *P. persimilis–*

T. urticae and *T. pyri–P. ulmi* interactions, however, it is not clear that the dynamics were stable, even at large spatial scales. At moderate spatial scales, the *P. persimilis–T. urticae* interaction was certainly unstable (Laing and Huffaker, 1969; Takafuji, 1977; Burnett, 1979; Takafuji *et al.*, 1983). Only at very large spatial scales did predator and prey persist for substantial periods of time, and even in this study the large oscillations present at the spatial scale of an entire greenhouse suggest that the interaction was not stable (Nachman, 1981, 1991). Similarly, *P. ulmi* was not persistent at moderate spatial scales and there is some evidence that *P. ulmi* may go extinct even at the spatial scale of whole orchards (Walde, 1994).

The prediction that increased local stability should enhance metapopulation persistence (Reeve, 1988) is generally supported by available empirical evidence. Local interactions involving all three acarine predators are unstable, but to different degrees and for varying reasons. The *P. persimilis–T. urticae* interaction is more unstable than the *T. occidentalis–T. urticae* interaction. *Phytoseiulus persimilis* is most voracious, has the highest reproductive rate, and aggregates most effectively in predator patches (Laing and Huffaker, 1969; Sabelis, 1985b; Zhang *et al.*, 1992). The source of the instability of the interaction is overexploitation of the prey by the predator, followed by predator extinction due to the absence of prey. Local dynamics are further destabilized by an unstable prey–host plant interaction. The source of the instability of local *T. occidentalis–T. urticae* dynamics is similar, but the overexploitation occurs at a slower rate. In addition, *Typhlodromus* species are much more cannibalistic than *P. persimilis*, which may enhance persistence when prey densities are low. We do see that persistence at the metapopulation level was easier to achieve for *T. occidentalis*, suggesting that predator characteristics affecting local stability really are important.

Overexploitation by the predator also occurs in the *T. pyri–P. ulmi* interaction, although at an even slower rate, but a major difference lies in the end point of the unstable interaction. For the *P. persimilis–T. urticae* interaction, the end point appears to be either elimination of the prey followed by disappearance of the predator or near-elimination of the prey followed by disappearance of the predator and resurgence of the prey. For the *T. pyri–P. ulmi* interaction the end point is elimination or near-elimination of the prey, but maintenance of the predator at moderate densities. The principal factor contributing to this difference is the generalist food habit of *T. pyri*, allowing the predator to survive and reproduce even in the absence of its preferred prey.

Is persistence favored by high migration rates as suggested by most cell-occupancy models or by low rates of migration as suggested by most models with internal dynamics? Here the empirical evidence is not consistent. Persistence was enhanced by lower migration rates (*P. persimilis–T. kanzawai*, Takafuji, 1977), by higher migration rates (*T. pyri–P. ulmi*, this study), by higher prey and lower predator migration rates (*T. occidentalis–E. sexmaculatus*, Huffaker, 1958), or was not affected by migration rate (*P. persimilis–T. urticae*, Takafuji *et al.*, 1983; *T. occidentalis–T. urticae*, van de Klashorst *et al.*, 1992). The greater persistence of the *T. occidentalis–T. urticae* interaction as compared with *P. persimilis–T. urticae* suggests that lower predator migration rates may enhance persistence.

The nature of the internal dynamics, and in particular the within-cell effectiveness of the predators, affects the relationship between migration rate and persistence. Lower migration rates, particularly lower migration by the predator, favor persistence when the source of the local dynamic instability is rapid overexploitation of the prey by a specialist predator, or one with no available alternate food. Higher migration rates, particularly higher migration by the prey, favors persistence when the instability is caused by the generalist habit of the predator. In both cases, persistence is favored by the migration pattern that tends to increase prey density, that is, the pattern that reduces predator effectiveness. This is true for both systems despite the fact that *P. persimilis* appears to disperse more readily than its prey (Sabelis and van der Meer, 1986), whereas *P. ulmi* disperses far more readily than does *T. pyri*. These results appear to be in conflict with some theoretical expectations, where high prey dispersal has been found to be more destabilizing than high predator dispersal (Nachman, 1987a; Reeve, 1988).

IV. Conclusions

At some spatial scale, spider mite–phytoseiid interactions must be persistent, as we certainly do not see global extinction of these species. However, the interactions at a local scale are very unstable, although for quite different reasons in different systems. We find a tendency toward increased persistence at larger spatial scales, and it is clear that dispersal plays a critical role in this increased level of stability. To that extent the presence of a metapopulation structure does tend to stabilize these predator–prey interactions.

Different dynamical outcomes emerged from metapopulation studies conducted on different spider mite–phytoseiid predator pairs. The three principal dynamic outcomes were: an unstable interaction even at the largest scale tested, where both predator and prey typically go extinct (*P. persimilis–T. urticae*), a possibly stable interaction at moderate to large spatial scales (*T. occidentalis–T. urticae*), and an unstable interaction that results in elimination of the prey but not the predator at moderate to large spatial scales (*T. pyri–P. ulmi*). These systems all have unstable predator–prey dynamics at the local population level, but internal dynamics vary as a result of differences in the biology of the predator and prey species, and this ultimately leads to different patterns at larger spatial scales.

The prediction that enhancement of local dynamic stability will increase metapopulation persistence seems to be borne out by the empirical evidence, both that obtained within studies and in comparisons across studies. The result that Reeve (1988) generated by simply varying the degree of aggregation in a negative binomial model may thus be quite general. The relationship between migration rate and persistence is complex, and depends on the details of the internal dynamics of the local populations. It has already been pointed out that the relationship is expected to depend on the effectiveness of the predators within local populations (Taylor, 1988), and the empirical evidence considered here supports this prediction. Thus basic aspects of the predator–prey interaction at the local scale, such as functional and numerical responses, will be critical to understanding the dynamics produced by a metapopulation structure. In addition, the inclusion of host plant dynamics will be essential for predictions in some systems, but be of lesser importance in others.

Finally, differences in internal dynamics generated by differences in environmental conditions are also likely to affect the pattern of dynamics observed. The difference between *T. pyri* dynamics and that of the other predators was attributed mainly to its generalist habit, but this aspect was evident only because these experiments were conducted in the field where alternate food was available. Thus breadth of diet and availability of alternate prey are also likely critical to predicting metapopulation dynamics. In addition, as shown in the large-scale *T. pyri–P. ulmi* experiments, conditions such as initial prey density can have a major effect on how important a metapopulation structure is to observed dynamics.

There are other predictions made by metapopulation theory that cannot as yet be evaluated using available data. For example, Reeve (1988) found that lower prey growth rates should enhance persistence. This has not been manipulated within a study, and there are no appropriate cross-study comparisons. Similarly, he suggests that with high prey migration rates, local populations should be better buffered, and predator numerical responses should be evident only at the metapopulation level. Data are not available to test these predictions, and it would be useful to aim further tests of metapopulation theory at specific predictions of general interest, rather than merely trying to establish whether we have metapopulation dynamics.

Finally, I conclude with a plea for the inclusion of additional types of internal dynamics in general metapopulation models. A review of the scanty empirical evidence available, coupled with a bit of deductive reasoning, suggests many links between various aspects of predator–prey interactions at the local level and dynamic patterns at the metapopulation level. However, some theoretical exploration would let us know if we are indeed on the right track.

Acknowledgments

I would like to thank G. Belovsky, N. Cappuccino, I. Hanski, A. Hunter, and D. Ruzzante for helpful comments on previous versions of this chapter. A. Lawson, C. Kennedy, S. Matthews, and G. Sidhu assisted with sample collection and processing. Many thanks also go to Frank and Dot Robinson for tolerating my idiosyncratic orchard on their land, to J. M. Hardman for permission to work on an Agriculture Canada orchard, and to M. Rogers for cutting and removing the trees. This work was supported by research grants from the Natural Sciences and Engineering Research Council of Canada.

References

Aliniazee, M. T., and Cranham, J. E. (1980). Effect of four synthetic pyrethroids on a predatory mite, *Typhlodromus pyri* and its prey. *Panonychus ulmi*, on apples in southeast England. *Environ. Entomol.* **9**, 436–439.

Bernstein, C. (1984). Prey and predator emigration responses in the acarine system *Tetranychus urticae–Phytoseiulus persimilis. Oecologia* **61**, 134–142.

Burnett, T. (1979). An acarine predator–prey population infesting roses. *Res. Popul. Ecol.* **20**, 227–234.

Clements, D. R., and Harmsen, R. (1990). Predatory behaviour and prey-stage preferences of stigmaeid and phytoseiid mites and their potential compatibility in biological control. *Can. Entomol.* **122**, 321–328.

Crowley, P. H. (1981). Dispersal and the stability of predator–prey interactions. *Am. Nat.* **118**, 673–701.

Dicke, M. (1988). Prey preference of the phytoseiid mite *Typhlodromus pyri*. 1. Response to volatile kairomones. *Exp. Appl. Acarol.* **4**, 1–13.

Dicke, M., and DeJong, M. (1988). Prey preference of the phytoseiid mite *Typhlodromus pyri*. 2. Electrophoretic diet analysis. *Exp. Appl. Acarol.* **4**, 15–25.

Dunley, J. E., and Croft, B. A. (1990). Dispersal between and colonization of apple by *Metaseiulus occidentalis* and *Typhlodromus pyri* (Acarina: Phytoseiidae). *Exp. Appl. Acarol.* **10**, 137–149.

Duso, C., and Camporese, P. (1991). Developmental times and oviposition rates of predatory mites *Typhlodromus pyri* and *Amblyseius andersoni* (Acari: Phytoseiidae) reared on different foods. *Exp. Appl. Acarol.* **13**, 117–128.

Fujita, K., Inoue, T., and Takafuji, A. (1979). Systems analysis of an acarine predator–prey system. *Res. Popul. Ecol.* **21**, 105–119.

Gilpin, M., and Hanski, I., eds. (1991). "Metapopulation Dynamics: Empirical and Theoretical Investigations." Academic Press, London.

Godfray, H. C. J., and Pacala, S. W. (1992). Aggregation and the population dynamics of parasitoids and predators. *Am. Nat.* **140**, 30–40.

Gurney, W. S. C., and Nisbet, R. M. (1978a). Single species population fluctuations in patchy environments. *Am. Nat.* **112**, 1075–1090.

Gurney, W. S. C., and Nisbet, R. M. (1978b). Predator–prey fluctuations in patchy environments. *J. Anim. Ecol.* **47**, 85–102.

Hanski, I. (1989). Metapopulation dynamics: Does it help to have more of the same? *Trends Ecol. Evol.* **4**, 113–114.

Hanski, I. (1991). Single-species metapopulation dynamics: Concepts, models and observations. *Biol. J. Linn. Soc.* **42**, 17–38.

Harrison, S. (1994). Metapopulations and conservation. *In* "Large-Scale Ecology and Conservation Biology" (P. J. Edwards, R. May, and N. R. Webb, eds.), pp. 111–128. Blackwell, Oxford.

Hassell, M. P. (1978). "The Dynamics of Arthropod Predator–Prey Systems." Princeton Univ. Press, Princeton, NJ.

Hastings, A. (1977). Spatial heterogeneity and the stability of predator–prey systems. *Theor. Popul. Biol.* **12**, 37–48.

Hastings, A. (1991). Structured models of metapopulation dynamics. *Biol. J. Linn. Soc.* **42**, 57–71.

Hastings, A. (1993). Complex interactions between dispersal and dynamics: Lessons from coupled logistic equations. *Ecology* **74**, 1362–1372.

Herbert, H. J. (1970). Limits of each stage in populations of the European red mite, *Panonychus ulmi. Can. Entomol.* **102**, 64–68.

Huffaker, C. B. (1958). Experimental studies on predation: Dispersion factors and the predator–prey oscillation. *Hilgardia* **27**, 343–383.

Jansen, V. A. A., and Sabelis, M. W. (1992). Prey dispersal and predator persistence. *Exp. Appl. Acarol.* **14**, 215–231.

Johnson, D. L., and Wellington, W. G. (1984). Simulation of the interactions of predatory *Typhlodromus* mites with the European red mite, *Panonychus ulmi* (Koch). *Res. Popul. Ecol.* **26**, 30–50.

Johnson, D. T., and Croft, B. A. (1976). Laboratory study of the dispersal behavior of *Amblyseius fallacis* (Acarina: Phytoseiidae). *Ann. Entomol. Soc. Am.* **69**, 1019–1023.

Kennedy, G. G., and Smitley, D. R. (1985). Dispersal. *In* "Spider Mites: Their Biology, Natural Enemies and Control" (W. Helle, and M. W. Sabelis, eds.), Vol. 1A, pp. 233–242. Elsevier, Amsterdam.

Laing, J. E., and Huffaker, C. B. (1969). Comparative studies of predation by *Phytoseiulus persimilis* Athias-Henriot and *Metaseiulus occidentalis* (Nesbitt) (Acarina: Phytoseiidae) on populations of *Tetranychus urticae* Koch (Acarina: Tetranychidae). *Res. Popul. Ecol.* **11**, 105–126.

Lawson, A. B., and Walde, S. J. (1993). Comparison of the responses of two predaceous mites, *Typhlodromus pyri* and *Zetzellia mali,* to variation in prey density. *Exp. Appl. Acarol.* **17**, 811–821.

Levins, R. (1969). Some demographic and genetic consequences of environmental heterogeneity for biological control. *Bull. Entomol. Soc. Am.* **15**, 237–240.

Murdoch, W. W., and Stewart-Oaten, A. (1989). Aggregation by parasitoids and predators: Effects on equilibrium and stability. *Am. Nat.* **134**, 288–310.

Murdoch, W. W., Chesson, J., and Chesson, P. L. (1985). Biological control in theory and practice. *Am. Nat.* **125**, 344–366.

Murdoch, W. W., Briggs, C. J., Nisbet, R. M., Gurney, W. S., and Stewart-Oaten, A. (1992). Aggregation and stability in metapopulation models. *Am. Nat.* **140**, 41–58.

Nachman, G. (1981). Temporal and spatial dynamics of an acarine predator–prey system. *J. Anim. Ecol.* **50**, 435–451.

Nachman, G. (1987a). Systems analysis of acarine predator–prey interactions. I. A stochastic simulation model of spatial processes. *J. Anim. Ecol.* **56**, 247–265.

Nachman, G. (1987b). Systems analysis of acarine predator–prey interactions. II. The role of spatial processes in system stability. *J. Anim. Ecol.* **56**, 267–281.

Nachman, G. (1991). An acarine predator–prey metapopulation system inhabiting greenhouse cucumbers. *Biol. J. Linn. Soc.* **42**, 285–303.

Nyrop, J. P. (1988). Spatial dynamics of an acarine predator–prey system: *Typhlodromus pyri* (Acarina: Phytoseiidae) preying on *Panonychus ulmi* (Acarina: Tetranychidae). *Environ. Entomol.* **17**, 1019–1031.

Oatman, E. R. (1973). An ecological study of arthropod populations on apple in northeastern Wisconsin: Population dynamics of mite species on the foliage. *Ann. Am. Soc. Entomol.* **66**, 122–131.

Quinn, J. F., and Hastings, A. (1987). Extinction in subdivided habitats. *Conserv. Biol.* **1**, 198–208.

Reeve, J. D. (1988). Environmental variability, migration and persistence in host–parasitoid systems. *Am. Nat.* **132**, 810–836.

Roff, D. A. (1974). Spatial heterogeneity and the persistence of populations. *Oecologia* **15**, 245–258.

Sabelis, M. W. (1985a). Reproductive strategies. *In* "Spider Mites: Their Biology, Natural Enemies and Control" (W. Helle and M. W. Sabelis, eds.), Vol. 1A, pp. 265–277. Elsevier, Amsterdam.

Sabelis, M. W. (1985b). Capacity for population increase. *In* "Spider Mites: Their Biology, Natural Enemies and Control" (W. Helle and M. W. Sabelis, eds.), Vol. 1B, pp. 35–42. Elsevier, Amsterdam.

Sabelis, M. W. (1985c). Reproduction. *In* "Spider Mites: Their Biology, Natural Enemies and Control" (W. Helle and M. W. Sabelis, eds.), Vol. 1B, pp. 73–82. Elsevier, Amsterdam.

Sabelis, M. W., and Diekmann, O. (1988). Overall population stability despite local extinction: The stabilizing influence of prey dispersal from predator-invaded patches. *Theor. Popul. Biol.* **34**, 169–176.

Sabelis, M. W., and Laane, W. E. M. (1986). Regional dynamics of spider-mite populations that become extinct locally because of food source depletion and predation by phytoseiid mites (Acarina: Tetranychidae, Phytoseiidae). *Lect. Notes Biomath.* **68**, 346–376.

Sabelis, M. W., and van der Meer, J. (1986). Local dynamics of the interaction between predator mites and two-spotted spiders. *Lect. Notes Biomath.* **68**, 322–344.

Sabelis, M. W., Diekmann, O., and Jansen, V. A. A. (1991). Metapopulation persistence despite local extinction: Predator–prey patch models of the Lotka–Volterra type. *Biol. J. Linn. Soc.* **42**, 267–283.

Stewart-Oaten, A., Murdoch, W. W., and Walde, S. J. (1995). Estimation of temporal variability in populations. *Am. Nat.* (in press).

Strickler, K., Cushing, N., Whalon, M., and Croft, B. A. (1987). Mite (Acari) species composition in Michigan apple orchards. *Environ. Entomol.* **16**, 30–36.

Takafuji, A. (1977). The effect of the rate of successful dispersal of a phytoseiid mite,

Phytoseiulus persimilis Athias-Henriot (Acarina: Phytoseiidae), on the persistence in the interaction system between the predator and its prey. *Res. Popul. Ecol.* **18**, 210–222.

Takafuji, A., Tsuda, T., and Miki, T. (1983). System behaviour in predator–prey interaction, with special reference to acarine predator–prey system. *Res. Popul. Ecol., Suppl.* **3**, 75–92.

Taylor, A. D. (1988). Large-scale spatial structure and population dynamics in arthropod predator–prey systems. *Ann. Zool. Fenn.* **25**, 63–74.

Taylor, A. D. (1991). Studying metapopulation effects in predator–prey systems. *Biol. J. Linn. Soc.* **42**, 305–323.

van de Klashorst, G., Readshaw, J. L., Sabelis, M. W., and Lingeman, R. (1992). A demonstration of asynchronous local cycles in an acarine predator–prey system. *Exp. Appl. Acarol.* **14**, 185–199.

Vandermeer, J. M. (1973). On the regional stabilization of locally unstable predator–prey relationships. *J. Theor. Biol.* **41**, 161–170.

Walde, S. J. (1991). Patch dynamics of a phytophagous mite population: Effect of number of subpopulations. *Ecology* **72**, 1591–1598.

Walde, S. J. (1994). Immigration and the dynamics of a predator–prey interaction in biological control. *J. Anim. Ecol.* **63**, 337–346.

Walde, S. J. (1995). How quality of host plant affects a predator–prey interaction in biological control. *Ecology* (*in press*).

Walde, S. J., Nyrop, J. P., and Hardman, J. M. (1992). Dynamics of *Panonychus ulmi* and *Typhlodromus pyri:* Factors contributing to persistence. *Exp. App. Acarol.* **14**, 261–292.

Walde, S. J., Magagula, C., and Morton, M. L. (1995). Feeding preference of *Zetzellia mali:* Does absolute or relative abundance of prey matter more? *Exp. Appl. Acarol.* (*in press*).

Wanibuchi, K., and Saito, T. (1983). The process of population increase and patterns of resource utilization of two spider mites, *Oligonychus ununguis* (Jacobi) and *Panonychus citri* (McGregor), under experimental conditions (Acari: Tetranychidae). *Res. Popul. Ecol.* **25**, 116–129.

Zhang, Z.-Q., Sanderson, S. P., and Nyrop, J. P. (1992). Foraging time and spatial patterns of predation in experimental populations. *Oecologia* **90**, 185–196.

Chapter 10

Herbivore–Natural Enemy Interactions in Fragmented and Continuous Forests

Jens Roland and Philip D. Taylor

I. Introduction

Studies of forest insect population dynamics have generally taken two approaches, using either long-term life-table data from a single forest stand or data from several widely separated sites where each site is treated as a replicate population and a composite picture of the dynamics is inferred from the multiple sties. Examples of the former include studies of the winter moth, *Operophtera brumata,* in Wytham Wood (Varley and Gradwell, 1968) and larch casebearer, *Coleophora laricella* (Ryan, 1990); examples of the latter include studies of the gypsy moth, *Lymantria dispar* (Campbell, 1973), and the spruce budworm, *Choristoneura fumiferana* (Mott, 1963; Royama, 1984).

By considering dynamics at only a single site, or at multiple but distant sites, these studies eliminate potentially critical elements of animal population dynamics: the effect of dynamics at neighboring sites, movement among sites, and how the structure of intervening landscapes differentially influences this movement for the animals in the system (landscape connectivity; Taylor *et al.,* 1993). Fragmentation of forests is one change in landscape structure that has received considerable attention in its effects on animal populations (e.g., Wilcove *et al.,* 1986). Forest fragmentation alters the relative kinds and positions of resources in the landscape and introduces new elements into landscapes that may differentially influence animal movement, and hence population dynamics.

In the case of insect population dynamics, movement of either the herbivore or its natural enemies may be strongly influenced by local habitat structure. Local populations of herbivorous insects whose dynamics are driven by natural enemies will be affected by the dynamics of neighboring populations of herbivores and parasites, depending on the ability of each to move through intervening habitat. The fact that they *are* linked to a greater or lesser extent will therefore influence their regional dynamics. This type of process is evident in the analyses of spruce budworm dynamics in New Brunswick, Canada, where dispersal of budworm is implicated as an important element in determining local abundance (Royama, 1984). However, the large distances between the studied budworm populations made direct evaluation of the importance of movement impossible.

Empirical evidence exists for the importance of differential herbivore and natural enemy movement from studies done at very fine spatial scales. For example, coccinellid predators are unable to aggregate to local patches of high aphid abundance when habitat (monocultures of the composite *Solidago*) is fragmented into small clumps separated by 1-m clearings (Kareiva, 1987). However, when the habitat is continuous, predators readily move, aggregate, and suppress aphid outbreaks. Maintaining connectivity of the landscape through which the predators forage allows them to respond to and suppress local prey abundance.

There is also evidence that landscape structure at a large spatial scale alters predator–prey interactions. For example, the forest tent caterpillar *Malacosoma disstria* exhibits short outbreaks (2–4 years) in continuous forests and longer outbreaks (3–6 years) in fragmented forests (Fig. 1; Roland, 1993). Suppression of tent caterpillar outbreaks is associated with high mortality by natural enemies, especially by the sarcophagid fly *Sarcophaga* (=*Arachnidomyia*) *aldrichi* (Hodson, 1939), the tachinid flies *Leschenaultia exul* (Bess, 1936; Sippell, 1962) and *Patelloa pachypyga* (Sippell, 1962), and nuclear polyhedrosis virus epizootics (Clarke, 1958; Stairs, 1966; Myers, 1993). Longer outbreaks of tent caterpillar in fragmented boreal forests imply a reduced impact of natural enemies.

We hypothesize that forest fragmentation may reduce the impact of natural enemies on forest tent caterpillar dynamics directly by reducing their efficiency at the forest edge adjacent to clearings. For example, parasitoid behavior may differ near the forest edge, where temperatures and insolation are higher than in the forest interior, and where aggregation of the herbivore (e.g., Bellinger *et al.,* 1989) might "swamp" attacking parasitoids. Viral pathogens may also be deactivated along forest edges because of higher levels of ultraviolet radiation, thereby reducing their effect on tent caterpillar. Alternatively, forest fragmentation may reduce the impact of natural enemies through a differential effect of clearings on movement of parasites and pathogens relative to that of the tent caterpillar.

Under each hypothesis, the effect of fragmentation is to uncouple the forest tent caterpillar population from the parasite/pathogen community associated with population collapse. It is likely that fragmentation will influence parasites and pathogens differently from its effect on the herbivore, resulting in potentially quite complex dynamics. For example, clearings may inhibit the movement of some parasitoids or the transmission of virus. Alternatively, or in addition, clearings may enhance the absolute distances that moths disperse by causing a more strongly directional pattern of movement compared to their movement through a continuous forest canopy. Each of these scenarios results in the natural enemies lagging behind the herbivore in fragmented forests, when compared to continuous forests. Such a pattern of decreasing parasitism with increased fragmentation

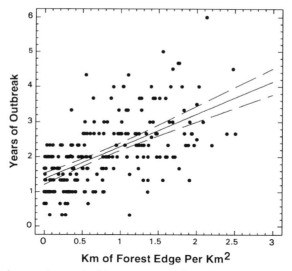

Figure 1. Mean duration (in years) of forest tent caterpillar outbreaks in northern Ontario townships as a function of the amount of forest edge in each township (after Roland, 1993).

has been implicated in fine-scale field studies of insect parasitism in habitat clumps of red clover (Kruess and Tscharntke, 1994). Consistent effects of landscape structure on fine-scale local movements of herbivores and their parasites and pathogens should produce large-scale changes in the population dynamics of these animals when there is extensive alteration of landscape structure such as forest fragmentation.

We have approached the study of forest tent caterpillar population dynamics by examining patterns of distribution of the herbivores and of the parasites at several spatial scales and simultaneously by examining how individual animals are influenced by different kinds of landscape structure. In this chapter, we present a broad outline of the spatial approach used to study this host and natural enemy system, and give an example of the details with a subset of our larger data set. This analysis shows, at two spatial scales, differential effects of fragmentation on forest tent caterpillar parasitism by two species of parasitic flies, *S. aldrichi* and *P. pachypyga*.

II. Methods

A. General Approach

We approach the study of tent caterpillar and its natural enemies at a spatial scale (420 km²) that is sufficiently large to detect variation in dynamics among subpopulations, but we do so with study sites arrayed at a sufficiently fine scale to permit linkage of local dynamics to those in adjacent stands. We feel that the associated problems of lack of independence among study sites is more than outweighed by the benefits of being able to identify the impact of dynamics at adjacent sites and of movement among sites. Spacing of study sites at a scale large enough to ensure independence would in effect obscure one of the most important factors driving the dynamics, that of movement.

1. Sampling Sites

Our main study area includes 127 population sample points (megagrid) spread across an area of 420 km^2 in the Cooking Lake and Ministik Hills area of Alberta, Canada (113° 00′ W, 53° 22′ N, Fig. 2). This area is at the southern edge of boreal forest and is dominated by trembling aspen (*Populus tremuloides*), the preferred host plant of forest tent caterpillar. Other tree species intermixed with aspen and in decreasing importance are balsam poplar (*Populus balsamifera*), paper birch (*Betula papyrifera*), and white spruce (*Picea glauca*). Clearing for agriculture has been variable in intensity throughout the study area, resulting in a broad range of forest fragmentation from highly fragmented to continuous forest (Fig. 2).

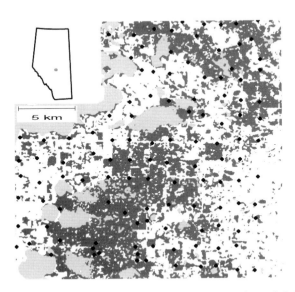

Figure 2. Map of large-scale study area (megagrid). Dark areas are forested, light gray areas are water, and white areas are pasture and cropland. Sample points of the megagrid are shown as black dots.

Each sample point of the megagrid is separated from adjacent points by, on average, 1.8 km. This large scale provides a proximity among sample points that is within the dispersal capabilities of both the forest tent caterpillar and its major parasitoids. Nested within this main study area is a smaller-scale grid (mesogrid) of 109 points with an average separation of 50 m, which permits us to examine the effects of fine-scale landscape structure on population dynamics in the absence of the larger-scale effects. At an even finer scale we sample from 14 small (10 × 10 m) isolated fragments of aspen forest (isolates) scattered throughout the main study area. These isolated patches allow us to examine the relative ability of individual tent caterpillars and parasitoids to colonize new sites, compared to sites in continuous forest.

2. Life Tables

At each sample point in the megagrid and mesogrid we estimate forest tent caterpillar abundance and several discrete mortality factors. These are not complete life tables (Varley and Gradwell, 1968), but sampling does include those stages acted on by the agents considered most important in causing collapse of forest tent caterpillar outbreaks (i.e., virus and late-larval and pupal parasitoids; Sippell, 1957; Hodson, 1939; Myers, 1993). A time-limited search for cocoons is used to estimate abundance; the time taken to find and collect 50 cocoons is recorded up to a maximum of 15 minutes. For samples that required less than 15 minutes, the count of 50 is multiplied by 15 minutes and divided by the actual time taken. In this way, an estimate is made of the number of cocoons that would have been collected if continued for a full 15 minutes. By monitoring the phenology of herbivore pupal development and of parasitism (Sippell, 1957), the two tachinid parasitoids *L. exul* and *P. pachypyga* can be recovered from cocoons, as well as the slightly later sarcophagid *S. aldrichi*. Parasitism is estimated for each by rearing the herbivore pupae until emergence of the parasitoid. Subtle differences in phenology of each parasitoid species (and its effect on absolute level of parasitism) is not considered a major problem because it is the relative rates of parasitism in the two landscapes for each species that is of greatest interest to us here.

Other data collected at each site, but not presented here, are (1) fecundity of tent caterpillar estimated from abdominal width of female pupal cases and the known relationship between abdominal width and egg mass size (J. Roland and R. Bourchier, unpublished) and (2) the presence/absence of nuclear polyhedrosis virus using a DNA probe for tent caterpillar virus (Kaupp and Ebling, 1993) on pupal and larval exuviae and dead bodies from each site.

3. Outbreak Spread

Each year we acquire false-color infrared aerial photography of the entire study area. The spatial pattern of defoliation provides a detailed record of caterpillar abundance, from which we can measure the spread of increasing populations and the contraction of declining populations. Life-table data at each point can then be related to life-table data at the same point in previous years, life-table data at adjacent points in the same or previous years, and the pattern of population spread or contraction around the site.

4. Individual Movement

We use two methods to assess movement of tent caterpillar and parasitoids, one direct and one indirect. First, mark–release–recapture of parasitoids is used to estimate movement through continuous forest and through fragmented forest. Second, we indirectly assess the difference in the ability of moths and parasitic flies to move across clearings of various sizes by examining their relative rates of "discovery" of isolated patches of the host plant, trembling aspen (*Populus tremuloides*).

B. Detailed Methods

1. Small-Scale Studies

Fourteen isolated fragments (50 to 450 m from large stands of aspen forest) were surveyed to estimate the relative ability of each insect to "discover" fragments at increasing distance from infested continuous stands. These were contrasted with 14 control sites in continuous forest. Presence of tent caterpillar cocoons and of parasitoids attacking the cocoons was assessed with respect to distance to nearest forest and distance to known defoliation in the previous year (from the false-color infrared aerial photography). Isolated sites were compared to sites within continuous forests with respect to abundance of tent caterpillar and level of parasitism by each fly species. The prediction was that clearings would act as barriers to parasitoid movement, and that parasitism of cocoons in forest isolates should decline as distance from continuous forest increased.

2. Large-Scale Studies

At each sample point of the megagrid, percentage defoliation by tent caterpillar was classified on a nonlinear scale from 1 to 9. Using false-color infrared photographs from 1993, we estimated the degree of forest fragmentation around each point by counting the number of quarter-sections (800 × 800 m) around the point that contained greater than 50% forest cover, giving a crude estimate of fragmentation that ranged from complete (0) to no fragmentation (4). We also measured the distance to the closest area of tent caterpillar defoliation in the previous year (1993). Pupae were sampled at all sites over a 1-week period from late June to early July, 1994. We test for the effects of the estimated abundance of forest tent caterpillar, the distance to areas of high defoliation in the previous year, and the degree of forest fragmentation around the sample point, on the abundance of fly parasites, *S. aldrichi* and *P. pachypyga*, recovered from the collected pupae.

3. Statistical Models

Statistical models were fit using the generalized linear models procedure in Splus (StatSci). The goodness-of-fit of a given model was assessed by (1) plotting fitted and residuals versus predicted values, (2) computing the overall change in model deviance when terms of interest were dropped, and (3) considering the size of the estimated standard errors of parameter estimates (*t*-tests, maximum probabilities of type-I error: 5%).

For the isolate data, we fit logistic models (binomial errors; McCullagh and Nelder, 1989) to the odds of a cocoon being parasitized versus not being parasitized. The proportion of cocoons parasitized was fit to a model that included the main effects of species (*S. aldrichi* versus *P. pachypyga*), distance to defoliation in the previous year (hereafter simply called distance), whether the sample came from an isolate or not, and the two-way interactions among these. We then sequentially dropped terms from the model and tested for a significant change in deviance against a χ^2 distribution.

For the large-scale data (megagrid), the response variable was the count of the total number of parasites of both species at each point. We fit a model to the counts (Poisson errors) of fly parasites, including the main effects of distance, fragmentation and fly species, and all two-way interactions. We again assessed the importance of each term by sequentially dropping them from the model, and testing for a significant change in deviance against a χ^2 distribution.

III. Results

A. Isolates

Cocoons from isolated patches were less likely to be parasitized than were those from continuous forest (Isolate term, Table 1), and the effect was much stronger for parasitism by *P. pachypyga* than for *S. aldrichi* (Species × Isolate interaction). *Patelloa pachypyga* caused a lower rate of parasitism in isolates than did *S. aldrichi* (4.4 ± 0.76% versus 19.7 ± 1.3%, respectively); both caused similar rates of parasitism in continuous forest (15.7 ± 1.4% versus 18.5 ± 1.6%, respectively). Cocoons in both the isolates and continuous forest were more likely to be parasitized when closer to an area of high defoliation in the previous year (Table 1; Main effect of distance). The distances of the isolates in which *S. aldrichi* parasitized tent caterpillar cocoons ranged up to 400 m from the nearest continuous forest; the only cocoons that *P. pachypyga* parasitized were in isolates within 125 m of nearby forest.

TABLE 1
Analysis of Deviance Table for Logistic Models
Assessing the Odds Ratio of a Cocoon Being Parasitized
as a Function of Fly Species (*Sarcophaga aldrichi* versus
***Patelloa pachypyga*), Distance from the Previous Year's**
Defoliation, and Forest Isolates versus Continuous Forest

Term	df	D[a]	$p(\chi^2)$
Species	1	7.20	0.01
Distance	1	26.00	<0.001
Isolate	1	6.40	0.01
Species × Isolate	1	10.70	0.0011
Residuals	43	91.31	

[a]The change in deviance (D) when the term is dropped from the full model is shown.

B. Megagrid

The dominance of each parasitoid species depended on both the degree of forest fragmentation and distance from the previous year's outbreak (Table 2; Species × Distance; Species × Fragmentation interactions). Parasitism by *P. pachypyga* declined with increasing fragmentation (Fig. 3), whereas *S. aldrichi* maintained its presence irrespective of the level of fragmentation. Numbers of attacks by both fly species declined as the distance from areas of high defoliation in previous years increased, but *P. pachypyga* was more prevalent at points close to previous year's outbreaks than was *S. aldrichi* (Table 2; Species × Distance interaction). The mean count of *P. pachypyga* from cocoons at sites within the area of the previous year's outbreak was 48 ± 5.5 versus 36 ± 6.2 for *S. aldrichi*.

TABLE 2
Analysis of Deviance Table for Poisson (Count) Models
of the Abundance of *Sarcophaga aldrichi* and *Patelloa*
***pachypyga* as a Proportion of All Fly Parasitism at Sample**
Points in the Large-Scale Survey (Megagrid)

Term	df	D	$p(\chi^2)$
Distance[a]	1	6.43	0.0011
Species × Distance	1	12.26	0.0005
Species × Fragmentation	1	7.13	0.0007
Residuals	242	829.40	

[a]Distance is the distance from the site to the nearest defoliated forest in the previous years, and the degree of fragmentation is fit as an ordinal factor.

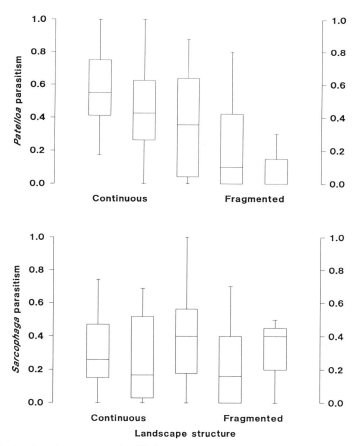

Figure 3. Boxplots showing mean (middle bar), interquartile range (inner bars), and range (outer bars) of parasitism rates for the two parasitoid species *Patelloa pachypyga* and *Sarcophaga aldrichi* across five levels of forest fragmentation.

IV. Discussion

A. Medium-Scale Dynamics of Tent Caterpillar

Two of the parasitic flies that attack forest tent caterpillar, *Sarcophaga aldrichi* and *Patelloa pachypyga,* are affected in different ways by forest fragmentation. Our results from both the isolate and megagrid studies show uniform parasitism by *Sarcophaga* at all sites, supporting the idea that clearings do not act as barriers to their movement. By contrast, the reduction in parasitism by *Patelloa* in forest isolates suggests there is limited movement of this fly across clearings. In preliminary mark–recapture studies, we have observed movement by *Sarcophaga* across clearings more than 300 m wide, over a period of several days. We have not yet done mark–recapture studies of *Patelloa* and so cannot unequivocally say that reduction in *Patelloa* efficacy in fragments is due to reduced movement. Results are consistent with a differential effect of landscape structure on movement by the two fly species, although other effects of fragmentation could explain the observed pattern. Parasitoid efficiency is reduced at forest edges (J. Roland and P. Taylor, unpublished), but that effect would not be expected to be related to distance from continuous forest unless movement was involved.

The spatial distribution of the herbivore and parasites across the megagrid is consistent with the results from the forest isolates; tent caterpillar readily colonize isolated and fragmented stands, *Sarcophaga* readily discover and parasitize tent caterpillar at all sites within 400 m of continuous forests, and *Patelloa* is unable to effectively attack tent caterpillar in fragmented forests where the fragments are more than 125 m from continuous stands. This pattern suggests that forest structure differentially affects the rates of movement of the three insects (herbivore and two parasitoids). Moths appear to move more readily than do flies, since sites that are far from the previous year's outbreak had moths but no flies (Fig. 4). Our sampling of herbivore cocoons to detect presence of parasitoids requires that the moths do colonize fragments first, a fact that biases the detection of greater movement by the tent caterpillar. The high abundance of cocoons at some of the distant sites (Fig. 4), however, suggests that tent caterpillar have been there for more than one year (although not presently at outbreak levels) and remain undiscovered by flies. Similar patterns of reduced parasitism with distance from main population sources for other insect species (Landis and Haas, 1992; Kruess and Tscharntke, 1994) have been attributed, in part, to reduced movement of parasitoids across clearings.

B. Large-Scale Dynamics of Tent Caterpillar

The large-scale pattern of prolonged outbreaks of tent caterpillar in fragmented boreal forests in Ontario (Roland, 1993) could result from a variety of mechanisms, including the reduced efficiency of parasitoids and a reduction in parasitoid movement in response to herbivore abundance in fragmented forests. *Patelloa pachypyga* is limited in fragmented forests because of either its limited movement across clearings or its reduced efficiency in fragments. Data presented here suggest that one parasitoid species, *S. aldrichi,* is affected little by forest fragmentation. Many other parasitoid species attack tent caterpillar (Witter and Kulman, 1979), several of which will likely to hindered by forest fragmentation. Reduction in efficiency of those species dominant during tent caterpillar decline (*P. pachypyga* is among these) would result in prolonged outbreak. Reduction in

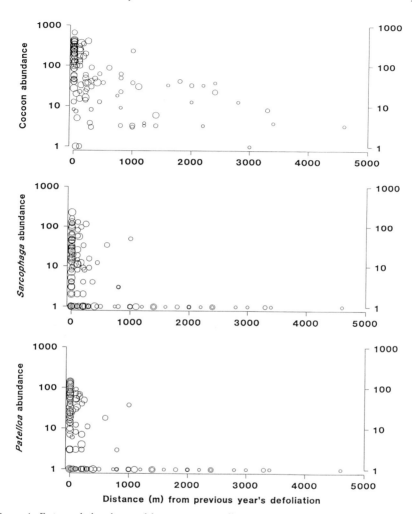

Figure 4. Estimated abundance of forest tent caterpillar cocoons, *Sarcophaga aldrichi*, and *Patelloa pachypyga* as a function of distance from areas of high defoliation in the previous year. Level of fragmentation is indicated for each site by the size of the circle: large circles = continuous forest, small circles = highly fragmented forest.

efficiency of those parasites that are thought to maintain low tent caterpillar densities (e.g., the braconid wasp *Aleiodes malacosomatus;* Sippell, 1962) would shorten the interval between outbreaks. Regulation of tent caterpillar at low density, however, is not known.

Moths move readily across clearings and easily colonize fragmented, isolated sites (sites indicated by small circles in Fig. 4). Fragmentation does not appear to inhibit moth movement and may actually enhance movement. Forest fragments could act as "stepping stones" for more rapid and more strongly directional dispersal compared to that through a continuous forest canopy. Both fly species parasitize tent caterpillar at up to 1000 m from the previous year's outbreak (Fig. 4), but if the intervening forest is fragmented, *Patelloa* drops out (Fig. 3). On the basis of the isolate data, we suspect that *Sarcophaga* moves readily through fragmented forest, and that *Patelloa* may move readily through continuous forest; the two forest types are approximately equally represented across our megagrid sites (Fig. 2). Similar overall distances moved by both fly species for the entire study area (Fig. 4) are therefore not surprising; it is the differential movement in each of the two landscapes that is of most interest here.

Parasitism rates across the megagrid do show a differential effect of forest fragmentation on the two fly species. *Sarcophaga* contributes about 35% of the total fly parasitism at all sites in the megagrid, regardless of level of fragmentation (Fig. 3). This pattern suggests that *Sarcophaga,* as a component of the parasitoid community, is able to keep up with changing spatial patterns of herbivore abundance. *Patelloa,* on the other hand, contributes 60% of the fly-caused parasitism in continuous-forest sites and contributes virtually nothing to parasitism at the most fragmented sites (Fig. 3). If the reduced parasitism in fragments is due to reduced discovery, it implies that *Patelloa* is less able to keep up with the spread of tent caterpillar populations in fragmented forest. Data from the forest isolates support the idea that differential movement of the two flies is important; *Sarcophaga* successfully parasitize tent caterpillar in forest isolates that are much farther from nearby forest (400 m) compared to the maximum distance at which *Patelloa* is able to parasitize (125 m).

In this chapter, we have shown that intervening habitat structure can affect the impact of parasitoids that are dominant during suppression of one of the most important defoliators of boreal forest. We have also shown that the impact of landscape structure differs between parasitoid species. If both flies had been negatively affected by forest fragmentation, then the pattern may simply have arisen from an effect of fragmentation on the herbivore (e.g., higher herbivore density in fragments resulting in "swamping" of all parasitoids), and not because of an effect of fragmentation on the flies themselves. The fact that one fly species was negatively affected by forest fragmentation and one species was not confirms that fly distribution is not simply a consequence of effects of forest structure on the tent caterpillar.

The interaction of landscape structure and population dynamics likely occurs for other insects, but the effect may not be the same as seen for tent caterpillar. For example, outbreaks of eastern spruce budworm tend to be more severe and longer lasting in continuous forests than in fragmented forests (Mott, 1963), a pattern opposite to that seen for tent caterpillar. The mechanism causing the difference in pattern of budworm outbreak in the two landscape types is not known. Dispersal of spruce budworm is on the order of tens or hundreds of kilometers (Royama, 1984) and many of its parasitoids are generalists with more than one generation per year, requiring alternate hosts later in the summer. Therefore, budworm are more loosely coupled with its parasitoid fauna than is forest tent caterpillar. Because of the looser linkage between budworm and its parasites, fragmentation may have little effect on further decoupling the interaction. In this situation, large continuous areas of host plants of the budworm (white spruce, black spruce, and balsam fir) may in fact promote longer outbreaks compared to those in small stands of these trees. Tight linkage between forest tent caterpillar and its parasites combined with shorter dispersal distances by the moths would make the effects of forest structure on natural enemies much more apparent.

In the study of population dynamics and conservation ecology it is becoming increasingly clear that ecological processes are affected by landscape structure at many spatial scales (Edwards *et al.,* 1993). It is also clear that new insights into the functioning of the systems emerge when data are collected at several spatial scales. To us this means that ultimately population dynamics studies will need to be placed within the landscape framework for dynamics to be fully understood.

Acknowledgments

Field studies were assisted by Kirsty Ward, Norine Ambrose, and Stacey Rasmussen. Helpful comments on the manuscript were provided by MaryCarol Rossiter and Naomi Cappuccino. This project is funded by the Natural Sciences and Engineering Research Council (J.R.), Green Plan Canada through the Canadian Forest Service (J.R.), Canadian Circumpolar Institute (P.D.T.), and the Central Research Fund, University of Alberta (P.D.T.).

References

Bellinger, R. G., Ravlin, F. W., and McManus, M. L. (1989). Forest edge effects and their influence on gypsy moth (Lepidoptera: Lymantriidae) egg mass distributions. *Environ. Entomol.* **18,** 840–843.

Bess, H. A. (1936). The biology of *Leschenaultia exul* Townsend, a tachinid parasite of *Malacosoma americana* Fibricius and *Malacosoma disstria* Hübner. *Ann. Entomol. Soc. Am.* **29,** 593–613.

Campbell, R. W. (1973). Numerical behavior of a gypsy moth population system. *For. Sci.* **19,** 162–167.

Clarke, E. C. (1958). Ecology of the polyhedrosis of tent caterpillars. *Ecology* **39,** 132–139.

Edwards, P. J., May, R. M., and Webb, N. R. (1993). "Large-Scale Ecology and Conservation Biology." Blackwell, Oxford.

Hodson, A. C. (1939). *Sarcophaga aldrichi* Parker as a parasite of *Malacosoma disstria* Hbn. *J. Econ. Entomol.* **32**, 396–401.

Kareiva, P. (1987). Habitat fragmentation and the stability of predator–prey interactions. *Nature (London)* **326**, 388–390.

Kaupp, W. J., and Ebling, P. M. (1993). Horseradish peroxidase-labelled probes and enhanced chemiluminescence to detect baculoviruses in gypsy moth and eastern spruce budworm larvae. *J. Virol. Methods* **44**, 89–98.

Kruess, A., and Tscharntke, T. (1994). Habitat fragmentation, species loss, and biological control. *Science* **264**, 1581–1584.

Landis, D. A., and Haas, M. J. (1992). Influence of landscape structure on abundance and within-field distribution of European corn borer (Lepidoptera: Pyralidae) larval parasitoids in Michigan. *Environ. Entomol.* **21**, 409–416.

McCullagh, P., and Nelder, G. A. (1989). "Generalized Linear Models." Chapman & Hall, London.

Mott, D. G. (1963). The forest and the spruce budworm. *Mem. Entomol. Soc. Can.* **31**, 189–202.

Myers, J. H. (1993). Population outbreaks in forest Lepidoptera. *Am. Sci.* **81**, 240–251.

Roland, J. (1993). Large-scale forest fragmentation increases the duration of forest tent caterpillar outbreak. *Oecologia* **93**, 25–30.

Royama, T. (1984). Population dynamics of the spruce budworm *Choristoneura fumiferana. Ecol. Monogr.* **54**, 429–462.

Ryan, R. B. (1990). Evaluation of biological control: Introduced parasites of larch casebearer (Lepidoptera: Coleophoridae) in Oregon. *Environ. Entomol.* **19**, 1873–1881.

Sippell, W. L. (1957). A study of the forest tent caterpillar *Malacosoma disstria* Hbn., and its parasite comples on Ontario. Ph.D. Thesis, University of Michigan, Ann Arbor.

Sippell, W. L. (1962). Outbreaks of the forest tent caterpillar, *Malacosoma disstria* Hbn., a periodic defoliator of broad-leaved trees in Ontario. *Can. Entomol.* **94**, 408–416.

Stairs, G. R. (1966). Transmission of virus in tent caterpillar populations. *Can. Entomol.* **98**, 1100–1104.

Taylor, P. D., Fahrig, L., Henein, K., and Merriam, G. (1993). Connectivity is a vital element of landscape structure. *Oikos* **68**, 571–573.

Varley, G. G., and Gradwell, G. R. (1968). Populations models for the winter moth. *In* "Insect Abundance" (T. R. E. Southwood, ed.), pp. 132–142. Blackwell, Oxford.

Wilcove, D. S., McLellan, C. H., and Dobson, A. P. (1986). Habitat fragmentation in the temperate zone. *In* (M. E. Soulé, ed.) "Conservation Biology: The Science of Scarcity and Diversity" Sinauer Assoc., Sunderland, MA.

Witter, J. A., and Kulman, H. M. (1979). The parasite complex of the forest tent caterpillar in northern Minnesota. *Environ. Entomol.* **8**, 723–731.

Chapter 11

Simple Models and Complex Interactions

Greg Dwyer

I. Introduction

The application of formal mathematical theory to interspecific interactions has a well-known history, dating from the work of Lotka and Volterra in the 1920s (Kingsland, 1985). The usefulness of such classical mathematical theory for understanding the population dynamics of herbivorous insects, however, has at times been questioned (Onstad, 1991; Strong, 1986). This chapter is intended to demonstrate that theory can be useful both qualitatively and quantitatively. I consider first a variety of mathematical models that have been used to reach qualitative conclusions, and then I describe my own work using a mathematical model to make quantitative predictions.

Theoreticians using differential equation models usually focus on either stability or what is known in mathematics as complex dynamics (Drazin, 1992), which means oscillatory behavior that includes limit cycles and chaos. The question of biological interest is, typically, under what conditions will populations of the species in a model be stable, or alternatively show complex dynamics? Whether mathematical stability properties are ecologically meaningful has been controversial (Murdoch *et al.*, 1985); in fact, stability (Connell and Sousa, 1983), limit cycles (Gilbert, 1984), and chaos (Hassell *et al.*, 1976; Berryman and Millstein, 1989) have all been questioned for their relevance to real ecological systems. Work with long time series of a wide variety of different animals, however, has suggested that many animal populations, at least, experience complex dynamics (Turchin and Taylor, 1992), in turn reemphasizing a need for simple mechanistic models.

One of the root causes of criticism of mathematical models by field ecologists is model simplicity (Onstad, 1991). That is, the models often consider only a small fraction of the biological detail that field ecologists believe is important, and this is sometimes used as an argument in favor of complex simulation models (Logan, 1994). Simple models, however, have the advantage that they allow the relationship between biological mechanism and population dynamics to be much

POPULATION DYNAMICS

more easily understood. Moreover, the idea of simplifying a situation is a standard research strategy, and is just as often used when designing models as when designing experiments. In Section II, an attempt is made to make clear the value of even the simplest models by briefly reviewing nonintuitive qualitative results from a variety of models of interspecific interactions among insects. In Section III, the aim is to show that simplicity is not necessarily a barrier to quantitative accuracy.

II. Simple Mathematical Models with Nonintuitive Results

Of the many messages that may be derived from mathematical models, perhaps one of the most important is that complicated population dynamic behavior may result from simple biology. Although the best-known example of this is the occurrence of chaotic dynamics in the discrete logistic model of population growth (May, 1976), in fact the message of the original Lotka–Volterra predator–prey models is similar, in that Volterra's work was an attempt to find a simple biological explanation for the dramatic fluctuations observed in Mediterranean fisheries (Braun, 1975). These two examples loosely motivate what follows. Four different kinds of interspecific interactions that can potentially affect the population dynamics of herbivorous insects are considered; host–parasitoid interactions, host–pathogen interactions, induced plant defenses, and maternal effects. My intent is not to review so much as to show how models of two-way interactions can lead to surprising conclusions.

A. Host–Parasitoid Interactions

Difference equation models of the interactions between insect parasitoids and their hosts represent some of the earliest theoretical work on interspecific interactions, dating from the well-known Nicholson–Bailey model (Hassell, 1978). The basis for this model and many of its derivatives is that adult female parasitoids are extremely mobile, but all other life stages are essentially sessile while developing inside the parasitized host. Female parasitoids are assumed to produce some constant number of offspring per infected host. The original model was formulated in discrete time to mimic the nonoverlapping generations of many parasitoids and their hosts, and partly for this reason it was unstable for all parameter values. Instability here implies that the parasitoid or the parasitoid and its host go extinct by means of unstable oscillations, essentially because the parasitoid population overshoots the host population by an increasing amount in each generation. Part of what is interesting about this first model is precisely that this behavior is unrealistic; that is, if there ever were a parasitoid–host interaction

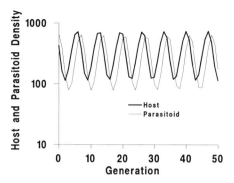

Figure 1. Dynamics of an insect host–parasitoid model, due originally to Hassell and Varley (1969). The model is

$$N_{t+1} = \lambda N_t e^{-QP_t^{1-m}} \tag{1}$$

$$P_{t+1} = N_t(1 - e^{-QP_t^{1-m}}), \tag{2}$$

where N_t is the host population at time t, P_t is the parasitoid population at time t, λ is the reproductive rate of the parasitoid, and Q and m are interference constants. Here the parameter values are $Q = 0.1$, $m = 0.5$, $\lambda = 5.0$. The figure depicts classic stable, multigenerational cycles in host and parasitoid.

that exactly fit the model, it has probably already gone extinct. Subsequent work has focused on a variety of different stability mechanisms; density-dependent reproduction in the host, for example, can lead to a point equilibrium or to stable cycles. In particular, a Nicholson–Bailey model that includes interference among parasitoids can lead to limit cycles (Hassell and Varley, 1969; see Fig. 1). This inspired Varley *et al.*'s (1973) argument that such models may provide an explanation for the population cycles seen in many forest-defoliating insects.

A thorough review of models of host–parasitoid interactions is beyond the scope of this chapter (see instead Hassell, 1978, or Hassell and Waage, 1984). Here I consider only spatial heterogeneity in parasitoid attack rates relative to host densities, which is perhaps the most intensively studied mechanism that is capable of stabilizing host–parasitoid interactions. The history of this particular mechanism illustrates how field ecology can beneficially affect theory. The original Nicholson–Bailey model assumes that parasitoid attack rates are essentially random and that parasitoids are distributed evenly, assumptions that are probably biologically unrealistic. An alternative model structure assumes that hosts are instead distributed patchily, and that parasitoid attacks are aggregated with respect to host density, so that host patches of higher density are subjected to higher attack rates (Hassell, 1978). Early models that incorporated such nonrandom attack rates, however, were too complex to be easily understood, prompting the development of a simpler model that was more phenomenological but far easier to understand (May, 1978). The simpler model demonstrated that aggregation strongly promoted stability, leading to the interpretation that aggregation to higher host densities is stabilizing. A further interpretation of this result was that parasitism rates should be positively correlated with host densities, so that the parasitoid has a density-dependent effect on its host. Empirical studies of host–parasitoid systems, however, indicated that parasitism rates are often either inversely density dependent or have no relationship to density and that in general stability may not be a useful concept for understanding real host–pathogen systems (Murdoch *et al.*, 1984, 1985).

In response to these observations, more detailed analysis of the May (1978) model pointed out that, in that particular model and in general, aggregation does not necessarily have to be related to host density for the parasitoid to stably regulate its host (Chesson and Murdoch, 1986). Moreover, aggregation to host patches of lower density is as effective in creating stability as is aggregation to host patches of higher density, which in turn implies the counterintuitive result that inverse density dependence can stabilize host–parasitoid interactions. Further empirical work suggested that aggregation is indeed often inversely density dependent (Stiling, 1987; Walde, and Murdoch, 1988). Further theoretical work showed that heterogeneity in parasitoid attack sufficient to create stability can arise from mechanisms other than just spatial patchiness (Hassell *et al.*, 1991) and that empirical levels of heterogeneity are often sufficient to lead to stability (Pacala and Hassell, 1991). An important caveat is that the degree to which the Nicholson–Bailey approach is relevant to real host–parasitoid systems continues to be controversial, which has prompted elaborations of continuous-time host–parasitoid models that show very different relationships between heterogeneity and stability (Murdoch and Stewart-Oaten, 1989; Murdoch *et al.*, 1992; Ives, 1992; Rohani *et al.*, 1994). The salient point, however, is that pressure from empirical observations has led to theoretical conclusions that are not at all intuitive.

B. Host–Pathogen Interactions

Although empirical work with infectious diseases of insects dates at least from the nineteenth century (Steinhaus, 1975), the first simple model of insect host–pathogen interactions was presented by Anderson and May (1980, 1981). They used their model to suggest that infectious diseases are responsible for the cyclic population dynamics observed in many forest-defoliating insects, following Varley *et al.*'s (1973) pioneering attempts to explain such cycles with host–parastoid models. Outbreaks of forest defoliators are often ended by epidemics of infectious diseases that destroy much of the host population, and many of the diseases in question have infectious stages that are capable of surviving outside of the host for decades (Thompson and Scott, 1979). Anderson and May's approach was to add this long-lived infectious stage to an otherwise standard epidemiological model, an addition that changes the long-term model behavior from a stable equilibrium to stable cycles for at least some parameter values. The externally surviving infectious stage, a rare feature in a pathogen, is thus the critical feature of the interaction between host and pathogen. The cycles arise because the disease is able to survive for long periods after the insect population has crashed. In the interoutbreak periods, the population of infectious pathogens outside of any hosts slowly declines, while the insect population grows slowly but exponentially (Fig. 2). When the insect population exceeds some threshold, the disease begins to increase in frequency and density at an accelerating rate, finally outstripping the insect population and causing it to crash. The other critical feature of the interaction is thus that the reproductive rate of the insect has to be slow enough to ensure that the pathogen does not bounce back too rapidly, leading to a stable equilibrium or damped oscillations.

As with host–parasitoid models, modifications of this basic model can lead to at least modestly different conclusions about dynamics. An immune host class, representing adult insects that typically do not become infected, lead to an apparently reduced chance of stable cycles (Brown, 1984). Time delays and more elaborate stage structure can give rise to cycles with a period of one year or less (Briggs and Godfray, 1995a). Including a refuge for the pathogen can lead to a stable point equilibrium (Hochberg, 1989), and adding density-dependent host reproduction also has a stabilizing effect (Bowers *et al.*, 1993). Nearly all of these models, however, are formulated in continuous time, whereas most temperate insects that are afflicted with pathogens have one or at most a few generations per year (Evans and Entwistle, 1987). Empirical studies of temperate insect host–pathogen systems have suggested that a more realistic scenario is one in which the pathogen has several generations during the host's larval period, followed by host reproduction and overwintering (Woods and Elkinton, 1987; Dwyer and Elkinton, 1993). For such conditions, however, the original model structure, in which transmission is a linear function of pathogen density, gives unstable oscillations leading to pathogen and/or host extinction for all parameter values, much as in the original Nicholson–Bailey model (Briggs and Godfray, 1995b). Including nonlinearity in transmission in this model leads to a stable point equilibrium, but long-term oscillations, including chaos, occur with the additional details of (1) sublethal infections in which some hosts survive infection but have a reduced reproductive rate or (2) a reservoir for the pathogen (Briggs and Godfray, 1995b). As is discussed later in this chapter, there is increasing empirical support for nonlinearity in disease transmission. Field observations and experiments have thus again pushed the theory into new and interesting avenues of development.

One of the most interesting aspects of this body of theory is that, in cases

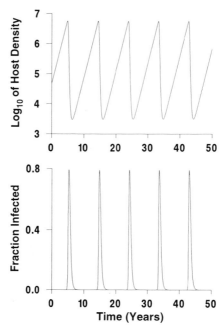

Figure 2. Dynamics of Anderson and May's (1980, 1981) invertebrate host–pathogen model. The model is

$$\frac{dS}{dt} = r(S + I) - vSP \qquad (3)$$

$$\frac{dI}{dt} = vSP - (\alpha + b)I \qquad (4)$$

$$\frac{dP}{dt} = \lambda I - (\mu + v(S + I))P, \qquad (5)$$

where S is the density of susceptible hosts, I is the density of infected hosts, P is the density of pathogen particles, r is the reproductive rate of the host, v is the transmission coefficient, α is the rate of disease-induced mortality, b is the rate of nondisease mortality, λ is the rate of production of pathogen particles by infected hosts, μ is the decay rate of the virus, and t is time. The upper graph plots the total host population, $S + I$, and the lower graph plots the fraction of the population that is infected with the disease, $I/(S + I)$. Here $r = 1.0$ year^{-1}, $v = 1 \times 10^{-10}$ year^{-1}, $\lambda = 1 \times 10^6$ year^{-1}, $\alpha = 14$ year^{-1}, $b = 3.3$ yr^{-1}, $\mu = 3.0$ year^{-1}. Note that the model shows stable cycles with about 11 years between host outbreaks, and that in the interoutbreak periods disease incidence is extremely low.

where long-term cycles are observed, most of the models in question predict that disease incidence in most years will be low (Anderson and May, 1981; Bowers *et al.*, 1993; Dwyer, 1994). In other words, pathogens can cause cycles even if in most years there is almost no disease infection. Although this observation attracts little comment, it is both counterintuitive and apparently robust.

C. Inducible Plant Defenses

Changes in host plant quality induced by defoliation also have been suggested as a possible mechanism driving complex dynamics among insect herbivores (Benz, 1974; Fischlin and Baltensweiler, 1979; Fox and Bryant, 1984). Although there is a vast empirical literature documenting such changes in any of a number of plant–herbivore systems (see Karban and Myers, 1989, for a review), models that attempt to understand how inducible defenses can affect herbivore population dynamics are rare (Edelstein-Keshet, 1986; Edelstein-Keshet and Rausher, 1989; Alder and Karban, 1994). Adler and Karban concentrate on evolutionary issues by examining conditions that favor in turn inducible, constitutive, and "moving target" defenses. Because of this emphasis, for the most part they assume that the herbivore population is constant. In contrast, Edelstein-Keshet and Rausher are concerned with herbivore population dynamics. The critical assumptions made in their paper are that herbivores are not selective, so that all plant biomass is fed upon equally, that herbivores affect plant defenses rather than plant abundance, that these defenses increase with defoliation to some asymptote, that the herbivore growth rate declines as the level of plant defenses increases, and that the herbivores reproduce continuously. Interestingly, the model that results, or rather the suite of models, produces cycles only for a very particular set of assumptions. This lack of cycles holds true for virtually *any* relationship between defoliation and defenses, as long as there is a maximum level of defenses, and for a broad range of herbivore growth rates, including density dependence. Cycles are only possible if the herbivore growth rate displays Allee dynamics, so that there is a density of herbivores at which the growth rate is a maximum, and the growth rate is lower for densities above and below the maximum (Fig. 3). Even in such a case, cycles are possible only for a limited range of parameter values.

Part of the reason why this result is a surprise is that the original intuition that induced defenses could cause cycles comes from experience with the Lotka–Volterra predator–prey equations. Perhaps the biggest difference between that model and the present one, however, is that here plant defenses increase to an asymptote, whereas for the predator–prey equations there is no a priori limit to either the prey or predator populations. This means that the induced defense, which plays a role equivalent to that of a predator in predator–prey models, can never overshoot the herbivore population. This is essentially the difference that prevents cycles. In contrast, in an appendix, Adler and Karban consider a model with a fluctuating herbivore population. This model, which has a structure similar to that of Lotka–Volterra models, is apparently likely to oscillate.

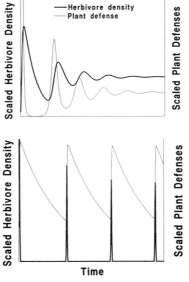

Figure 3. Dynamics of inducible defenses models of Edelstein-Keshet and Rausher (1989). In the upper graph, the model is

$$\frac{d\bar{Q}}{dt} = K_{max}h/(k_n + h) - a\bar{Q} \tag{6}$$

$$\frac{dh}{dt} = hr_0(1 - \bar{Q}/q_c), \tag{7}$$

where $\bar{Q}$ is the (mean) level of plant defenses, h is the density of herbivores, K_{max} is the maximal rate at which the defenses are induced, k_n is the level of herbivory that stimulates induction at one-half the maximal rate, a is the rate at which the defenses decay, q_c is the critical level of plant defenses at which the herbivore reproductive rate is zero, and r_0 is the maximum reproductive rate of the herbivore (its intrinsic rate of increase). The axes depict the scaled herbivore density, h/k_n, and the scaled plant defenses, $\bar{Q}/q_c$. For this model, plant defenses and herbivore density always reach a point equilibrium; here, this equilibrium is reached through damped oscillations. In the lower graph, the model includes Allee herbivore dynamics, so that the model is

$$\frac{d\bar{Q}}{dt} = K_{max}h/(k_n + h) - a\bar{Q} \tag{8}$$

$$\frac{dh}{dt} = hr_0[(1 - \bar{Q}/q_c) - \mu(1 - h/H_0)^2], \tag{9}$$

where the new parameters are μ, a dimensionless Allee parameter, and H_0, the maximal herbivore carrying capacity in the absence of defenses. Now the scaled herbivore density is h/H_0 and the scaled plant defense variable is $\bar{Q}/q_c$. This model can show a variety of behaviors, ranging from a point equilibrium through damped oscillations to the stable limit cycle shown here.

The lack of attention that the Edelstein-Keshet and Rausher paper has received is surprising (six citations between the beginning of 1990 and the end of 1993). Even if the lack of cycles has deflected attention, the paper includes an extensive discussion of the conditions under which plant defenses may regulate herbivore populations. Presumably empirical work related to the theory could drive further elaborations of the theory; for one thing, it remains to be seen whether cycles would still occur in herbivores that have discrete generations.

D. Maternal Effects

One of the newer models of complex dynamics in herbivorous insects has to do with the effect of maternal "quality" upon offspring survival and fecundity (Ginzburg and Taneyhill, 1994). "Quality" in this case means the amount of resource passed on from a female insect to her offspring in the egg, in the form of, for example, storage proteins (Leonard, 1970), which can affect the performance of those offspring in terms of their survival or fecundity. Because quality in this case is presumably related to some kind of biotic resource, I am taking this model to fall under the rubric of interspecific interactions, although strictly speaking the model keeps track of only one population. Although the notion of maternal effects was first proposed by Wellington (1957), it has lately received increased attention (Myers, 1990). Rossiter (1991), for example, has shown that the performance of gypsy moth larvae can depend on the species of tree that their mothers had fed upon as larvae (although Rossiter's data suggest positive feedbacks).

Inspired by these observations, Ginzburg and Taneyhill constructed a model to explore the consequences of maternal effects on population dynamics. Their model keeps track of both population density and population quality, such that population growth rate increases with population quality to some asymptote, while population quality declines from its maximum as population density rises. If the rates of increase in quality and density are high enough, the model shows undamped cycles (Fig. 4). Such cycles may be either neutrally stable or truly stable, depending on the details of the responses of density and quality to each other. In the tradition of Varley *et al.* (1973) and Anderson and May (1981), the authors fit their model to a variety of time series of forest defoliators and showed that their model reproduces the complex dynamics exhibited by many forest-defoliating insects better at least than a time-delayed discrete logistic model. The notion that maternal effects are a likely explanation for such oscillations has already been challenged, however. For example, one of the conclusions of Ginzburg and Taneyhill's model is that the time between outbreaks will be longer in insects with low growth rates, yet Berryman (1995) presents data showing that the time between outbreaks for many forest Lepidoptera is longer in insects with higher growth rates. More directly, Myers and Kukan (1995) present data show-

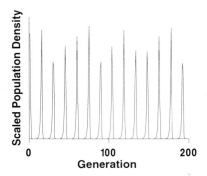

Figure 4. Dynamics of maternal effects model of Ginzburg and Taneyhill (1994). The model is

$$N_{t+1} = N_t \ \frac{Rx_{t+1}}{1 + x_t} \tag{10}$$

$$x_{t+1} = x_t \ \frac{M}{1 + N_{t+1}}, \tag{11}$$

where N_t is the density of the insect at time t, x_t is the quality of the insect population at time t, R is the maximum reproductive rate, and M is the maximum possible increase in quality. In fact, density and quality here have been scaled relative to parameters that control rates of increase to asymptotes of density and quality. The figure shows the dynamics of the scaled density variable. Note the neutrally stable cycles with long intervals between outbreaks. The parameter values here are $R = 3.0$ and $M = 10$.

ing that, in populations of western tent caterpillars (*Malacosoma californicum pluviale*) in various locations in British Columbia, Canada, densities vary over several orders of magnitude, but are only slightly related to fecundity; similar relationships hold for gypsy moth, *Lymantria dispar* (J. S. Elkinton, personal communication.

E. Overview

Hopefully, by now it is clear that several different interspecific interactions provide possible mechanisms that can cause complex dynamics in insect populations, as demonstrated by efforts to apply very different models to the same data sets on forest-defoliating insects. It remains to be seen whether any one mechanism can explain any or all of the cases of cycling forest defoliators, which is an issue that ultimately can only be settled empirically. More generally, it is well known that oscillations can be driven only by some kind of delayed density dependence (Berryman, 1978), and in every case reviewed here there is indeed some kind of delay or lag between the time when the herbivore reaches high density and the time when the controlling agent causes the highest mortality in the herbivore or the greatest reduction in the herbivore's growth rate. Part of what is interesting about the models that I have focused on is that in every case the oscillations arise in a "natural" way; that is, the lag in the density-dependent factor is not assumed a priori (for explicit delays, see Kuang, 1993, or Murray, 1989), but rather arises from the interaction between different species, or between density and quality. An important point, however, is that such time lags per se do not necessarily lead to oscillations in ecological models. For example, Hastings (1983) showed that introducing age structure into Lotka–Volterra predator–prey models has complex effects. Depending on assumptions about the model form describing maturation, as the time delay between birth and maturation increases, the predator–prey dynamics are first unstable, then stable, and then again unstable. Levin and Goodyear (1980) similarly showed that adding a time delay in the form of age structure to the Ricker fisheries model can give complicated results; as year-to-year survival is increased, stability is at first more likely and then less likely.

1. With No Theory, What Would We Get Wrong?

Here I summarize the nonintuitive results of the theory that I have covered, focusing on conclusions that may have been utterly missed without the theory.

1. Insect parasitoids can stably regulate their host populations even if parasitism is inversely density dependent. Important caveats are that aggregation may not be stabilizing for overlapping generations, and may be less stabilizing for multiple rounds of redistribution within each discrete generation.

2. Insect pathogens can drive oscillations in their hosts even if disease incidence is unmeasureably low in most years. The conditions under which such oscillations are stable, however, are more complex for insects with discrete generations.

3. If there is a maximum level of an induced plant defense, then the induced defense alone is unlikely to cause oscillations in specialist insects feeding on plants. This conclusion may or may not hold up for herbivores with discrete generations.

4. Maternal effects alone can produce oscillations. It remains to be seen, however, whether such effects are at all widespread.

III. Using a Simple Model to Make Quantitative Predictions: Gypsy Moth and Its Virus

In the examples given in the preceding section, models are used as tools for reaching qualitative conclusions about how the biology of a species interaction can affect the population dynamics of that interaction. Efforts to use simple models in a more quantitative fashion are rarer (but see, for example, Kareiva and Odell, 1987; Morris, 1993). As discussed earlier, part of the reason for this rarity is the popular notion that real ecological systems are so complex that quantitative predictions require complex simulation models (Onstad, 1991; Logan, 1994). In fact, however, in the few cases in which a simple differential equation model and a more complex simulation model were compared, the simpler model did at least as good a job as the more complex model at reproducing a given data set or ecological phenomenon (Ludwig *et al.,* 1978; Gillman and Crawley, 1990). This at least suggests that simple models may provide sufficient explanations for a least some ecological phenomena. What follows is a description of my own efforts, in collaboration with Joseph S. Elkinton, to use simple differential equations models as testable, quantitative hypotheses about the mechanisms driving virus epizootics in insect populations.

The host–pathogen models described earlier consider host–pathogen dynamics over many host generations and pathogen epidemics. Practical use of these models for univoltine insects, however, requires modification of the usual model structure to consider single-season epizootics. Moreover, qualitative theoretical conclusions usually rely on long-term model behavior, and thus often require estimates of only a few or even none of a model's parameters. Using a model to make quantitative predictions, however, requires estimates of each model parameter.

With these caveats in mind, we attempted to use a conventional insect host–pathogen model to understand the factors determining the timing and intensity of epidemics of nuclear polyhedrosis virus (NPV) within the larval period of gypsy moth (*Lymantria dispar*) populations (Dwyer and Elkinton, 1993). Virus epizootics within a gypsy moth larval period begin when larvae hatch in the spring from eggs that were laid the previous fall in contact with virus from infectious cadavers (Murray and Elkinton, 1989, 1990). If gypsy moth densities are high enough, and enough larvae hatch out infected, these initially infected larvae will start an epidemic. The modified model thus extrapolates from the densities of healthy and infected larvae at the beginning of the season to predict the fraction of larvae that are infected at each sampling period during the season. Our intent was to use the model as a hypothesis about virus spread, particularly the assumption that the instantaneous rate of virus transmission is linearly dependent on the densities of host and pathogen, and that this rate depends on nothing else. In fact, laboratory data suggest that transmission may be affected by a wide variety of different factors (Dwyer, 1991) and, as mentioned earlier, theory suggests that nonlinearity can be critical to stability. To estimate the model's transmission parameter under the simple assumptions of the model, we created small-scale virus epizootics by enclosing infected larval cadavers and healthy larvae together on red oak foliage in the field. The cadavers release virus that is available for the initially healthy larvae to consume and so to become infected. In the field, the virus causes death at about 14 days. After 7 days, a period short enough to prevent secondary mortality, the initially healthy larvae are reared individually in the lab to see if they are infected, so that the experiment consists of a single round of transmission. The resulting data give the number of larvae that become infected as a function of the initial densities of healthy larvae and infectious cadavers that began the experiment. We then fit the model to these data to estimate the transmission parameter, and the remaining parameters can be estimated from literature data.

Another consequence of using the model to understand short-term rather than long-term disease spread is that, over the short term, the initial densities of host and pathogen inevitably have a large effect (long-term dynamics are generally affected only by initial conditions in cases of neutral stability or chaos). Consequently, comparing the model predictions to data requires both initial densities of larvae and virus and a season-long time series of virus infection. Earlier work in our lab had produced just such a data set for eight distinct gypsy moth populations (Woods and Elkinton, 1987; Woods *et al.*, 1991). The fit of the model to the data is shown in Fig. 5. There are two critical points to keep in mind when assessing the fit of the model. First, the model is extrapolating from transmission data collected over a short temporal scale (7 days) at a very small spatial scale (approximately 0.05 m² of foliage) to data from populations on 4- to 9-hectare plots over entire seasons. Second, the model is based on experimental data that are completely independent of the observational test data. Given these caveats, the model does a surprisingly good job of predicting the time course of virus infection in the eight populations. Nevertheless, there are some significant discrepancies between model and data; specifically, the model is very close to the data at high density (plot 1, 1983; plot 10, 1984), not too far from the data at moderate density (plot 8, 1983; plot 5, 1985; plot 16, 1985), and very far from the data at low density (plot 3, 1983; plot 5, 1983; plot 1, 1985). In other words, under some conditions the model does an excellent job of quantitatively predicting virus epizootics. Nevertheless, it appears that the critical assumption of the model, that transmission depends only on the linear densities of host and pathogen, is incorrect; either densities should somehow be represented nonlinearly in the model, or the model neglects some other aspect of transmission. Further small-scale transmission experiments have suggested that this nonlinearity can be observed at a small scale (D'Amico *et al.*, 1995), which has in turn inspired further development of the theory (Dwyer and Elkinton, 1995).

The method that we have adopted, of using a model to make short-term predictions rather than to generate qualitative understanding, taught us that there is a gap in our understanding of the virus, and this gap was not clear from more conventional analysis of the field data (Woods *et al.*, 1991). We have also tried to understand the field data using the Gypsy Moth Life System Model (GMLSM), a vastly more complicated computer simulation that attempts to include nearly all

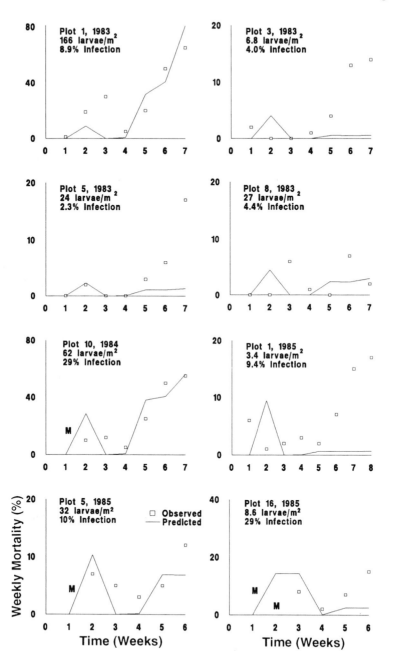

known aspects of gypsy moth biology. Unfortunately, even after we had considerably simplified the GMLSM, there were still parameters that could not be estimated, so that it was not possible to use it to make predictions. Nevertheless, adjusting two of the transmission parameters to fit the data gave us a slightly better fit to the data than did the Dwyer and Elkinton model. That is, for a single set of parameters, the GMLSM was able to fit all eight plots reasonably well, whereas there is no one set of parameters that will do the same for the simpler model. Adding even a single adjustable parameter to the simple model, however, gives as good a fit as the GMLSM. In other words, greater detail per se is not the answer; what we need to add is the *correct* detail, although we have not yet identified the biological basis for this additional parameter. Since a very slight modification to the model greatly improves the model fit, either a simple model is sufficient or, if there *are* a plethora of factors that affect transmission, we may need a vastly larger data set to detect them. For all practical purposes, then, disease spread may be a far simpler process than is widely believed. In short, using the model to make short-term predictions has been extremely informative, even though we have *not* attempted to explain the long-term dynamics of gypsy moth and its virus.

Two points have hopefully emerged from this description of our efforts to use theory quantitatively. The first is that quantitative comparisons between theory and data can expose inadequacies in the theory clearly and rapidly, in turn exposing inadequacies in our understanding of an organism's ecology. The second point has to do with how population dynamic theories can be rigorously

---

Figure 5. Within-season dynamics of the Dwyer and Elkinton model, modified from the Anderson and May model in Fig. 2. The modifications are that here there is no reproduction, there is now a fixed time between infection and death, and the loss of virus due to consumption by larvae is assumed to be offset by loss of foliage area. The model is

$$\frac{dS}{dt} = -\nu SP \qquad (12)$$

$$\frac{dI}{dt} = \nu PS - \nu P(t - \tau)S(t - \tau) \qquad (13)$$

$$\frac{dP}{dt} = \Lambda \nu P(t - \tau)S(t - \tau) - \mu P, \qquad (14)$$

where S is the density of susceptible hosts, I is the density of infected hosts, P is the density of the pathogen in the environment, ν is the transmission constant, τ is the time between infection and death of the host, Λ is the number of pathogen particles produced by an infected larva, μ is the decay rate of the pathogen, and t is time. The lines are the model predictions, and the squares are data from Woods and Elkinton (1987). For each plot, larvae per m^2 is the initial density of larvae, and percent infection is the percent of larvae that hatched out infected at the beginning of the season. The parameter values are $\nu = 1.45 \, m^2 \, day^{-1}$, $\tau = 14 \, days$, $\mu = 3.0 \times 10^{-3} \, day^{-1}$, and $\Lambda = 2 \times 10^9$.

tested. Classic ecological experimentation typically relies on qualitative perturbations of species densities, such as species removals or additions (Bender *et al.*, 1984). As the comparison of model with data in Fig. 5 indicates, quantitative variation in initial densities can be very effective for testing theory. In other work with the gypsy moth virus, we have used this approach experimentally to further test our theoretical understanding of disease spread (Dwyer and Elkinton, 1995).

References

Adler, F. R., and Karban, R. (1994). Defended fortresses or moving targets? Another model of inducible defenses inspired by military metaphors. *Am. Nat.* **144,** 813–832.

Anderson, R. M., and May, R. M. (1980). Infectious diseases and population cycles of forest insects. *Science* **210,** 658–661.

Anderson, R. M., and May, R. M. (1981). The population dynamics of microparasites and their invertebrate hosts. *Philos. Trans. R. Soc. London, Ser. B.* **291,** 451–524.

Bender, E. A., Case, T. J., and Gilpin, M. E. (1984). Perturbation experiments in community ecology: Theory and practice. *Ecology* **65,** 1–13.

Benz, G. (1974). Negative feedback by competition for food and space, and by cyclic induced changes in the nutritional base as regulatory principles in the population dynamics of the larch budmoth, *Zeiraphera diniana*. *Z. Angew. Entomol.* **76,** 196–228.

Berryman, A. A. (1978). Population cycles of the Douglas-fir tussock moth: The time-delay hypothesis. *Can. Entomol.* **110,** 513–518.

Berryman, A. A. (1987). Equilibrium or nonequilibrium: Is that the question? *Bull. Ecol. Soc. Am.* **68,** 500–502.

Berryman, A. A. (1995). Population cycles: A critique of the maternal and allomeric hypotheses. *J. Anim. Ecol.* **62,** 290–348.

Berryman, A. A., and Millstein, J. A. (1989). Are ecological systems chaotic—And if not, why not? *Trends Ecol. Evol.* **4,** 26–28.

Bowers, R. G., Begon, M., and Hodgkinson, D. E. (1993). Host–pathogen population cycles in forest insects? Lessons from simple models reconsidered. *Oikos* **67.**

Braun, M. (1975). "Differential Equations and Their Applications." Springer-Verlag, New York.

Briggs, C. J., and Godfray, H. C. J. (1995a). The dynamics of insect–pathogen interactions in stage-structured populations. *Am. Nat. (in press).*

Briggs, C. J., and Godfray, H. C. J. (1995b). The dynamics of insect–pathogen interactions in seasonal environments. *J. Theor. Biol. (in press).*

Brown, G. C. (1984). Stability in an insect–pathogen model incorporating age-dependent immunity and seasonal host reproduction. *Bull. Math. Biol.* **46,** 139–153.

Chesson, P. L., and Murdoch, W. W. (1986). Aggregation of risk: Relationships among host–parasitoid models. *Am. Nat.* **127,** 696–715.

Connell, J. H., and Sousa, W. P. (1983). On the evidence needed to judge ecological stability or persistence. *Am. Nat.* **121,** 789–824.

D'Amico, V., Elkinton, J. S., Dwyer, G., and Burand, J. P. (1995). Disease transmission in an insect virus is not a simple mass-action process. *Ecology* (in press).

Drazin, P. G. (1992). "Nonlinear Systems." Cambridge Univ. Press, New York.

Dwyer, G. (1991). The roles of density, stage and patchiness in the transmission of an insect virus. *Ecology* **72**, 559–574.

Dwyer, G. (1994). Density-dependence and spatial structure in the dynamics of insect pathogens. *Am. Nat. (in press)*.

Dwyer, G., and Elkinton, J. S. (1993). Using simple models to predict virus epizootics in gypsy moth populations. *J. Anim. Ecol.* **62**, 1–11.

Dwyer, G., and Elkinton, J. S. (1995). Host dispersal and the spatial spread of insect pathogens. *Ecology (in press)*.

Edelstein-Keshet, L. (1986). Mathematical theory for plant–herbivore systems. *J. Math. Biol.* **24**, 25–58.

Edelstein-Keshet, L., and Rauscher, M. D. (1989). The effects of inducible plant defenses on herbivore populations. I. Mobile herbivores in continuous time. *Am. Nat.* **133**, 787–810.

Evans, H. F., and Entwistle, P. F. (1987). Viral diseases. *In* "Epizootiology of Insect Diseases" (J. R. Fuxa and Y. Tanada, eds.), Wiley, New York.

Fischlin, A., and Baltensweiler, W. (1979). Systems analysis of the larch budworm system. Part I. The larch–larch budworm relationship. *In* "Dispersal of Forest Insects: Evaluation, Theory and Management Implications" (V. Delucchi and W. Baltensweiler, eds.), Proc. IUFRO Conf., 1978, pp. 273–289.

Fox, J. F., and Bryant, J. P. (1984). Instability of the snowshoe hare and woody plant interaction. *Oecologia* **63**, 128–135.

Gilbert, N. (1984). What they didn't tell you about limit cycles. *Oecologia* **65**, 112–113.

Gillman, M. P., and Crawley, M. J. (1990). A comparative evaluation of models of cinnabar moth dynamics. *Oecologia* **82**, 437–445.

Ginzburg, L. R., and Taneyhill, D. E. (1994). Population cycles of forest Lepidoptera: A maternal effect hypothesis. *J. Anim. Ecol.* **63**, 79–92.

Hassell, M. P. (1978). "The Dynamics of Arthropod Predator–prey Systems." Princeton Univ. Press, Princeton, NJ.

Hassell, M. P., and Varley, G. C. (1969). New inductive population model for insect parasites and its bearing on biological control. *Nature (London)* **223**, 1133–1136.

Hassell, M. P., and Waage, J. K. (1984). Host–parasitoid population interactions. *Nature (London)* **223**, 1133–1136.

Hassell, M. P., Lawton, J. H., and May, R. M. (1976). Patterns of dynamical behavior in single species population models. *J. Anim. Ecol.* **45**, 471–486.

Hassell, M. P., May, R. M., Pacala, S. W., and Chesson, P. L. (1991). The persistence of host–parasitoid associations in patchy environments. I. A general criterion. *Am. Nat.* **138**, 568–583.

Hastings, A. (1983). Age-dependent predation is not a simple process. I. Continuous time models. *Theor. Popul. Biol.* **23**, 347–362.

Hochberg, M. E. (1989). The potential role of pathogens in biological control. *Nature (London)* **337**, 262–264.

Ives, A. R. (1992). Continuous-time models of host–parasitoid interactions. *Am. Nat.* **140**, 1–29.

Karban, R., and Myers, J. H. (1989). Induced plant responses to herbivory. *Ann. Rev. Ecol. Syst.* **20**, 331–348.

Kareiva, P., and Odell, G. (1987). Swarms of predators exhibit 'preytaxis' if individual predators use area-restricted search. *Am. Nat.* **130**, 233–247.

Kingsland, S. E. (1985). "Modeling Nature: Episodes in the History of Population Ecology." Univ. of Chicago Press, Chicago.

Kuang, Y. (1993). "Delay Differential Equations with Applications in Population Dynamics." Academic Press, Boston.

Leonard, D. E. (1970). Intrinsic factors causing qualitative changes in populations of *Porthetria dispar* (Lep: Lymantriidae). *Can. Entomol.* **102**, 239–249.

Levin, S. A., and Goodyear, C. P. (1980). Analysis of an age-structured fishery model. *J. Math. Biol.* **9**, 245–274.

Logan, J. A. (1994). In defense of big ugly models. *Am. Entomol.* **40**, 202–206.

Ludwig, D., Jones, D. D., and Holling, C. S. (1978). Qualitative analysis of insect outbreak systems: The spruce budworm and forest. *J. Anim. Ecol.* **47**, 315–322.

May, R. M. (1976). Simple mathematical models with very complicated dynamics. *Nature (London)* **261**, 459–467.

May, R. M. (1978). Host–parasitoid systems in patchy environments: A phenomenological model. *J. Anim. Ecol.* **47**, 833–844.

Morris, W. F. (1993). Predicting the consequences of plant spacing and biased movement for pollen dispersal by honey bees. *Ecology* **74**, 493–500.

Murdoch, W. W., and Stewart-Oaten, A. (1989). Aggregation by parasitoids and predators: Effects on equilibrium and stability. *Am. Nat.* **134**, 288–310.

Murdoch, W. W., Reeve, J. D., Huffaker, C. E., and Kennet, C. E. (1984). Biological control of scale insects and ecological theory. *Am. Nat.* **123**, 371–392.

Murdoch, W. W., Chesson, J., and Chesson, P. L. (1985). Biological control in theory and practice. *Am. Nat.* **125**, 344–366.

Murdoch, W. W., Briggs, C. J., Nisbet, R. M., Gurney, W. S. C., and Stewart-Oaten, A. (1992). Aggregation and stability in metapopulation models. *Am. Nat.* **140**, 41–58.

Murray, J. D. (1989). "Mathematical Biology." Springer-Verlag, New York.

Murray, K. D., and Elkinton, J. S. (1989). Environmental contamination of egg masses as a major component of transgenerational transmission of gypsy moth nuclear polyhedrosis virus (LdMNPV). *J. Invertebr. Pathol.* **54**, 324–334.

Murray, K. D., and Elkinton, J. S. (1990). Transmission of nuclear polyhedrosis virus to gypsy moth (Lepidoptera: Lymantriidae) eggs via contaminated substrates. *Environ. Entomol.* **19**, 662–665.

Myers, J. H. (1990). Population cycles of western tent caterpillars: Experimental introductions and synchrony of fluctuations. *Ecology* **71**, 986–995.

Myers, J. H., and Kukan, B. (1995). Changes in the fecundity of tent caterpillars: A correlated character of disease resistance or sublethal effect of disease? *Oecologia (in press)*.

Onstad, D. W. (1991). Good models and immature theories. *Am. Entomol.* **72**, 202–204.

Pacala, S. W., and Hassell, M. P. (1991). The persistence of host–parasitoid associations in patchy environments. II. *Am. Nat.* **138**, 584–605.

Rohani, P., Godfray, H. C. J., and Hassell, M. P. (1994). Aggregation and the dynamics of host–parasitoid systems.: A discrete-generation model with within-generation distribution. *Am. Nat.* **144**, 491–509.

Rossiter, M. C. (1991). Environmentally-based maternal effects: A hidden force in insect population dynamics? *Oecologia* **87**, 288–294.

Steinhaus, E. A. (1975). "Disease in a Minor Chord." Ohio State Univ. Press, Columbus.

Stiling, P. D. (1987). The frequency of density dependence in insect host–parasitoid systems. *Ecology* **68**, 844–856.

Strong, D. R. (1986). Population theory and understanding pest outbreaks. *In* "Ecological Theory and IPM Practice" (M. Kogan, ed.), pp. 37–58. Wiley (Interscience), New York.

Thompson, C. G., and Scott, D. W. (1979). Production and persistence of the nuclear polyhedrosis virus of the Douglas-fir tussock moth. *Orgyia pseudotsugata* (Lepidoptera: Lymantriidae), in the forest ecosystem. *J. Invertebr. Pathol.* **33**, 57–65.

Turchin, P., and Taylor, A. (1992). Complex dynamics in ecological time series. *Ecology* **73**, 289–305.

Varley, G. C., Gradwell, G. R., and Hassell, M. P. (1973). "Insect Population Ecology: An Analytical Approach." Blackwell, Oxford.

Walde, S. J., and Murdoch, W. W. (1988). Spatial density dependence in parasitoids. *Annu. Rev. Entomol.* **33**, 441–466.

Wellington, W. G. (1957). Individual differences as a factor in population dynamics: The development of a problem. *Can. J. Zool.* **35**, 293–323.

Woods, S., and Elkinton, J. S. (1987). Bimodal patterns of mortality from nuclear polyhedrosis virus in gypsy moth (*Lymantria dispar*) populations. *J. Invertebr. Pathol.* **50**, 151–157.

Woods, S., Elkinton, J. S., Murray, K. D., Liebhold, A. M., Gould, J. R., and Podgwaite, J. D. (1991). Transmission dynamics of a nuclear polyhedrosis virus and predicting mortality in gypsy moth (Lepidoptera: Lymantriidae) populations. *J. Econ. Entomol.* **84**, 423–430.

Chapter 12

Field Experiments to Study Regulation of Fluctuating Populations

Judith H. Myers and Lorne D. Rothman

I. Population Regulation: A Critique of the Descriptive Approach

To Murdoch, "population regulation is a fundamental process related to most phenomena in ecology, including evolutionary ecology" (Murdoch, 1994). However, others think that population regulation remains a vague, poorly defined, and controversial concept (Strong, 1986; Murray, 1994). "What is population regulation?" can be the determining question on a Ph.D. qualifying examination. But in fact many ecologists are still searching for common agreement on the definition for density-dependent population regulation and on testable hypotheses arising from the concept (Murray, 1994).

In the simplest case, arguments have centered on whether numbers are normally regulated around an equilibrium, or if density is determined by independent phenomena pushing the population first in one direction and then in the other, within some limit of densities set by the physical environment. A useful definition of density-dependent population regulation is that proposed by Murray (1994): "a decline in the birthrate and increase in mortality rate with an increase in population density until mortality exceeds births and the population declines." In textbooks, and probably in the minds of many ecology students, the underlying model for density dependence is a linear or curvilinear increase in mortality and decrease in birth rate with increase in density (Figs. 1A and 1B). It is implied by these figures that the relationships between birth and death rates and density are consistent and follow similar trajectories with increasing and declining population densities. Therefore, looking for correlations between density and density-adjusted natality and mortality is a means to identify density dependence and therefore population regulation. This representation is likely to be an oversimplification for many field populations.

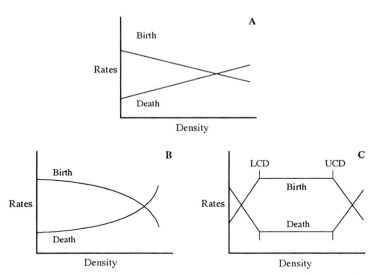

Figure 1. Diagrammatic relationships between birth and death rates and density from Murray (1994). UCD and LCD are upper and lower critical densities.

A more complex pattern for the relationship between birth and death rates and density was suggested by Murray (1994) and includes upper and lower critical thresholds at which relationships with density change (Fig. 1C). However, mortality and birth rates of natural populations can be influenced not only by current density, but by the history of previous density (Fig. 2A). For example, egg batches (i.e., average fecundity) of tent caterpillars, *Malacosoma pluviale*, were bigger as the population increased and remained smaller over several years of population decline (Fig. 2B). Similarly, survival of early-instar tent caterpillars increased with higher densities over part of the range (Fig. 2C). Delayed density dependence will lead to a spiral rather than a straight-line relationship between density and survival and fecundity. Indeed, simple relationships between density and population parameters may be the exception rather than the rule among natural populations.

A popular way of looking for regulation has been through the analysis of time-series density data (Pollard *et al.*, 1987; Turchin, 1990; Woiwod and Hanski, 1992). However, the validity of techniques for demonstrating statistically significant density dependence remains controversial (Wolda and Dennis, 1993; Holyoak, 1994). If the incidence of statistically significant density dependence increases with sampling errors in the data, and if patterns of annual rainfall are found to be density dependent (Wolda and Dennis, 1993), warning flags rise. Measurement errors, procedural problems, and insufficient data in time series are

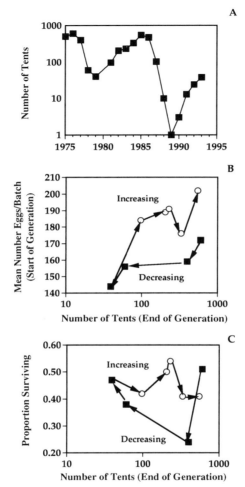

Figure 2. (A) Population trend (number of tents counted) for western tent caterpillars on Mandarte Island, British Columbia. (B) Mean number of eggs per batch for eggs collected at the beginning of the generation for which density is plotted, during an increase and decline phase of the population fluctuation (1974 to 1985). (C) Proportion of caterpillars surviving from hatch to late instars estimated from the sizes of the tents and egg batches (Myers 1990) during an increase and decline phase (1974 to 1985). Arrows indicate the temporal sequence. Open circles, increasing populations; solid squares, declining populations.

reasons for not relying heavily on statistical analyses of time-series data as a means for showing whether populations are "regulated." Lack of statistical significance certainly does not mean that density dependence does not occur for that population under some conditions. Similarly, the demonstration of statistically significant density dependence could be erroneous, and furthermore it does not identify regulatory mechanisms. Time-series analysis should not be viewed as an end in itself, although some might find it useful as a first cut leading to further work for identifying causal mechanisms.

Another popular approach has been to search for complex dynamics (e.g., limit cycles, chaos, and quasi-periodicity) in time-series data (Hassel *et al.*, 1976; Turchin and Taylor, 1992). Time series of densities of forest Lepidoptera showing periodic fluctuations have been used for statistical analyses in an attempt to identify delayed density dependence and to test if dynamics are simple or complex (examples in Turchin and Taylor, 1992; Ginzburg and Taneyhill, 1994). Methodological problems also exist for these analyses (Perry *et al.*, 1993; Holyoak, 1994). A particularly trendy approach has been the search for chaotic behavior in population data, but few (if any) sets of data are sufficiently long for rigorous analysis. Furthermore, it is unclear how the demonstration of chaotic behavior would contribute to understanding mechanisms or developing interesting testable hypotheses. "Cyclic" population dynamics have long been recognized to result from second-order, density-dependent processes (Hutchinson, 1948; Royama, 1977). Cyclic field populations will rarely show the perfectly uniform periodicity that characterizes the outcome of some population models. However, if one is interested in explaining the cyclic dynamics of field populations it is useful to look for processes that might lead to lags in recovery of populations after declines.

II. Population Regulation: The Experimental Approach

Experimental manipulation is one alternative to statistical analysis of population trends and the search for patterns between density and natality and mortality parameters. Table 1 lists feasible manipulations that might be done to test hypotheses. This section gives examples of how such experiments can improve our understanding of the dynamics of fluctuating populations of forest Lepidoptera.

A. Suppressing the Population Peak: Spray Programs

If populations are regulated by density-related mortality and fecundity, reductions of density should lead to compensation such that perturbed populations come back into synchrony with surrounding unperturbed populations. If populations are not tightly regulated, greatly reduced populations may remain low and out of phase with unmanipulated populations since their rate of increase will not be influenced by low density. A large-scale manipulation that has frequently been

TABLE 1
Hypotheses That Can Be Tested through Experimental Approaches to the Study of Population Regulation of Forest Lepidoptera[a]

HYPOTHESES TO BE TESTED BY PREVENTING OUTBREAKS

Insecticide-Spraying Programs

1. Sprayed populations of insect herbivores will become asynchronous with unsprayed populations
2. Density reduction and prevention of a viral epizootic will lead to resurgence of the herbivore population

Cropping Experiments

1. Protecting foliage quality will lead to resurgence of the population after release
2. Preventing crowding will lead to resurgence of the population after release

HYPOTHESES TO BE TESTED BY CREATING OUTBREAKS

Introductions of Exotic Species of Insect Herbivores

1. Introductions from outbreaking source populations will be more likely to become established in new sites than those from declining populations
2. Introduced parasitoids will reduce introduced populations

Experimental Introductions of Populations of Native Insect Herbivores

1. Generalist parasitoids can prevent outbreaks
2. Insect "quality" influences population dynamics
3. Populations will be resilient to perturbation of density and will show regional synchrony

HYPOTHESES TO BE TESTED BY MANIPULATING PROCESSES

Phenology of Leaves and Insects

1. Age of leaves will modify the growth of insect herbivores
2. Spring temperatures will influence growth and survival of insect herbivores

Introducing Virus

1. Spread of virus is determined by host density
2. Susceptibility of caterpillars to virus is related to leaf age at the time of ingestion
3. Sublethal effects of virus can influence insect quality, fecundity, and population growth

[a]Tests of some of these hypotheses are discussed in the text.

carried out is the suppression of populations of defoliating insects through spray programs. Unfortunately, insecticide-spraying programs have rarely been treated as large-scale experiments, and little quantitative information can be found in the literature to show the long-term impact of insecticide treatment on insect populations. Auer *et al.* (1981) sprayed a population of larch budmoth, *Zeiraphera diniana,* with Bt (*Bacillus thuringiensis*) and found that the population came back into synchrony with unsprayed populations almost immediately. This suggests regionwide synchrony and regulation of populations.

A particularly interesting result occurred following the spraying of a population of nun moth, *Lymantria fumida,* with chemical insecticide (Katagiri, 1969).

In this species, as with any other forest Lepidoptera, population outbreaks usually end with epizootics of nuclear polyhedral virus (NPV) (Myers, 1988). The spray program apparently reduced host density and prevented the viral epizootic. Without the continued influence of viral disease, the population almost immediately rebounded to outbreak density. Although other interpretations are possible, this suggests that NPV may be necessary for the "regulation" of nun moth. Without the production of the viral inoculum necessary for the development of an epizootic in the next generation, the typical population fluctuation was disrupted. This hypothesis might be tested experimentally by partially removing pathogens that contaminate the host environment as polyhedral inclusion bodies following an epizootic. Bleaching the food plants before budburst could potentially remove the virus from the environment and thus prolong outbreaks.

Spraying NPV on an increasing population of tussock moth, *Orgyia pseudotsugata,* in British Columbia, Canada, prevented the outbreak of the population without the premature resurgence seen in the nun moth (Shepherd *et al.,* 1984; Otvos *et al.,* 1987a,b). These experiments support the hypothesis that viral disease plays a role in both the decline and the delayed recovery of fluctuating populations of forest Lepidoptera (Myers, 1993). Although viral disease is not associated with all declines of tussock moth, it does seem to be associated with synchronous, periodic outbreaks that occur over a wide geographic area in more northern locations (Shepherd *et al.,* 1988).

It is possible that abundant unpublished data from insecticide treatment programs exist in the files of government agencies. If so, these could be analyzed to test whether premature suppression of population densities of defoliating insects maintains populations in outbreak phase or if sprayed populations follow the same long-term dynamics as unsprayed populations. Since data analysis is cheaper than data collection, students on tight research budgets might look for such data. Information on large-scale suppression of populations could be used to test hypotheses about population regulation, such as the impacts of preventing herbivore damage to foliage, eliminating viral epizootics, suppressing densities of parasitoids, or creating out-of-phase populations. A complication of this approach is that several or all of these factors may be influenced at the same time, making interpretation of the experiment difficult.

In conclusion, suppression of forest Lepidoptera early in the increase phase may stop the normal viral epizootic and cause resurgence of the population. Spraying with virus late in the outbreak may terminate a population outbreak without resurgence. Spraying with Bt or insecticides that do not cause an epizootic may lead to temporary asynchrony of the treated population.

B. Suppressing the Population Peak: Cropping Experiments

Population cropping is commonly used as a management tool, particularly for large vertebrates, and often appears to prolong high density populations

(Caughley, 1985). Cropping or reducing the density of herbivores to levels at which plants suffer little feeding damage is an experimental tool for testing hypotheses on damage-induced deterioration of food plant quality with increasing herbivore populations. Food plants that are not damaged should continue to be good quality and populations of herbivores should increase rapidly when released from cropping.

For several years we cropped an expanding population of western tent caterpillars on a small island, in the Haro Straits near Vancouver Island, to test the hypothesis that protecting foliage from feeding damage by caterpillars would maintain high foliage quality and lead to an outbreak when cropping stopped (Myers, 1993). Our prediction was not supported. For 2 years the population increased following experimental reduction. However, in the third year the population declined in synchrony with populations on surrounding islands. These results indicate that some populations are regulated by a mechanism that is resilient to experimentally reduced density and to the protection of foliage quality by preventing defoliation. By late in the phase of population increase the state of the population may already be determined such that decline will follow.

Unlike the nun moth described earlier, preventing the outbreak of tent caterpillars by removing insects in the early instars or as egg masses did not lead to an immediate population resurgence. Perhaps densities remained sufficiently high to support the spread of virus.

C. Attempting to Create an Outbreak

1. Introducing Populations: Gypsy Moth

Gypsy moth, *Lymantria dispar,* was initially introduced to North America in 1869 from Europe and has since spread throughout the eastern United States and Canada. A number of parasitoids have subsequently been introduced, but biological control has been considered to be a failure. Outbreaks have continued to occur as the species spreads to the west and the south. Campbell and Sloan (1977) concluded from life-table studies that parasitism was not a large or important source of mortality in northeastern North America.

Two large-scale experiments have evaluated the impact of some of the introduced parasitoids and have come to a different conclusion. In the first of these experiments, Liebhold and Elkinton (1989) increased the density of gypsy moth egg masses on study sites in Cape Cod, Massachusetts, from approximately 20 egg masses/ha to 8000 egg masses/ha. They used eggs collected from other field sites and laboratory-reared egg masses that were genetically modified to prevent the development of an F_2 generation. This experimental density of egg masses would be considered to be high in natural populations, but occurred only within a 1-ha plot in the center of a 9-ha experimental area. They monitored densities of gypsy moth caterpillars over the summer. An introduced wasp parasitoid, *Cotesia melanoscela* (Ratzeburg), parasitized 40 to 48% of the late-instar caterpillars in these experimental populations and 7 to 21% in control plots. Parasitism by an introduced tachinid, *Compsilura concinnata* (Meigen), was also higher in the experimentally augmented plots. In the end, both the experimental and control populations were reduced to very low densities, and no outbreak was created. Therefore, on a small scale the introduced parasitoids were capable of preventing an increase in the gypsy moth population.

In the second experiment, Gould *et al.* (1990) created a range of gypsy moth densities from 174 egg masses/ha (approximately 44,000 larvae) to 4600 egg masses/ha (approximately 1.14 million larvae) in eight 1-ha experimental plots in western Massachusetts. Gypsy moth did not occur in these plots immediately prior to the experiment. Over 5 weeks, densities of gypsy moth declined in all plots to levels below that at the beginning of the experiment, and fewer egg masses occurred at the end of the generation on the initially high-density plots than in the lower-density plots. Parasitism by the introduced tachinid *C. concinnata* was highly correlated with the density of gypsy moth on the plots. In addition, loss of early-instar larvae, perhaps associated with dispersal by ballooning, was also related to the density in the plots. Another parasitoid, *Parasetigena silvestris,* was initially inversely related to the density of the gypsy moth, but later in the season it too was positively related to density. *Compsilura concinnata,* also introduced as a biological control agent, is a generalist parasitoid that has up to four generations a year. Therefore, it can be maintained by alternative hosts and can respond to spatial heterogeneity in the density of the gypsy moth.

Elkinton and colleagues point out that the experimental creation of a range of gypsy moth densities produced the density-related response of introduced, generalist parasitoids. These manipulations were done on a large scale with hectare-size plots and millions of eggs. Though they show the suppression of localized increases of caterpillars by parasitoids, this does not mean that parasitoids "regulate" or reduce gypsy moth populations following the escape to outbreak levels on a large geographic scale. The mechanisms underlying spatially density-dependent rates of parasitism or predation may vary with scale of observation (Rothman and Darling, 1991). The detection of density dependence at large spatial scales relative to the dispersal ability of natural enemies could reflect functional and numerical responses of natural enemies *in situ,* and therefore approximate temporal population regulation. Spatial density dependence at smaller scales could result from "optimal" foraging by natural enemies and may not necessarily be related to temporal density dependence if populations increase simultaneously on a large geographic scale.

Further research failed to demonstrate temporal density-dependent mortality by *C. concinnata* and *P. silvestris* at the sites initially studied by Gould *et al.* (1990). The next year (1988), gypsy moth eggs were again introduced to these sites, and the impact of the parasitoids was evaluated. There was no relationship between the density of either parasitoid in 1987 and its killing power in 1988 (Ferguson *et al.,* 1994).

These experiments show that while generalized parasitoids might help suppress outbreaks of gypsy moth at sites of new introductions, their role in explaining long periods of low density following an initial outbreak of gypsy moth populations in eastern North America is not confirmed and remains controversial (Berryman, 1991; Leibhold and Elkinton, 1991). Temporal density dependence was not found to occur.

2. Introducing Populations: Tent Caterpillars

The experimental increases of gypsy moth populations described here were both done in 1987 when existing populations were at low density, approximately 6 years after a widespread outbreak in 1981. To determine the fate of introductions made during a period of regional increase and peak population density of a forest Lepidoptera, we carried out similar introductions of western tent caterpillar egg masses to sites with low or no natural populations in southwestern British Columbia in 1984 and 1986 (Myers, 1990, 1993). These latter experiments were made on a smaller scale than the gypsy moth experiments, but also showed the resilience of the population dynamics within a region. Of five populations initiated from egg masses introduced during the phase of population increase, two increased and declined in synchrony with control populations, whereas three of the populations increased and declined a year after the control populations. Two populations initiated with eggs from populations at peak density declined in synchrony with the source populations. In the year of population decline, the causes of late-instar larval and pupal mortality varied considerably among populations. In addition, introduced populations tended to survive better than source populations. Nevertheless, populations declined in synchrony, and by 1988 all populations were at low density and the fecundity of moths was reduced compared with that at peak densities (Myers, 1990).

This observed variation in the apparent causes and timing of mortality among populations declining in synchrony indicates why searches for simple patterns of density dependence in population parameters frequently fail. These experiments, however, show that the future fate of the population is carried with the egg masses, at least at the beginning of the decline, and could be related to their quality (Roissiter, 1991), or to disease contamination. Processes acting regionally appear to synchronize population dynamics at least some of the time.

In conclusion, introduction experiments show that parasitoids can respond to locally increased populations, and that egg masses may carry the "history" of their population with them. Regional population dynamics can be resilient.

D. Biological Control of Introduced Forest Lepidoptera: Experimental Evaluation of Parasitoids

Accidental introductions of phytophagous insects to new areas are experiments awaiting analysis. Introduced insects often reach high densities and spread following their establishment in exotic areas. These situations allow analysis of the impacts of introduced parasitoids and of population dynamics of hosts in new areas.

1. Gypsy Moth

Because the introduction of gypsy moth to the West Coast of North America is of concern, pheromone trapping for male moths has been carried out since 1978 in British Columbia. Although not all introductions are identified by pheromone trapping, the data do indicate interesting patterns. From this record we have identified 76 geographically widespread introductions that are likely to have been independent events, and for which spray programs were not initially carried out. Of these, 27 persisted in subsequent years, as indicated by increasing numbers of catches of male moths in traps at the site (Fig. 3). This level of establishment of one-third success is about half that for phytophagous insects intentionally introduced in biological control of weeds programs (Myers, 1989). The pattern of introductions and establishment can be compared with the years of high gypsy moth density in eastern North America (Fig. 3). The number of introductions has been increasing with the spread of the gypsy moth in eastern North America, but introductions were also more numerous in years following outbreaks in the East (1981 to 1982, 1985 to 1986, and 1991 to 1992). Establishment is highest, 45–50%, for introductions discovered within one or two years of peak densities in the East (1983, 1987, and 1992) as compared to 0 to 33% in other years. Heterogeneity in establishment success could be associated with the

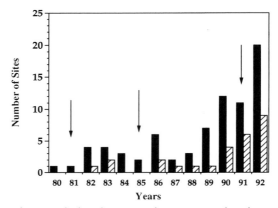

Figure 3. Number of sites at which male gypsy moths were captured in pheromone traps in British Columbia (solid bars) and number of sites at which gypsy moths became established as indicted by increasing numbers of males caught in traps in subsequent years (hatched bars). Vertical arrows indicate years of high populations in eastern North America, namely, 1981, and 1991 in the northeastern United States and 1985 and 1991 in Ontario, Canada. Data summarized from Forest Insect and Disease Survey, British Columbia.

quantity or quality of the introduced insects that probably arrive as egg masses on vehicles and transported goods.

Two of the parasitoids, *C. melanoscela* and *C. cincinnata,* which responded to the experimentally augmented populations of gypsy moth in Massachusetts described earlier, have been introduced from Europe to British Columbia in a previous biological control program against the Satin moth, *Leucoma salicis* (Forbes and Ross, 1971). They are therefore already part of the generalist parasitoid fauna in areas to which gypsy moths are being continuously accidentally introduced. These parasitoids could potentially help to suppress local populations of gypsy moth as they did in the experiments of Elkinton and colleagues in Massachusetts.

Whether or not the introduced parasitoids were involved in suppressing and eliminating the introduced populations that went extinct is unlikely, but not known. However, the lower rate of establishment of gypsy moth compared with intentionally introduced phytophagous biological control agents is possibly associated with the introduced generalist parasitoids responding to patches of elevated gypsy moth density. Experimental release of eggs that produce sterile adults would allow evaluation of the effectiveness of the parasitoids in British Columbia. It is not known how numerous these parasitoids are or the range of native hosts that they use.

2. Winter Moth

Introduced populations of winter moth, *Operophtera brumata,* provide another experiment in population regulation. In initial studies of this species in Britain, Varley *et al.* (1973) concluded that winter moth was regulated by density-related predation of pupae in the soil. In the 19 years of life-table data collected by Varley and Gradwell, mortality from the fly parasitoid *Cyzenis albicans* was not density dependent and, therefore, this species was not considered to be involved in regulation of the population. Even so, following the accidental introduction of winter moth to North America, *C. albicans* was imported as a biological control agent, and within 6 years winter moth population densities declined (review in Roland and Embree, 1995). Following the introduction of winter moth to the West Coast, probably in the late 1960s, *C. albicans* again was also introduced. Within 5 years parasitism by *C. albicans* was high, and the winter moth population declined.

In North America, unexplained increases in predation of winter moth pupae in the soil have preceded the increased parasitization by *C. albicans* and the decline of winter moth. As in Britain, predation by soil predators on winter moth pupae in North America is density dependent, and parasitism contributes to further, but variable, levels of mortality. Without the additional mortality from the introduced parasitoid, winter moth would probably fluctuate at higher densities and continue to defoliate trees (Roland, 1994).

The importance of mortality from *C. albicans* was not fully appreciated by Varley and Gradwell because, although their study continued for 19 years with minor outbreaks occurring at 6 to 8 years, major outbreaks of winter moth occurred only immediately prior and shortly after their study (Horgan *et al.*, 1995). In both North America and Britain, parasitization by *C. albicans* is greatest during the decline following high host densities. Neither generalist soil predators nor *C. albicans* could prevent the initial outbreak of winter moth when it spread into the Vancouver area of British Columbia from the initial introduction on Vancouver Island. However, mortality from both the predators in the soil (beetles, ants, and small mammals) and the introduced parasitoids was associated with the decline in the population (Horgan *et al.*, 1995) and should maintain densities at lower levels in the future (Roland, 1994).

Another interesting aspect of the winter moth introductions to British Columbia is that the population on Vancouver Island had a resurgence in 1991 when the initial outbreak occurred on the lower mainland. Small numbers of winter moth had been captured in traps on the mainland since the early 1980s, but densities sufficient to cause defoliation did not occur until 1989–1991. Similarly, in eastern Canada minor eruptions of winter moth have continued in synchrony with other geometrids, such as fall cankerworm and Bruce spanworm (Roland and Embree, 1995). This suggests that regionwide conditions can influence the population dynamics of introduced insects.

The studies of the introduced winter moth show the potential pitfalls of an observational study carried out without spatial replication or manipulative experimentation (Krebs, 1991). The accidental introduction of winter moth to North America is certainly a large-scale manipulation that has caused continuing damage to trees in urban areas and apple orchards. But it did allow tests of the predictions of an observational, life-table study. Ground predators are density dependent in both situations and parasitation by *Cyzanis* follows high host density.

In conclusion, biological control introductions allow the evaluation of parasitoids on host population dynamics in ways that are not possible in native habitats. Experiments have shown parasitoids to be more influential than was previously thought, but the lack of controls weakens conclusions from biological control programs. No sites were monitored which had winter moth but no introduced parasitoids.

E. Evaluating Geographic Variation in Population Dynamics

Naturally occurring variation provides experiments without the need for manipulation. Comparison of the dynamics of viral disease in populations of tussock moth that do not have periodic outbreaks with those with cyclic dynamics has been suggested by Shepherd *et al.* (1988). This approach has been used for small mammals. Comparing cyclic populations in northern Sweden with noncyclic populations in southern Sweden has led to hypotheses about the impacts of generalist and specialist predators on the dynamics of voles and lemmings (Hansson, 1987; Hanski *et al.*, 1991). Periodic outbreaks of forest Lepidoptera are also more pronounced in northern populations, and the prevalence of viral disease should be compared between northern and southern populations.

F. Large-Scale Environmental Variables as Experimental Perturbations

Experimental manipulation of weather is nearly impossible, however, the impact of natural disasters on insect populations can sometimes be treated as experimental perturbations. For example, the dust cloud associated with the eruption of Mt. St. Helens in Oregon is thought to have suppressed the cyclic peak in density of tussock moth in northern Idaho for two years (Berryman *et al.*, 1990). Replication of this experiment fortunately has not occurred.

Fluctuations in sunspot activity are another type of natural variation that might influence insect populations. Population trends of four well-studied forest Lepidoptera in North America and Europe show tendencies to reach peak densities within a year or two of each other and of the trough of sunspot activity (Fig. 4). In 5 of the last 60 years, temperatures in the Northern Hemisphere between March and June, the season of caterpillar development, have been particularly cool. Four of these 5 cold years have been during the trough of

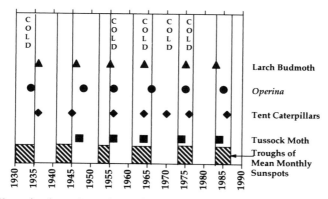

Figure 4. Years of peak population density for four species of forest Lepidoptera for which long-term data are available: larch budmoth (Baltensweiler and Fischlin, 1988), *Oporinia autumnata* (Haukioja *et al.*, 1988), tussock moth (A. Berryman, personal communication), and tent caterpillar (Myers, 1990). Years of low mean monthly sunspot activity, indicated by hatching, are based on data from A. Sinclair (personal communication and U.S. Department of Commerce). "COLD" years are deviations from the mean of surface air temperatures for the Northern Hemisphere from the reference period (1951–1970) for months of March to June combined (Jones *et al.*, 1986; Jones, 1988).

sunspot numbers, and within a year of peak densities of the four species of forest Lepidoptera. The anomalous outbreak of tent caterpillars in British Columbia in 1970 followed a low peak in sunspot activity during the late 1960s and early 1970s, and a cool spring in 1971. Although it is difficult to extract cause and effect from patterns, global weather variation could be viewed as an experimental variable and the population trends of different species can serve as replicates for interpreting the impacts of large-scale perturbations. The population processes will vary with the histories of individual populations, but large-scale conditions could influence the synchrony of populations.

The potential robustness of associations between sunspots and population dynamics is suggested by the observation that fluctuating populations of snowshoe hares also appear to be associated with sunspot activity (Sinclair *et al.* 1993).

G. Experiments to Identify Mechanisms Affecting Population Density

1. Effects of Varying the Synchrony between Caterpillar and Leaf Phenology

One aspect of the interactions between weather and insect population dynamics that can be perturbed is the synchrony between egg hatch and leaf development. The potential influence of temporal changes in the quality of leaves on the growth and survival of caterpillars was recognized over 20 years ago in the work of Feeny (1970) on winter moth. The absolute and relative concentrations of proteins, sugars, phenolics, resins, and alkaloids in leaves change from the earliest stages of expansion in the spring to senescence in the autumn. This in turn influences their nutritive value and toughness. The synchrony between hatching of caterpillar eggs and leaf development can influence the growth, development, and dispersal of caterpillars, and possibly also the growth of the population (Hunter, 1993). Some studies have found relationships between leaf phenology and defoliation (Hunter, 1992); whereas others have not (Crawley and Akhteruzzaman, 1988; Watt and McFarlane, 1991). Quality of leaves could influence the growth, potential fecundity, and susceptibility of caterpillars to virus, and variation in the time of egg hatch could modify the exposure of caterpillars to parasitoids.

Egg hatch can be delayed by storing egg masses at low temperatures or accelerated by placing them at higher temperatures and thus can be easily manipulated. Western tent caterpillar eggs normally hatch when leaves of their most common food plant, red alder, *Alnus rubra*, are just expanding. To determine the impact of delayed egg hatch, we deployed tent caterpillar egg masses at the beginning of a year of peak density and initial decline at sites that had not previously supported a high-density population (Myers, 1992). The egg masses had been collected from several high-density populations and kept cool. They were deployed at several times, which resulted in hatching dates from approximately 7 April, the time of natural egg hatch, to 15 May. Survival of caterpillars can be estimated from the sizes of tents formed by late instars (Myers, 1990), and this indicates that survival among early- and late-hatching caterpillars was similar. Parasitization of caterpillars in areas in which a high proportion of the caterpillars were delayed in hatching was similar to that of control sites. Warmer weather as spring progresses accelerates caterpillar growth, and although hatching as much as 6 weeks behind the normal hatch date, caterpillars from introduced and delayed eggs pupated at approximately the same time as control caterpillars.

These experiments show the resilience of tent caterpillars to variation in the phenology of egg hatch and leaf development. They suggest that phenological synchrony might have only weak influences on population dynamics of forest Lepidoptera.

2. Effects of Phenology on Susceptibility to Nuclear Polyhedral Virus and Parasitization

As described earlier, the dynamics of viral disease in populations of forest Lepidoptera have been investigated in large-scale experiments by spraying with virus or by reducing host populations with insecticide sprays to prevent the viral epizootic. An exciting approach has been to use small-scale experiments to explore the details of the transmission and impact of NPV (Dwyer, 1991, 1992; Dwyer and Elkinton, 1993). In these small-scale experiments one has good control over the numbers, ages, and condition of the experimental caterpillars, and these experiments can be very efficient for estimating parameters for describing host and disease dynamics. Three conclusions arise from these large- and small-scale experiments: (1) outbreaks can be suppressed by spraying virus, (2) preventing epizootics can result in immediate resurgence of the populations, and (3) the local spread of virus within a generation is not always accurately described by simple density-dependent models. Spread in low-density populations can be greater than predicted from models based on parameters arising from small-scale experiments (Dwyer and Elkinton, 1993), which suggests that some processes may have been overlooked or the models are oversimplified.

The gregarious behavior of tent caterpillars is advantageous for small-scale experiments involving the introduction and evaluation of NPV. Tent caterpillars occur on a gradient of over 340 m near Vancouver, and egg hatch varies over approximately 6 weeks from low to high elevations. Tent caterpillars were collected as discovered between early April and late May, 1992, from Cypress Mountain, just north of Vancouver. Even when densities are low and egg masses are nearly impossible to find, tents of first- or second-instar larvae can be spotted from a distance. Colonies of early instars were collected and transferred to

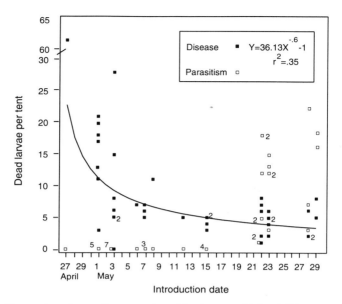

Figure 5. The number of caterpillars per colony dying between third and fifth instars of disease or wasp parasitization on date of introduction of 10 viral-infected caterpillars to third-instar colonies (n = 40). The date at which the colony reached the third instar was determined by the time of egg hatch, with eggs collected from higher elevations being later to hatch. Numbers on the graph indicate the number of overlapping data points.

the campus of the University of British Columbia onto red alder trees that had leafed out at the end of March. When caterpillars reached the third instar, 10 individuals from each colony were collected and brought into the laboratory. Colonies were randomly assigned to either control groups (10 caterpillars fed leaf discs dipped in water) or virus treatment (10 caterpillars fed leaf discs dipped in a high concentration of NPV). After 24–36 hours in the laboratory, the caterpillars were returned to their original tents. Caterpillars dying on or in the vicinity of tents were counted until caterpillars reached the fifth instar and began to disperse. Deaths from parasitoids and from disease were scored.

Overall mortality was significantly increased by the addition of infected individuals to the tents, and 33% of treated colonies failed to produce any fifth instars whereas only 7% of control colonies failed to do so ($P < 0.001$). Levels of parasitism were similar in both treated and control groups. However, both the number of deaths from virus and from parasitization varied with the date of introduction of viral-treated caterpillars to the third-instar colony (Fig. 5). The susceptibility to virus was apparently greater early in the spring than later, and the impact of ichneumonid and braconid parasitoids was higher later in the spring.

Total mortality remained similar over the duration of the experiment, and therefore variation in mortality from disease and parasitism could buffer the influences of early and late springs on the population dynamics of tent caterpillars.

Several interpretations of these results could be tested with further studies. First, the apparently low susceptibility to virus later in the spring could be caused by parasitization killing the caterpillars before the virus does. Second, the susceptibility of caterpillars to virus may be influenced by changes in leaf quality associated with age (Schultz and Foster, 1990). Third, the "quality" of individuals from different sites could influence their susceptibility to virus or parasitoids. Fourth, faster development of caterpillars transferred later in the spring when it is warmer could influence the expression of virus. The important result in terms of population regulation is that mortality agents may vary in their phenology and buffer the impact of variation in egg hatch from year to year or place to place. This would help keep populations synchronized.

In conclusion, experimental manipulation of the timing of egg hatch is a good technique for exposing the potential influences of early and late springs on insect population dynamics, and for investigating the impacts of various mortality agents that vary over the developmental period of the insect. Experiments show the resilience of tent caterpillars to phenological variation in egg hatch and larval development and also demonstrate the influence of compensatory mortality. It is surprising that experimental manipulation of the phenology of egg hatch has not been used more widely as an experimental tool.

3. Effect of Insect Quality on Population Dynamics

Nutritionally based maternal effects have been found in several species of forest Lepidoptera with outbreak dynamics (Rossiter, 1991, and Chapter 13, this volume). In western tent caterpillars, the first eggs laid in the egg mass are reported to have greater nutrient stores and to yield more active larvae than the last-laid eggs (Wellington, 1965). To determine if this potential variation influenced the survival of caterpillars in the field, egg masses from a peak population were cut in half to form colonies in which all caterpillars were from two top halves of egg masses and colonies that were from two bottom halves (Myers, 1978). Caterpillars from first-laid eggs survived as well as those from last-laid eggs. Populations in two experimental sites declined in synchrony with the general pattern for the area (Myers, 1978). Although this experiment failed to demonstrate an impact of egg quality on population dynamics, the approach deserves further attention. Particularly valuable would be the simultaneous introduction of eggs from an increasing population, which are expected to have high quality, and those from a declining population, which should be poor quality. Whether or not maternal effects are sufficiently robust to influence population dynamics of field populations remains to be determined by experimental analysis.

III. Conclusions

The purely descriptive approach, in which one looks for statistically significant density dependence in time-series data, appears to be bogged down in methodological problems. It is also unclear what a field ecologist would do differently if population dynamics were best described as being damped oscillations, limit cycles, chaotic, quasi-periodic, or exponentially stable. For fluctuating populations of forest Lepidoptera, the relationships between density, mortality, and natality rates will not be the same in increasing and declining populations. Analysis of density relationships must consider the phase of the population fluctuation.

The resilience of populations to perturbation provides good evidence for regulation. This does not mean that environmental conditions will not sometimes have dominant impacts on populations. These could cause some populations to go out of synchrony, as may have occurred with the tussock moth following the eruption of Mt. St. Helens, or to become synchronized, as with the cool springs in early 1950s and synchrony of outbreaks of forest Lepidoptera in the Northern Hemisphere.

The scale of field experiments will always be a problem for interpretation. We do not know if the suppression of high densities of gypsy moth that occurred on 1-ha plots would also occur on 2- or 10- or 20-ha plots. We cannot be sure that the experiments with tent caterpillars were not influenced by influx of parasitoids and even tent caterpillars from other areas. However, when the total budget for research in population ecology of the National Science Foundation is only $2 million per year, it is unlikely that funding will ever be available to carry out experiments on appropriate geographical scales. Financial support is frequently available for insecticide-spraying programs, and population ecology could be served by analyzing these as large-scale perturbations. The frequency of introductions of exotic species is probably going to continue to be high with the considerable movement of people and goods. These introductions can also be viewed as large-scale manipulations for which the impacts of parasitoids can be evaluated.

Small-scale manipulations can be used to study mechanisms that influence populations, such as the phenological mismatch between egg hatch and leaf development that may affect growth, parasitization, or susceptibility of caterpillars to virus. Small-scale experiments that describe and quantify mechanisms are valuable tools for developing models that can be scaled up to explore population dynamics (e.g., Dwyer and Elkinton, 1993). This combination of small-scale experiments and models will help population ecology progress even when budgets are tight.

Several important conclusions arise from field experiments with fluctuating populations of forest Lepidoptera: (1) dynamics are resilient to local perturbations; (2) different mortality agents act at different stages of the fluctuation, and agents that suppress populations at low densities will not necessarily be those that reduce high-density populations; (3) different parasitoids can play the same role in different geographic areas; (4) populations of different species can be synchronized over large geographic areas and major perturbations such as volcanic eruptions can disrupt synchrony; and (5) introductions from increasing populations appear to become established and increase to a greater degree than those from declining populations.

If there is frustration about what might be perceived as a lack of progress and understanding in population ecology, it might be because there has been a tendency to oversimplify and look for consistent density relationships. Experimental study of mechanisms causing population change has had an influential past, and can have a bright future even in an era of restrictive budgets.

Acknowledgments

We thank Jamie Smith, Greg Dwyer, MaryCarol Rossiter, Joe Elkinton, and Dennis Chitty for comments on this manuscript and acknowledge the Natural Sciences and Engineering Research Council of Canada for providing support for our work.

References

Auer, P. C., Roques, A., Goussard, F., and Charles, P. J. (1981). Effects de l'accroissement provoqué du niveau de population de la tordeuse du mélèze *Zeiraphera diniana* Guénée (Lep. Tortricidae) au cours de la phase de régression dans un massif forestier du Briançonnais. *Z. Angew. Entomol.* **92**, 286–303.

Baltensweiler, W., and Fischlin, A. (1988). The larch budmoth in the Alps. *In* "Dynamics of Forest Insect Populations: Patterns, Causes, Implications" (A. Berryman, ed.), pp. 332–353. Plenum, New York.

Berryman, A. A. (1991). Gypsy moth dynamics. Reply. *Trends Ecol. Evol.* **6**, 264.

Berryman, A. A., Millstein, J. A., and Mason, R. R. (1990). Modelling Douglas-fir tussock moth population dynamics: The case for simple theoretical models. *In* "Population Dynamics of Forest Insects" (A. D. Watt, S. R. Leather, M. D. Hunter, and N. A. C. Kidd, eds.), pp. 369–380. Intercept, Andover, UK.

Campbell, R. W., and Sloan, R. J. (1977). Natural regulation of innocuous gypsy moth populations. *Environ. Entomol.* **6**, 315–322.

Caughley, G. (1985). Harvesting of wildlife: Past, present and future. *In* "Game Harvest Management," (S. L. Beason and S. F. Roberson, eds.), pp. 3–14. Caesar Kleberg Wildlife Research Institute, Kingsville, TX.

Crawley, M. J., and Akhteruzzaman, M. (1988). Individual variation in the phenology of oak trees and its consequences for herbivorous insects. *Funct. Ecol.* **2**, 409–415.

Dwyer, G. (1991). The roles of density, stage and patchiness in the transmission of an insect virus. *Ecology*, **72**, 559–574.

Dwyer, G. (1992). On the spatial spread of insect pathogens: Theory and experiment. *Ecology*, **73**, 479–484.

Dwyer, G., and Elkinton, J. S. (1993). Using simple models to predict virus epizootics in gypsy moth populations. *J. Anim. Ecol.* **62**, 1–11.

Feeny, P. (1970). Seasonal changes in oak leaf tannins and nutrients as a cause of spring feeding by winter moth caterpillars. *Ecology* **51**, 565–581.

Ferguson, C. S., Elkinton, J. S., Gould, J. R., and Wallner, W. E. (1994). Population regulation of gypsy moth (Lepidoptera: Lymantriidae) by parasitoids: Does spatial density dependence lead to temporal density dependence? *Environ. Entomol.* **23**, 1155–1164.

Forbes, R. S., and Ross, D. A. (1971). *Stilpnotia salicis* (L.), satin moth (Lepidoptera: Liparidae). *In* "Biological Control Programmes Against Insects and Weeds in Canada 1959–1969" (J. S. Kelleher and M. A. Hulme, eds.), pp. 205–212. CAB International, Wallingford.

Ginzburg L. R., and Taneyhill, D. E. (1994). Population cycles of forest Lepidoptera: A maternal effect hypothesis. *J. Anim. Ecol.* **63**, 79–92.

Gould, J. R., Elkinton, J. S., and Wallner, W. E. (1990). Density-dependent suppression of experimentally created gypsy moth, *Lymantria dispar* (Lepidoptera: Lymantriidae) populations by natural enemies. *J. Anim. Ecol.* **59**, 213–233.

Hanski, I., Hansson, L., and Henttonen, H. (1991). Specialist predators, generalist predators, and microtine rodent cycle. *J. Anim. Ecol.* **60**, 353–367.

Hansson, L. (1987). An interpretation of rodent dynamics as due to trophic interactions. *Oikos,* **50**, 308–318.

Hassell, M. P., Lawton, J. H., and May, R. M. (1976). Patterns of dynamical behaviour in single species populations. *J. Anim. Ecol.* **45**, 471–486.

Haukioja, E., Neuvonen, S., Hanhimaki, S., and Niemela, P. (1988). The autumnal moth in Fennoscandia. *In* "Dynamics of Forest Insect Populations: Patterns, Causes, Implications" (A. Berryman, ed.), pp. 163–178. Plenum, New York.

Holyoak, M. (1994). Identifying delayed density dependence in time-series data. *Oikos,* **70**, 296–304.

Horgan, F., Myers, J. H., and VanMeel, R. (1995). *Cyzenis albicans* (Fall.) does not prevent the outbreak of introduced winter moth (*Operophtera brumata* (L.)) in the lower mainland of British Columbia, Canada. In manuscript.

Hunter, A. F. (1993). Gypsy moth population sizes and the window of opportunity in spring. *Oikos* **68**, 531–538.

Hunter, M. D. (1992). A variable insect–plant interaction: The relationship between tree budburst phenology and population levels of insect herbivores among trees. *Ecol. Entomol.* **16**, 91–95.

Hutchinson, G. E. (1948). Circular causal systems in ecology. *Ann. N.Y. Acad. Sci.* **50**, 221–246.

Jones, P. D. (1988). Hemispheric surface air temperature variations: Recent trends and an update to 1987. *J. Clim.* **1**, 654–660.

Jones, P. D., Raper, S. C. B., Bradley, R. S., Diaz, H. F., Kelly, P. M., and Wigley, R. M. L. (1986). Northern Hemisphere surface air temperature variations: 1851–1984. *J. Clim. Appl. Meteorol.* **25**, 161–179.

Katagiri, K. (1969). Review on microbial control of insect pests in forests in Japan. *Entomophaga* **14**, 203–214.

Krebs, C. J. (1991). The experimental paradigm and long-term population studies. *Ibis* **133**, 3–8.

Liebhold, A. M., and Elkinton, J. S. (1989). Elevated parasitism in artificially augmented populations of *Lymantria dispar* (Lepidoptera: Lymantriidae). *Environ. Entomol.* **18,** 986–995.

Liebhold, A. M., and Elkinton, J. S. (1991). Gypsy moth dynamics. *Trends Ecol. Evol.* **6,** 263.

Murdoch, W. W. (1994). Population regulation in theory and practice. *Ecology* **75,** 271–287.

Murray, B. G. (1994). On density dependence. *Oikos* **69,** 520–523.

Myers, J. H. (1978). A search for behavioural variation in first and last laid eggs of western tent caterpillar and an attempt to prevent a population decline. *Can. J. Zool.* **56,** 2359–2363.

Myers, J. H. (1988). Can a general hypothesis explain population cycles of forest Lepidoptera? *Adv. Ecol. Res.* **18,** 179–242.

Myers, J. H. (1989). The ability of plants to compensate for insect attack: Why biological control of weeds with insects is so difficult. *Proc. Int. Symp. Biol. Control Weeds, 7th, 1988,* pp. 67–73.

Myers, J. H. (1990). Population cycles of western tent caterpillars: Experimental introductions and synchrony of populations. *Ecology* **71,** 986–995.

Myers, J. H. (1992). Experimental manipulation of the phenology of egg hatch in cyclic populations of tent caterpillars. *Can. Entomol.* **124,** 737–742.

Myers, J. H. (1993). Population outbreaks in forest Lepidoptera. *Am. Sci.* **81,** 240–251.

Otvos, I. S., Cunningham, J. C., and Friskie, L. M. (1987a). Aerial application of nuclear polyhedrosis virus against Douglas-fir tussock moth, *Orgyia pseudotsugata* (McDunnough) (Lepidoptera: Lymantriidae). I. Impact in the year of application. *Can. Entomol.* **119,** 697–706.

Otvos, I. S., Cunningham, J. C., and Alfaro, R. I. (1987b). Aerial application of nuclear polyhedrosis virus against Douglas-fir tussock moth, *Orgyia pseudotsugata* (McDunnough) (Lepidoptera: Lymantriidae). II. Impact 1 and 2 years after application. *Can. Entomol.* **119,** 707–715.

Perry, J., Woiwod, I. P., and Hanski, I. (1993). Using response-surface methodology to detect chaos in ecological time series. *Oikos* **68,** 329–339.

Pollard, E., Lakhani, K. H., and Rothery, P. (1987). The detection of density dependence from a series of annual censuses. *Ecology* **68,** 2046–2055.

Roland, J. (1994). After the decline: What maintains low winter moth density after successful biological control? *J. Anim. Ecol.* **63,** 192–198.

Roland, J., and Embree, D. G. (1995). Biological control of the winter moth. *Annu. Rev. Entomol.* **40,** 475–492.

Rossiter, M. (1991). Environmentally-based maternal effects: A hidden force in insect population dynamics? *Oecologia* **87,** 288–294.

Rothman, L. D., and Darling, D. C. (1991). Spatial density dependence: Effects of scale, host spatial pattern and parasitoid reproductive strategy. *Oikos,* **62,** 221–230.

Royama, T. (1977). Population persistence and density dependence. *Ecol. Monogr.* **47,** 1–35.

Schultz, J. C., and Foster, M. A. (1990). Host plant-mediated impacts of baculovirus on gypsy moth populations. *In* "Population Dynamics of Forest Insects" (A. D. Watt, S. R. Leather, M. D. Hunter, and N. A. C. Kidd, eds.), pp. 303–313. Intercept, Andover, UK.

Shepherd, R., Otvos, I. S., Chorney, R. J., and Cunningham, J. C. (1984). Pest management of Douglas-fir tussock moth (Lepidoptera: Lymantridae): Prevention of an outbreak through early treatment with a nuclear polyhedrosis virus by ground and aerial application. *Can. Entomol.* **116,** 1533–1542.

Shepherd, R., Bennett, D. D., Dale, J. W., Tunnock, S., Dolph, R. E., and Their, R. W. (1988). Evidence of synchronized cycles in outbreak patterns of Douglas-fir tussock moth, *Orgyia pseudotsugata* (McDunnough) (Lepidoptera: Lymantriidae). *Mem. Entomol. Soc. Can.* **146,** 107–121.

Sinclair, A. R. E., Gosline, J. M., Holdsworth, G., Krebs, C. J., Boutin, S., Smith, J. N. M., Boonstra, R., and Dale, M. (1993). Can the solar cycle and climate synchronize the snowshoe hare cycle in Canada? Evidence from tree rings and ice cores. *Am. Nat.* **141,** 173–198.

Strong, D. R. (1986). Density-vague population change. *Trends Ecol. Evol.* **1,** 39–42.

Turchin, P. (1990). Rarity of density dependence or population regulation with lags? *Nature (London)* **334,** 660–663.

Turchin, P., and Taylor, A. D. (1992). Complex dynamics in ecological time series. *Ecology,* **73,** 289–305.

U.S. Department of Commerce. (1993). "Solar–Geophysical Data." U.S. Dept. of Commerce, Boulder, CO.

Varley, G. C., Gradwell, G. R., and Hassell, M. P. (1973). "Insect Population Ecology: An Analytical Approach." Blackwell, Oxford.

Watt, A. D., and McFarlane, A. M. (1991). Winter moth on Sitka spruce: Synchrony of egg hatch and budburst, and its effect on larval survival. *Ecol. Entomol.* **16,** 387–390.

Wellington, W. G. (1965). Some maternal influences on progeny quality in the western tent caterpillar, *Malacosoma pluviale* (Dyar). *Can. Entomol.* **97,** 1–14.

Woiwod, I. P., and Hanski, I. (1992). Patterns of density dependence in moths and aphids. *J. Anim. Ecol.* **61,** 619–629.

Wolda, H., and Dennis, B. (1993). Density-dependence tests, are they? *Oecologia* **95,** 581–591.

Chapter 13

Impact of Life-History Evolution on Population Dynamics: Predicting the Presence of Maternal Effects

MaryCarol Rossiter

I. Importance of Population Quality Parameters

If we want to predict, or even understand population growth and decline, it is critical to consider the variation in quality among individuals of a population. Historically, herbivore population growth has been described by the change in number over time. Recently, the inclusion of spatial components of change in number over time has greatly improved the realism of population growth models as well as our understanding of what factors play a role in population maintenance, outbreak, or extinction. Similarly, inclusion of population quality variables in population growth models can improve their utility whenever population quality is variable over space or time. Variation in herbivore population quality over space or time can adjust population growth potential or the outcome of interactions with extrinsic agents of natality or mortality. Current models do not account for this reality and the omission will be critical when prediction, rather than description, of population growth is the goal of the model (Rossiter, 1992).

For many species, variation in population quality is far less noted than variation in population size. Perhaps this is the outcome of a perspective that favors quantitative over qualitative measures. In any case, the first challenge is to determine if some variable aspect of population quality influences population growth. Quality traits are those that have a marked effect on survival to reproduction and reproductive output itself, such as toxin tolerance, desiccation tolerance, readiness to enter or leave diapause, or tendency to disperse. Accumulated differences in individual quality can be described collectively for the population in terms of a mean and variance.

One particularly fascinating aspect of population quality comes from environmentally based maternal effects, in which the quality of the individual, and collectively the population, is influenced by the environment of the previous generation. In other words, the impact of the environment is realized on a time lag. This is important because time-lagged effects are known to promote fluctuations in population growth (Hutchinson, 1948; May, 1974). Environmentally based maternal effects should really be called environmentally based parental effects so as to include paternal effects. However, in deference to the recognizability of the term "maternal effects" and the dominance of the mother as the source of such nongenetic, lineage effects, I will use the term maternal effects with the understanding that the phenomena can extend to the much less studied paternal and grandparental effects.

For most young, the earliest environmental experiences are a function of parental quality, a fact easily imagined for humans and mammals in general, but less obvious for fish, insects, rotifers, and the like, where parental care is more subtle. For organisms with limited parental care, environmentally based maternal effects come in the form of a packed lunch that can be nutritional, endocrine, defensive, or even regulatory (e.g., Mousseau and Dingle, 1991a; Giesel, 1988). It has been hypothesized that environmentally based maternal and paternal effects offer some insurance against the loss of parental control over offspring well-being, particularly when environmental quality for offspring is unpredictable (e.g., Kaplan and Cooper, 1984).

A species for which there is much data on environmentally based maternal effects is the gypsy moth, *Lymantria dispar*. For the gypsy moth, for example, food quality in the parental generation influences population quality immediately, adjusting both the probability of survival and total reproductive output of parents, and population quality in the subsequent generation, adjusting the probability of survival to age of reproduction and fecundity of offspring through maternal effects (Rossiter, 1991a, 1994). This cross-generational transmission of environmentally mediated population quality effects occurs via the egg. Consequently, the transmission can appear to be genetic in origin. So, in maternal effect studies, it is important to distinguish genetic from time-lagged maternal effects. I have used this partitioning approach extensively with the gypsy moth to show that parental environmental experience influences the following offspring traits: egg quality (Rossiter *et al.*, 1988, 1993; Rossiter, 1991b), hatch phenology (Rossiter, 1991b), neonate starvation resistance (Rossiter, 1994), length of the prefeeding/dispersal stage (Rossiter, 1991a), toxin resistance (Rossiter *et al.*, 1990), larval development time, pupal weight, and fecundity (Rossiter, 1991a,b). Each of these traits contributes to offspring survival or reproductive output.

Besides this one species, there is documentation for environmentally based maternal effects in other insect taxa. Mousseau and Dingle (1991a) compiled a list of such species, as seen in Table 1, which is based on their review. Nine insect orders are represented, ranging from springtails to parasitoids, with environmentally based maternal effects influencing many offspring traits. Given the

TABLE 1
Phylogenetic Distribution of Documented Maternal Effects in Insects[a]

Order	Number of genera	Offspring traits affected	Examples
Collembola	1	Development, weight, fecundity	Springtails
Orthoptera	6	Diapause, color, behavior, lipid reserves, development time	Crickets, locusts
Psocoptera	3	Diapause	Psocids
Hemiptera	6	Diapause, development time, wing polymorphism	True bugs
Homoptera	5	Diapause	Hoppers
		Wing polymorphism, sex ratio	Aphids
Coleoptera	4	Diapause, dispersal, body size, growth rate	Beetles
Lepidoptera	5	Diapause, growth, survival, toxin resistance	Moths
Diptera	12	Diapause, toxin resistance, survival	Mosquitos, midges
		Body size, development time	Flies
Hymenoptera	8	Diapause	Parasites

[a]Based on information presented in a review by Mousseau and Dingle (1991a).

phylogenetic span of maternal effects, it is conceivable that environmentally based maternal effects are ubiquitous, but vary chiefly in their strength of expression.

II. Maternal Effects Hypothesis of Herbivore Outbreak

From my experimental work on maternal effects, I developed a hypothesis that posits that if maternal effects deliver a sufficiently strong adjustment in population quality, fluctuations in population size will follow, given favorable environmental conditions (Rossiter, 1992, 1994). There was already a good theoretical foundation for the mathematical aspects of this hypothesis (e.g., Berryman, 1981). What was novel was the description of how the biology and ecology of an herbivorous species might generate some of the time-lagged effects that have been detected in the population dynamics for a diversity of species (Turchin and Taylor, 1992).

In its simplest form, the maternal effects hypothesis of herbivore outbreak states that across some equable environmental range, population quality influences population size directly in the current generation and subsequently in

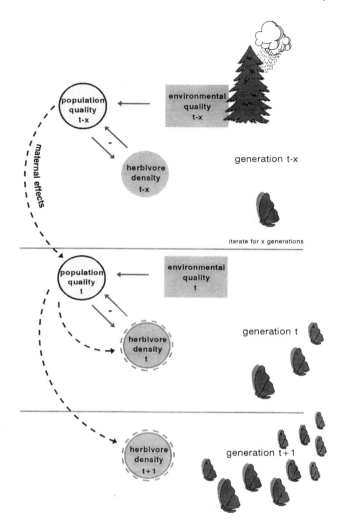

Figure 1. Maternal effects adjust population quality in subsequent generations through a change in offspring survival (arrow a) and offspring fecundity (arrow b), thereby affecting the density (+ or −) of at least two successive generations. Here, the maternal effects hypothesis of outbreak is shown under the condition that maternal effects expression is sensitive to environmental quality. Environmental quality can influence population quality in each generation so as to enhance or diminish the magnitude or direction of the maternal effect. In this example, environmental quality has a density-independent effect on the expression of the maternal effect, and there is a density-dependent interaction between herbivore population quality and density, although such an interaction is not a requisite for the hypothesis (based on Rossiter, 1994).

offspring generations via maternal effects (Fig 1). If the environmentally based maternal effect is great enough, the result will be an adjustment in population size in the next few generations through enhancement or reduction in offspring survival (arrow a) and through enhancement or reduction in offspring fecundity (arrow b), the latter of which adjusts population size two generations later. Whether such adjustments in population size, mediated by maternal effects, influence population dynamics will depend entirely on the relative strength of other ecological effects, such as the impact of weather or habitat composition and natural enemy presence on survival and quality of reproductive output (see Rossiter, 1994, for a complete discussion of this hypothesis).

Although this hypothesis was set forth in its earliest form several years ago (see Rossiter, 1992), it remains to be tested in the field. The magnitude of such a test is enormous, involving large-scale field manipulations as well as pertinent description of undisturbed populations on extensive spatial and temporal scales. Given the scope of the work, it is a good idea to explore other avenues to determine if such a large-scale effort would be valuable. One avenue is to develop a mathematical model with a time-lagged population quality term to see if it makes realistic predictions of population dynamics for outbreak species. Ginzburg and Taneyhill (1994) developed such a model and the model did very well for describing the dynamics of six forest outbreak species.

Another avenue is to develop predictions about which species are most likely to express maternal effects that are strong enough to influence population dynamics. This is a complicated task for several reasons. At present, relatively little is known about the occurrence and expression of maternal effects in phytophagous insects. What is clear from the work, however, is that the strength of maternal effects expression is sensitive to environmental conditions. Consequently, we must consider that the expression of maternal effects within a species may be intermittent or variable in strength. Beyond this, the impact of maternal effects on herbivore population growth potential must also be considered in light of community interactions. Despite the extensive context within which maternal effects must ultimately be evaluated, and the limited data that address the subject directly, we must begin somewhere. And so I will use an indirect approach to develop a set of predictions about the strength of maternal effect expression for herbivorous insects. Then, I will relate these expectations to what is known about the dynamics of the species under consideration.

Though it is true that maternal effects generate time lags, they are only one of several good mechanisms. Furthermore, not all outbreak species will express strong maternal effects and not all species that express strong maternal effects will experience significant fluctuations in population size because the effect of other community interactions may override the influence of maternal effects.

III. Predicting the Presence of Maternal Effects

The goal of this chapter is to develop criteria for predicting the propensity for maternal effects in herbivorous insect species, to make predictions for a taxonomically diverse group of species based on these criteria, and to relate the prediction of maternal effects expression (=the maternal effects score) to population dynamics for each species considered. In the absence of data on direct tests for presence or absence of maternal effects, I will use pertinent life-history data, based on the selection criteria discussed in the following, to infer which species are most likely to express maternal effects.

First, I developed a set of criteria based on evolutionary scenarios to predict which life-history traits would most likely favor maternal effects expression in a species. Then, I gathered data from the literature on the life-history traits of interest for nearly 100 lepidopteran species. With these data I calculated a maternal effects score (ME score), which predicts the probability of expression of significant maternal effects. Finally, I sorted the species on the basis of their reported population dynamics and looked at how the maternal effects scores were distributed among the three classes of population dynamics. Price *et al.* (1990) have argued that there is an evolved basis for differences in population dynamics among herbivores that goes beyond extrinsic factors, and that we would do well to consider life-history evolution while trying to make some sense of population dynamics.

IV. Criteria for Choice of Life-History Traits for ME Score

From existing work on the ecology, behavior, and evolution of herbivores, I constructed three selection criteria that favored greater probability of survival or increased fecundity, but in doing so, increased the unpredictability of the environment faced by offspring at hatch—thus favoring the development or enhanced expression of maternal effects (i.e., the lunchbox option).

Scenario a: Natural selection favors escape from mortality agents during the relatively immobile, less defended reproductive states: pupal, mating/oviposition stage, or egg stages (e.g., see Rossiter, 1987). As a consequence, oviposition occurs physically or temporally away from the host or the highest-quality host tissues. Here, the cost for improved survival of parents or prehatch eggs can be unpredictability in the quality of the environment faced by offspring at hatch. In compensation, neonate dispersal is favored. Under this scenario, the expression of neonate mobility predicts the presence of maternal effects that increase the likelihood that offspring live long enough to locate acceptable food.

Scenario b: Natural selection favors the use of high-quality food or feeding in enemy-free space as would be achieved by being an early-season feeder (e.g., see Holmes *et al.*, 1979; Nothnagle and Schultz, 1987). For most early-season feeders, the overwintering stage is that of the egg (Hunter, 1991), a condition that reduces mothers' ability to position offspring to their best nutritional advantage given the unpredictability of budburst the following spring. Maternal control over the external environment of young is further reduced by the long temporal separation between oviposition and first feeding. Consequently, early spring hatch favors the provision of reserves to enable offspring to avoid starvation until acceptable food is available. Under this scenario, the expression of early spring feeding in neonates predicts the presence of maternal effects that can act to increase the chance of survival by providing metabolic requirements through the period of overwintering and unpredictable budburst.

Scenario c: Natural selection favors increased fecundity by redistribution of energy from somatic to reproductive tissue. A significant reduction in mobility or loss of flight in the female results, and eggs are deposited in one or several groups. With less opportunity to place each egg in an optimal location, the ability of the mother to maximize the quality of the hatching environment faced by offspring is reduced. Under this scenario, the expression of egg clumping predicts the presence of maternal effects because egg provisions provide resources for location of high-quality environment (good food and not crowded). For colonial species (larvae live and feed together), the egg clumping trait had to precede the colonial trait. Consequently, there was the opportunity for evolution of maternal effects even though extant taxa remain near the clumped oviposition site.

In each scenario some advantage is gained: (a) improved odds for parental and egg survival by ovipositing away from the site of neonate feeding, (b) access of neonates to the high-quality environment of early spring, and (c) increased fecundity. In every case, the cost of the advantage is a reduction in the chance of neonate survival due to the increased uncertainty in finding high-quality food or habitat in the earliest stages of development. Based on these scenarios, there are five life-history traits that predict the presence of maternal effects: overwintering as an egg (scenario a or b), early spring feeding as a neonate (scenario b), flightlessness in adult females (scenario c), eggs oviposited in clumps (scenario c), and active or passive neonate mobility (scenarios a, b, and c).

The presence of any of these traits predicts the evolution of maternal effects because maternal effects can act to compensate for the experience of environmental uncertainty. Although it is true that these traits can be co-correlated, they need not be. Since the presence of any single trait is not predicted on the presence of any other, I decided on a protocol for scoring the likelihood of maternal effect expression that treats each of the five traits independently. Within a species, the presence of any individual trait contributed one point to the maternal effects score (ME score), and ME scores could range from 0 to 5. There were two circum-

TABLE 2

Data on the Population Dynamics and Life-History Characteristics of Ninety-two Herbivorous Moth Species from Twenty-one Lepidopteran Families[a]

Family	Species	V1 dynm	V2 ME	V3 phen	V4 ovw	V5 femfly	V6 eggd	V7 nmob	V8 greg	V9 #gen	V10 orig
Oecophoridae	*Machimia tentoriferella*	1	0.0	4	5	.	.	.	1	1	1
Oecophoridae	*Psilocorsis reflexella*	1	0.0	4	5	.	.	.	1	1	1
Pyralidae	*Nephopterix subcaesiella*	1	0.0	3	5	3	.	.	.	.	1
Geometridae	*Semiothisa granitata*	1	0.0	4	5	.	2	.	1	.	1
Geometridae	*Synchlora aerata*	1	0.0	2	5	3	2	.	1	2	1
Mimallonidae	*Cicinnus melsheimeri*	1	0.0	3	5	3	.	1	.	.	1
Lasiocampidae	*Phyllodesma americana*	1	0.0	2	5	3	.	.	2	.	1
Saturniidae	*Automeris io*	1	0.0	3	5	3	.	.	2	.	1
Saturniidae	*Antheraea polyphemus*	1	0.0	3	5	3	.	.	.	.	1
Pyralidae	*Dioryctria ponderosae*	1	0.5	7	2	.	1	.	1	.	1
Gracilariidae	*Caloptila invariabli*	1	1.0	3	1	.	.	.	.	.	1
Tortricidae	*Argyrotaenia pinatubana*	1	1.0	1	5	.	.	.	1	2	1
Limacodidae	*Phobreton pithecium*	1	1.0	4	4	3	3	.	2	.	1
Geometridae	*Hydria undulata*	1	1.0	3	5	3	3	.	3	.	1
Saturniidae	*Callosamia promethea*	1	1.0	3	5	3	3	.	2	.	1
Arctiidae	*Halisidota tessellaris*	1	1.0	4	5	3	4	.	1	.	1
Noctuidae	*Acronycta americana*	1	1.0	3	5	3	.	3	.	.	1
Gelechidae	*Coleotechnites piceaella*	1	1.5	6	2	.	1	.	1	.	1
Lasiocampidae	*Tolype vellada*	1	2.0	3	1	.	3	.	.	.	1
Gelechidae	*Dichomeris ligulella*	2	0.0	2	6	3	2	1	1	1	1
Tortricidae	*Barbara colfaxiana*	2	0.0	2	5	3	1	1	1	.5–1	1
Megalopygidae	*Megalopyge opercularis*	2	0.0	3	4	.	.	5	2	2	1
Pyralidae	*Pantographa limata*	2	0.0	4	4	3	.	.	1	.	1
Pyralidae	*Actrix nyssaecolella*	2	0.0	4	5	.	.	.	1	1	1
Geometridae	*Epimecis hortaria*	2	0.0	3	5	3	.	5	1	1	1

Family	Species										
Noctuidae	*Lithophane antennata*	2	0.0	2	6	3	·	1	·	3	1	1
Argyresthiidae	*Argyresthia laricella*	2	1.0	6	4	·	1	·	3	1	1	1
Tortricidae	*Rhyacionia rigidana*	2	1.0	2	5	3	3	·	3	1	1	1
Tortricidae	*Petrova comstockia*	2	1.0	5	4	·	1	·	3	1	1	1
Tortricidae	*Epinotia aceriella*	2	1.0	4	5	3	2	3	3	1	1	2
Limacodidae	*Cnidocampa flavescens*	2	1.0	4	4	·	2	3	3	2	1	1
Pyralidae	*Tetralopha robustella*	2	1.0	4	4	3	3	1	1	3	1	1
Pyralidae	*Acrobasis indigenella*	2	1.0	5	3	3	2	·	1	3	1	1
Notodontidae	*Danata ministra*	2	1.0	3	5	3	4	5	5	3	1	1
Arctiidae	*Estigmene acrea*	2	1.0	2	5	3	4	5	5	1	2	1
Arctiidae	*Halisidota harrisii*	2	1.0	4	5	3	1	1	·	2	1	1
Gelechiidae	*Exoteleia pinifoliella*	2	1.5	6	2	·	·	1	1	1	1	1
Gelechiidae	*Battaristis vittella*	2	1.5	6	2	·	·	·	1	1	1	1
Tortricidae	*Spilonota ocellana*	2	1.5	6	2	·	·	·	1	1	2	2
Arctiidae	*Lophocampa caryae*	2	1.5	3	5	2	3	·	3	3	1	1
Yponomeutidae	*Yponomeuta multipunctella*	2	2.0	4	·	·	4	3	3	3	·	1
Tortricidae	*Zeiraphera canadensis*	2	2.0	2	1	3	2	4	4	3	1	2
Geometridae	*Phigalia titea*	2	2.0	2	5	1	4	·	·	1	1	1
Geometridae	*Sabulodes aegrotata*	2	2.0	3	1	·	3	4	1	1	2	1
Tortricidae	*Choristoneura rosaceana*	2	3.0	7	2	2	3	4	4	1	3	1
Tortricidae	*Archips cerasivorana*	2	3.0	2	1	·	4	2	2	3	1	1
Saturniidae	*Hemileuca maia*	2	4.0	1	7	3	4	4	4	3	1	1
Gelechiidae	*Coleotechnites milleri*	3	0.0	7	5	·	1	1	1	1	4	1
Geometridae	*Rheumaptera hastata*	3	0.0	3	5	3	2	·	2	2	1	1
Notodontidae	*Heterocampa guttivitta*	3	0.0	3	5	3	1	5	1	2	1	1
Notodontidae	*Lochmaeus manteo*	3	0.0	4	4	·	·	5	·	1	·	1
Tortricidae	*Epinotia meritana*	3	0.5	3	2	3	1	·	1	1	.5–1	1
Pyralidae	*Dioryctria reniculelloides*	3	0.5	7	2	·	·	·	1	1	1	1
Argyresthiidae	*Argyresthia thuiella*	3	1.0	6	4	·	1	·	·	1	1	1
Tortricidae	*Rhyacionia frustrana*	3	1.0	1	5	3	1	·	1	1	4	1
Geometridae	*Hydria prunivorata*	3	1.0	3	5	3	4	·	3	3	1	1

(continues)

TABLE 2 (continued)

Family	Species	V1 dynm	V2 ME	V3 phen	V4 ovw	V5 femfly	V6 eggd	V7 nmob	V8 greg	V9 #gen	V10 orig
Saturniidae	*Dryocampa rubicunda*	3	1.0	3	5	3	3	5	2	1	1
Saturniidae	*Anisota senatoria*	3	1.0	4	5	3	4	5	2	1	1
Notodontidae	*Clostera inclusa*	3	1.0	2	5	3	3	5	3	2	1
Notodontidae	*Danata integerrima*	3	1.0	3	5	3	4	5	3	1	1
Notodontidae	*Symmerista canicosta*	3	1.0	3	5	3	3	5	2	1	1
Notodontidae	*Schizura concinna*	3	1.0	4	4	3	4	5	3	1	1
Arctiidae	*Hyphantria cunea*	3	1.0	4	5	3	4	5	3	1	1
Noctuidae	*Spodoptera frugiperda*	3	1.0	3	5	3	3	5	.	2	1
Noctuidae	*Peridroma saucia*	3	1.0	3	5	3	4	5	1	2	1
Saturniidae	*Coloradia pandora*	3	1.5	5	2	3	3	.	2	.2–.5	1
Sphingidae	*Ceratomia catalpae*	3	1.5	4	2	3	4	.	2	2	1
Dioptidae	*Phryganidia californica*	3	1.5	7	2	3	2	2	1	2	1
Lymantriidae	*Dasychira plagiata*	3	1.5	6	2	3	2	.	.	1	1
Bucculatrigidae	*Bucculatrix ainsliella*	3	2.0	3	5	.	3	2	3	2	1
Tortricidae	*Endothenia albolineana*	3	2.0	5	4	.	3	5	2	.	1
Tortricidae	*Croesia semipurpurana*	3	2.0	1	1	3	.	.	1	.	1
Geometridae	*Lambdina fiscellaria*	3	2.0	3	1	3	2	4	1	1	1
Geometridae	*Cingilia catenaria*	3	2.0	3	1	3	1	4	1	1	1

Family	Species	V1	V2	V3	V4	V5	V6	V7	V8	V9	V10
Arctiidae	*Lophocampa argentata*	3	2.5	6	2	3	4	1	3	1	1
Lymantriidae	*Leucoma salicis*	3	2.5	6	2	3	4	5	1	1	2
Tortricidae	*Acleris variana*	3	3.0	1	1	3	1	4	1	1	1
Geometridae	*Phigalia plumogeraria*	3	3.0	1	5	1	4	·	·	1	1
Geometridae	*Paleacrita vernata*	3	3.0	2	4	1	3	4	3	1	1
Lasiocampidae	*Malacosoma disstria*	3	3.0	1	1	3	4	5	3	1	1
Lymantriidae	*Euproctis chrysorrhea*	3	3.5	5	2	3	4	4	3	1	2
Psychidae	*Thyridopteryx ephemeraeformis*	3	4.0	2	1	1	4	4	1	1	1
Tortricidae	*Choristoneura fumiferana*	3	4.0	6	2	2	3	2	2	1	1
Tortricidae	*Archips semiferana*	3	4.0	1	1	3	4	4	2	1	1
Geometridae	*Erannis tiliaria*	3	4.0	1	1	1	2	4	1	1	1
Geometridae	*Ennomos subsignaria*	3	4.0	1	1	3	3	4	1	1	1
Geometridae	*Operophtera bruceata*	3	4.0	1	1	1	2	2	1	1	1
Geometridae	*Alsophila pometaria*	3	5.0	1	1	1	4	4	1	1	1
Lymantriidae	*Orgyia vetusta*	3	5.0	1	1	1	4	4	1	1	1
Lymantriidae	*Orgyia pseudotsugata*	3	5.0	1	1	1	4	2	2	1	1
Lymantriidae	*Orgyia leucostigma*	3	5.0	1	1	1	4	4	1	2	1
Lymantriidae	*Lymantria dispar*	3	5.0	1	1	1	4	4	2	1	2

[a]The meaning of the character state value for each of 10 variables is described in Section VI. V1 = type of population dynamics; V2 = ME score based on the character state for variables V3 through V7; V3 = feeding phenology; V4 = overwintering stage; V5 = mother's flying ability; V6 = degree of egg clumping; V7 = neonate mobility; V8 = degree of gregarious behavior; V9 = voltinism; V10 = origin of species.

stances where a modified value of 0.5 was assigned during ME score tabulation: when female flight was present but very poor and when the overwintering stage was that of the hatched but very young larva.

Details on the data set used to calculate the ME scores for nearly 100 species of herbivore species are described in Sections V and VI. It is important to remember that the criteria used for the ME scores is not all-inclusive. There are certainly other circumstances where natural selection would favor the development of maternal effects (see Section X). The aim of the ME scoring presented here is to make a start in an uncharted area.

V. Criteria for Choice of Phytophagous Insect Species

The choice of species for this analysis was limited to tree-feeding lepidopterans found in North America. An effort was made to use a phylogenetic diversity of representatives from this group. Of the 92 moth species ultimately included, 42% are microlepidopterans and the rest are macrolepidopterans. In total, 21 families and 78 genera are represented. Two species from the same genus were permitted only if the two differed in dynamics class or exhibited significant differences in life history. Life-history data were derived from the following references: Berryman (1988), Brues (1946), Burgess and Sharpe (1981), Coulson and Witter (1984), Craighead (1950), Davidson and Peairs (1966), Doane et al. (1936), Essig (1926), Forbes (1948, 1954, 1960), Herrick (1935), Hodges (1983), Johnson and Lyon (1988), Keen (1952), Martineau (1984), Metcalf and Flint (1962), Peterson (1962), Rose and Lindquist (1973, 1977, 1980, 1982), U.S. Department of Agriculture (1979), and van der Geest and Evenhuis (1991).

Each species was also classified by its reported population dynamics into one of three categories: no significant fluctuations, significant fluctuations or very occasional outbreak, and frequent outbreaks. These reports are, for the most part, qualitative rather than quantitative assessments. Consequently, they are biased by the degree of economic stress caused by a "fluctuation." Recognizing this limitation, I made the assumption that this type of categorization would be robust enough for the type of assessment I wanted to make. Nearly half of the total species evaluated were designated as species with frequent outbreaks, whereas 30 and 21% were designated, respectively, as species with and without significant fluctuations.

VI. Protocol for Scoring Herbivore Characteristics

Data on the population dynamics and life-history characteristics of 92 herbivorous moth species are presented in Table 2. Here, species are sorted by the type of population dynamics they exhibit. The character state of this first variable (V1, dynm) is indicated as follows: 1 = no fluctuations, 2 = moderate fluctuations or occasional outbreak, and 3 = regular outbreak. Within each population dynamics class, species are sorted by their final maternal effect score (V2, ME), which ranges from 0 to 5 as described earlier. The next five variables (V3–V7) are those used to calculate the ME score. The last three variables (V8–V10) provide additional data pertinent to the final discussion. The character state values for variables 3–10 are as follows:

V3, phen: time of season when larval feeding begins; 1 = budbreak, 2 = late spring, 3 = early summer, 4 = late summer, 5 = summer and budbreak, 6 = fall and budbreak, 7 = fall and late spring; when initial feeding is split across two time periods, egg hatch occurs in the earliest time period and this may or may not include some feeding; when the character state of V3 was "1" or "6," a value of 1 was added to the ME score.

V4, ovw: overwintering stage; 1 = egg, 2 = early-stage larva, 3 = midstage larva, 4 = late-stage larva, 5 = pupa, 6 = adult, 7 = two stages; when the character state of V4 was "1" or "2," a value of 1 or 0.5, respectively, was added to the ME score.

V5, femfly: degree of adult female mobility; 1 = flightless, 2 = poor flight, 3 = good flight; when the character state of V5 was "1" or "2," a value of 1 or 0.5, respectively, was added to the ME score.

V6, eggd: egg deposition pattern; 1 = individually, 2 = one to several eggs, 3 = clusters, 4 = one mass; when the character state of V6 was "3" or "4," a value of 1 was added to the ME score.

V7, nmob: neonate mobility; 1 = no, 2 = yes, passive (through silking or dropping), 3 = yes, passive for both early and late larval stages, 4 = yes, passive for early larval stages and active mobility for late larval stages, 5 = no, only active mobility for late-stage larvae; when the character state of V7 was "2," "3," or "4," a value of 1 was added to the ME score.

V8, greg: degree of gregariousness; 1 = solitary, 2 = gregarious only during the early larval stages, 3 = gregarious for most to all of the larval period.

V9, #gen: maximum number of generations per year.

V10, orig: origin; 1 = native to North America, 2 = introduced.

VII. Predictions

Based on the character state of life-history traits V3 through V7, a maternal effects score was calculated for each species. The ME scores ranged from 0 to 5, where 5 predicts the greatest likelihood of the expression of significant maternal effects. Overall, 66% of the species have very low ME scores (≤ 1.5), whereas 21% have moderate scores (2.0–3.5) and 13% have very high scores (4.0–5.0).

TABLE 3
Frequency Distribution of Maternal Effects Scores Relative to Species' Population
Dynamics Classifications

ME score	Type of population fluctuations			Totals	
	Negligible	Moderate	Outbreak		
0.0–0.5	10 ⎫ 95%	7 ⎫ 75%	6 ⎫ 49%	23	25% ⎫ 66%
1.0–1.5	8 ⎭	14 ⎭	16 ⎭	38	41% ⎭
2.0–2.5	1 ⎫ 5%	4 ⎫ 21%	7 ⎫ 27%	12	13%
3.0–3.5	0 ⎭	2 ⎭	5 ⎭	7	8% ⎫ 34%
4.0–4.5	0 ⎫ 0%	1 ⎫ 4%	6 ⎫ 24%	7	8% ⎭
5.0	0 ⎭	0 ⎭	5 ⎭	5	5%
Totals	19	28	45	92	
	21%	30%	49%		100%

These scores suggest that 34% of the species studied are good prospects for the expression of maternal effects.

Table 3 presents the frequency distribution of ME scores relative to the species' population dynamics. The first thing to notice is that there is a relationship between the ME scores and the type of population dynamics. We can look at this in two ways. First, the proportion of species with a very low ME score (≤ 1.5) is related to the type of population dynamics: a very low ME score is found in 95% of the species with negligible fluctuation, 75% of the species with moderate fluctuation, and 49% of the species with regular outbreak. Second, the proportion of species with a very high ME score (4.0–5.0) is related to the type of population dynamics: a very high ME score is found in 0% of the species with negligible fluctuation, 4% of the species with moderate fluctuation, and 24% of the species with regular outbreak. These results show that as the proportion of life-history traits that favor the development of maternal effects increases (i.e., a high ME score), the likelihood of population fluctuations increases. However, the converse is not so; an absence of the life-history traits that favor the development of maternal effects (i.e., a low ME score) says nothing about the likelihood of population fluctuation.

Based on the criteria I used for the ME score, there is a low probability that maternal effects are involved in the fluctuations of those outbreak species with low ME scores (i.e., half of the outbreak species). The direct actions of extrinsic agents are more likely to be sources of extreme population fluctuation. The fact that half of the outbreak species had low ME scores might also indicate that population outbreak precedes the development of strong maternal effects. This is a complex topic and deserves further thought.

TABLE 4

Forest Lepidopteran Herbivore Species with Biological or Theoretical Evidence for Maternal Effects and Evidence for a Cross-Generational Shift in Population Quality[a]

Outbreak species	Common name	ME score	Evidence for:[a] Maternal effects Biological	Evidence for:[a] Maternal effects Theoretical	Cross-generation Pop. quality shift
Bupalus piniaria	Pine looper	1.0	?	6	5, 10
Hyphantria cunea	Fall webworm	1.0	13	6-yes, 15-no	13
Zeiraphera diniana	Larch budmoth	2.0	?	15	1, 2
Acleris variana	Black-headed budworm	3.0	?	6	?
Malacosoma pluviale	Western tent caterpillar	3.0	16, 17	?	18
Epirrata autumnal	Autumnal moth	4.0	9	6	8
Choristoneura fumiferana	Spruce budworm	4.0	3, 7	6-yes, 15-no	?
Orgyia pseudotsugata	Douglas fir tussock moth	5.0	?	15	12
Lymantria dispar	Gypsy moth	5.0	11, 14	6, 15	4

[a]References: (1) Baltensweiler (1971); (2) Baltensweiler *et al.* (1977); (3) Campbell (1962); (4) Campbell (1967); (5) Engel (1942); (6) Ginzburg and Taneyhill (1994); (7) Harvey (1985); (8) Haukioja and Neuvonen (1985); (9) Haukioja and Neuvonen (1987); (10) Klompf (1966); (11) Leonard (1970); (12) Mason and Wickman (1988); (13) Morris (1967); (14) Rossiter (1991a); (15) Turchin (1990); (16) Wellington (1965); (17) Wellington (1977); (18) Wellington (1964).

It is worth mentioning that additional criteria can be included in the prediction of maternal effects, such as diapause or phase polymorphism, traits known to be influenced by the parental environmental experience. The addition of these traits to the ME score may reduce the proportion of outbreak species that currently exhibit low ME scores. I did not include such traits in the calculation of the ME score for lack of sufficient documentation in the literature for the full complement of sample species in Table 2. But the relationship between population dynamics and the expression of these maternal effects-influenced traits is discussed near the end of the chapter.

Within the context of lepidopteran species, I compiled a list of outbreak species for which maternal effects have been studied, then ranked these species based on their maternal effects score in order to assess the reliability of the ME score (Table 4). The evidence for maternal effects arose from biological sources (investigations on the nature of the maternal effect) or theoretical sources (inference from models incorporating delayed density dependence and using time-series data). The model of Turchin (1990) includes time-lagged effects that arise from unspecified biological sources. The model of Ginzburg and Taneyhill (1994) specifies that the time lag arises from maternal effects based on the inclusion of a population quality term that has time-lagged consequences. The low ME scores for the pine looper and fall webworm stand in contrast to the referenced support for the presence of maternal effects (Table 4). This suggests two possibilities. First, the cross-generational shift in population quality documented for both the pine looper and the fall webworm is due to selection rather than maternal effect. Second, the criteria for the ME score must be expanded. As noted earlier, the criteria used for the ME scores were, in part, limited by the type of data available in the literature (but see Section X).

VIII. Comparison of Inferences with Other Studies

Several other researchers have noted an association between outbreak dynamics and some of same life-history traits I have used in calculating the ME score. Nothnagle and Schultz (1987) surveyed the status of 12 life-history traits for 29 outbreak species of Macrolepidoptera in an attempt to characterize pest species. Among other things, they found that half of the species overwinter as eggs and begin feeding at budbreak, one-third of the outbreak species had flightless females, and species with higher fecundities tend to cluster eggs. Still it was not clear if these patterns are common to Macrolepidoptera in general or particular to pest species, because the life-history patterns of nonoutbreak species were unavailable for comparison. Hunter (1991) clarified this issue with a comparison of life-history patterns for 313 tree-feeding macrolepidopteran species, the majority of which are nonoutbreak herbivores. The results indicated that two life-history patterns were disproportionately represented in outbreak species: over-wintering in the egg stage followed by feeding in early spring, and gregarious behavior combined with summer feeding. She also found that outbreak species cluster eggs more often than do nonoutbreak species and poor female flight occurs more often among outbreak species. So, in two other studies the occurrence of outbreak was seen in species more likely to express four of the character states from which one can predict the likelihood of maternal effects expression, based on several evolutionary scenarios.

I suggest that maternal effects are the mechanism responsible for the convergence of results in these studies. The result of Hunter that summer-feeding outbreak species are usually gregarious is of particular interest. I chose not to use sociality as a component of the ME score because it need not result in greater environmental uncertainty during early development, the latter being the ultimate criterion for inclusion in the ME score. However, the gregarious habit strongly suggests an evolutionary history that favored, at least for some period, the development of maternal effects because it is a trait that emerges from egg clustering.

IX. Gregarious Larval Behavior

Interestingly, the first extensive study of maternal effects in insects came from an outbreak species with gregarious habits and hatch at budbreak, the western tent caterpillar (Wellington, 1965). Wellington found variation among groups in size and average quality and these differences were generated by differences in parental quality (Wellington, 1977). Moreover, a simulation model for this species indicated persistence and growth of populations with variable quality among groups even in the face of high mortality from outside sources, so long as the high-quality groups were not preferentially removed (Wellington *et al.*, 1975).

Sociality need not predict the presence of maternal effects. There were data for 83 species on the degree of gregarious behavior and one-third of these exhibited some degree of gregarious behavior. For solitary species, 65% had low ME scores (≤ 1.5); for gregarious species, 66% had low ME scores. From this I conclude that the presence of gregarious behavior does not necessarily predict the presence of selective criteria favoring the evolution of maternal effects beyond that of egg clustering.

There are, however, some gregarious species with high ME scores. In this data set, 7 of the 16 fully gregarious species have a ME score of 2.0 or more, and all 7 exhibit some type of fluctuations (3 moderate, 4 outbreak). Since maternal effects can adjust population quality, their expression in a gregarious species may further amplify the variation in the population's spatial distribution. This is of interest because it has been hypothesized that a clumped spatial distribution can alter the impact of natural enemies (Hanski, 1987) and the response of host plant defense systems (e.g., Berryman, 1979).

At this point, it is of interest to consider another taxonomic group, the diprionid sawflies. Hanski (1987) found that among 11 sympatric species of European diprionids, outbreak and nonoutbreak species could be distinguished not by their host plant use patterns or distribution, but by their behavior: outbreak species were gregarious and nonoutbreak species were solitary. Larsson *et al.* (1993) expanded the data base and compared life-history traits of 27 diprionid sawfly species using canonical discrimination analysis. They found that outbreak species were significantly distinguished from the others by a gregarious life-style, capability of prolonged diapause, and feeding on young needles. Hanski (1987) hypothesized that gregarious behavior promoted outbreaks because it led to a patchy distribution, which in turn increased variation in the probability of escape from natural enemies, a condition that promotes instability. This conclusion is in keeping with that of Wellington (1957), who considered differential mortality to be a result of both colony size and quality, the latter of which was significantly influenced by maternal quality for the western tent caterpillar. In an assessment of life history relative to diprionid sawfly dynamics, Larsson *et al.* (1993) make a compelling argument that the degree of patchiness in herbivore distribution may be augmented by the impact of host quality on fecundity in the previous generation. Though there has been no work on maternal effects in diprionid sawflies, the results presented by Hanski and Larsson suggest that they may occur in sawflies and be strongly expressed in the outbreak species. From his work on a gregarious sawfly (*Neodiprion swainei*), Lyons (1962) suggested that an "internal factor," possibly of maternal origin, was responsible for patterns of differential mortality among colonies.

X. Other Criteria for Predicting Maternal Effects

The chief criterion for traits included in the ME score, generation of greater uncertainty for success during early development, ignored several other situations that would also favor the development of strong maternal effects: facultative diapause, telescoping generations, and presence of phase polymorphism. Facultative entry into diapause by eggs or early-stage juveniles is a trait shown to be under the control of environmentally based maternal effects for a number of insects (Mousseau and Dingle, 1991a). The sensitive stage of the maternal generation that experiences the environmental influence can happen as early as the mother's egg stage (e.g., *Bombyx mori,* the silkworm; Fukuda, 1951), but more typically it occurs closer to the time of reproduction. The environmentally based maternal influence on initiation of diapause is strongest when diapause occurs in egg or early larval stages, but can also occur in species that diapause in later developmental stages (Mousseau and Dingle, 1991a). In cases where the environmental control is from an unpredictable source (e.g., rainfall, temperature, or host quality during the parental life, rather than photoperiod), the addition of a facultative diapause trait to the calculation of the ME score would boost the score for some species. For example, in a locust species (*Locusta pardalina*), only half of the eggs from the smaller, dark, aggregating mothers enter diapause compared to all of the eggs from the larger, light, solitary mothers (Matthée, 1951).

Facultative diapause was not included as a variable in the prediction of maternal effects only because there are too little data available. However, it is of interest to consider the distribution of voltinism among the scored species, relative to their population dynamics. Forty-one percent of the 22 species with very low ME scores and outbreak dynamics are multivoltine (i.e., number of generations per year is not fixed at 1; see Table 2). By comparison, only 13% of the 39 species with very low ME scores and nonoutbreak dynamics are multivoltine.

In aphids, oocytes for as many as three subsequent generations are already formed by the large embryo stage. This telescoping of generations offers extensive opportunity for the input of the maternal environment on offspring phenotype (Lees, 1983). The impact of such environmentally based maternal effects on offspring phenotype may be an important component of population dynamics of aphid species. A shift in sex ratio and the proportion of dispersing forms are known to be correlated with density in previous generations for several aphid species (see Mousseau and Dingle, 1991b).

In a study of population fluctuation in seven indigenous tree-feeding aphid species of Britain, Dixon (1990) concluded that each species fluctuates in a strongly density-dependent manner around its own species-specific density from one year to the next. However, within years, the succession of parthenogenic generations experiences overcompensating density dependence (i.e., high density in the spring is followed by low density in the fall and vice versa). Using several experimental approaches, he concludes that natural enemies are not critical to the within-year population dynamics for four of these species. He does note a delayed density-dependent shift in aphid quality in parthenogenic generations. After ruling out a delayed aphid-induced shift in host quality as the source of the aphid quality shift, Dixon suggests that intraspecific competition may be responsible for within-year fluctuations. It is also possible that maternal effects underlie all of this, increasing the magnitude of fluctuations because of the time-lagged nature of their delivery of the effects of density-dependent intraspecific competition.

An adjustment of phenotype in response to the environment is called plasticity. Barbosa and Baltensweiler (1987) compiled many examples of density-associated changes in phenotype for insect species, thereby providing a list of good candidates for investigation of environmentally based maternal effects, the latter being a form of time-lagged phenotypic plasticity. Only a few of the traits discussed by Barbosa and Baltensweiler have been investigated as to whether the source of plasticity is environmentally based maternal effects versus the immediate environment. One interesting example is that of phase polymorphism in locust species with fluctuating dynamics: as density increases, the frequency of the dark, gregarious morph increases (Uvarov, 1961). For *Locusta migratoria,* the African migratory locust, Ellis (1959) showed that the offspring of crowded parents tend to be darker and march more vigorously that those from isolated parents, and later work demonstrated the presence of maternal effects in this species as well as in *Schistocerca gregaria,* the desert locust (Hardie and Lees, 1985). Dark-phase individuals produce eggs of higher quality (better able to withstand starvation at hatch) and offspring with more ovarioles (the total number is already determined at hatch) compared to light-phase mothers (Uvarov, 1966). In the favored habitat type of the light, solitary form, prolonged rainfall or soil water availability for several successive locust generations can shift the balance of the phase polymorphism toward the dark, gregarious form. This shift occurs before an outbreak is evident. So, although the origin of outbreak in locust species is related to a previous period of prolonged rainfall (White, 1976), maternal effects may be the essential mechanism for the initiation and maintenance of gregarization.

XI. Closing Remarks

In summary, it is important to mention that the approach I have taken here aims to bring us one step closer to a widespread and useful assessment of the role of maternal effects in population dynamics. This approach, for good and for bad, avoids some complications of reality, such as the undoubtedly important contribution of the environment to the strength of maternal effects expression. In my own work with the gypsy moth, I find that the use of one host plant species over another can markedly change the contribution of maternal effects to phenotype (unpublished data). So, I suggest that the best place to study the interaction between maternal effects, environmental quality, and herbivore population dynamics is with species known to express maternal effects and whose dynamics vary geographically. For example, geographic variation in population dynamics is known for several species. The black-headed budworm (*Acleris variana*) experiences cycling dynamics in the Canadian maritime provinces but stays in a low and stable state elsewhere (Miller, 1966). The gypsy moth (*L. dispar*) maintains a history of outbreak only where oak species dominate (Giese and Schneider, 1979). The larch budmoth (*Zeiraphera diniana*) has outbreaks at high but not low altitudes, although even there populations exhibit cyclical fluctuations of much smaller amplitude (Baltensweiler and Fischlin, 1988). The autumnal moth (*Epirrata autumnal*) has outbreaks in the higher latitudes but not the lower latitudes of its range (Tenow, 1972). The Douglas fir tussock moth (*Orgyia pseudostugata*) has outbreaks in high-density stands of its host plants in Idaho, Washington, and Oregon but not in California (Stoszek and Mika, 1978), although outbreaks do occur there (Dahlsten *et al.*, 1990). If maternal effects can be documented for at least one population of any species, then it is likely that the capability is pervasive and variation in its function is premised on environmental quality. Variable strength in the expression of maternal effects across an herbivore's geographic range is one possible cause of geographic variation in outbreak dynamics within a species.

Acknowledgments

I thank Greg Dwyer, Judy Myers, and Lorne Rothman for comments on the manuscript, and Naomi Cappuccino and Peter Price for their efforts to encourage new approaches to old problems. This work was supported by an NRI/USDA Competitive Research Grant (#91-37302-6292) and a grant from the National Science and Engineering Research Council of Canada (NSERC #OGP0121386).

References

Baltensweiler, W. (1971). The relevance of changes in the composition of larch budmoth populations for the dynamics of its numbers. *Proc. Adv. Study Inst. Dyn. Numbers Popul., 1970*, pp. 208–219.

Baltensweiler, W., and Fischlin, A. (1988). The larch budmoth in the Alps. *In* "Dynamics of Forest Insect Populations: Patterns, Causes, and Implications" (A. A. Berryman, ed.), pp. 331–351. Plenum, New York.

Baltensweiler, W., Benz, G., Bovey, P., and DeLuicchi, V. (1977). Dynamics of larch budmoth populations. *Annu. Rev. Entomol.* **22**, 79–100.

Barbosa, P., and Baltensweiler, W. (1987). Phenotypic plasticity and herbivore outbreaks. *In* "Insect Outbreaks" (P. Barbosa and J. Shultz, eds.), pp. 469–503. Academic Press, Orlando, FL.

Berryman, A. A. (1979). Dynamics of bark beetle populations: Analysis of dispersal and redistribution. *Mitt. Schweiz. Entomol. Ges.* **52**, 227–234.

Berryman, A. A. (1981). "Population Systems: A General Introduction." Plenum, New York.

Berryman, A. A., ed. (1988). "Dynamics of Forest Insect Populations: Patterns, Causes, and Implications." Plenum, New York.

Brues, C. T. (1946). "Insect Dietary." Harvard Univ. Press, Cambridge, MA.

Burgess, R. L., and Sharpe, D. M. (1981). "Forest and Stand Dynamics in Man-Dominated Landscapes," Ecol. Stud. Ser., Vol. 41. Springer-Verlag, New York.

Campbell, I. M. (1962). Reproductive capacity in the genus *Choristoneura* Led. (Lepidoptera: Tortricidae). I. Quantitative inheritance and genes as controllers of rates. *Can. J. Genet. Cytol.* **4**, 272–288.

272   MaryCarol Rossiter

Campbell, R. W. (1967). "The Analysis of Numerical Change in Gypsy Moth Populations," For. Sci. Monogr., Vol. 15. Society of American Foresters, Bethesda, MD.

Coulson, R. N., and Witter, J. A. (1984). "Forest Entomology." Wiley, New York.

Craighead, F. C. (1950). Insect enemies of eastern forests. *Miscellaneous Publ.—U.S., Dep. Agric.* **657.**

Dahlsten, D. L., Rowney, D. L., Copper, W. A., Tait, S. M., and Wenz, J. M. (1990). Long-term population studies of the douglas-fir tussock moth. *In* "Population Dynamics of Forest Insects" (A. D. Watt, S. R. Leather, M. D. Hunter, and N. C. Kidd, eds.), pp. 45–58. Intercept, Andover, UK.

Davidson, R. H., and Peairs, L. M. (1966). "Insect Pests of Farm, Garden, and Orchard." Wiley, New York.

Dixon, A. F. G. (1990). Population dynamics and abundance of deciduous tree-dwelling aphids. *In* "Population Dynamics of Forest Insects" (A. D. Watt, S. R. Leather, M. D. Hunter, and N. C. Kidd, eds.), pp. 11–23. Intercept, Andover, UK.

Doane, R. W., Van Dyke, E. C., Chamberlin, W. J., and Burke, H. E. (1936). "Forest Insects." McGraw-Hill, New York.

Ellis, P. E. (1959). Some factors influencing phase characters in the nymphs of the locust, *Locusta migratoria migratorioides* (R. and F.). *Insectes Soc.* **6,** 21–39.

Engel, H. (1942). Uber die populationsbewegung des kiefernspanners (*Bupalus piniarius*) (L.) in verschiedenen Bestandstypen. *Z. Angew. Entomol.* **29,** 116–163.

Essig, E. O. (1926). "Insects of Western North America." Macmillan, New York.

Forbes, W. T. M. (1948). Lepidoptera of New York and neighboring states. Part II. Geometridae, Sphingidae, Notodontidae, Lymantriidae. *Mem.—N.Y., Agric. Exp. Stn. (Ithaca)* **274.**

Forbes, W. T. M. (1954). Lepidoptera of New York and neighboring states. Part III. Noctuidae. *Mem.—N.Y., Agric. Exp. Stn. (Ithaca)* **329.**

Forbes, W. T. M. (1960). Lepidoptera of New York and neighboring states. Part IV. Agaristidae through Nymphalidae. *Mem.—N.Y., Agric. Exp. Stn. (Ithaca)* **371.**

Fukuda, S. (1951). Factors determining the production of nondiapause eggs in the silkworm. *Proc. Jpn. Acad.* **27,** 582–586.

Giese, R. L., and Schneider, M. L. (1979). Cartographic comparisons of Eurasian gypsy moth distribution. *Entomol. News* **90,** 1–16.

Giesel, J. T. (1988). Effects of parental photoperiod on development time and density sensitivity of progeny of *Drosophila melanogaster*. *Evolution (Lawrence, Kans.)* **42,** 1348–1350.

Ginzburg, L. R., and Taneyhill, D. E. (1994). Population cycles of forest Lepidoptera: A maternal effect hypothesis. *J. Anim. Ecol.* **63,** 79–92.

Hanski, I. (1987). Pine sawfly population dynamics: Patterns, processes, problems. *Oikos* **50,** 327–335.

Hardie, J., and Lees, A. D. (1985). Endocrine control of polymorphism and polyphenism. *In* "Comprehensive Insect Physiology, Biochemistry and Pharmacology" (G. A. Kerkut, and L. I. Gilbert, eds.), Vol. 8, pp. 441–490. Pergamon, Oxford.

Harvey, G. T. L. I. (1985). Egg weight as a factor in the overwintering survival of spruce budworm (Lepidoptera: Tortricidae) larvae. *Can. Entomol.* **17,** 1451–1461.

Haukioja, E., and Neuvonen, S. (1985). The relationship between male size and reproductive potential in *Epirrita autumnata* (Lep., Geometridae). *Ecol. Entomol.* **10,** 267–270.

Haukioja, E., and Neuvonen, S. (1987). Insect population dynamics and induction of plant resistance: The test of hypotheses. *In* "Insect Outbreaks" (P. Barbosa and J. C. Schultz, eds.), pp. 411–432. Academic Press, New York.

Herrick, G. W. (1935). "Insect Enemies of Shade Trees." Cornell Univ. Press (Comstock), Ithaca.

Hodges, R. W. (1983). "Checklist of the Lepidoptera North of Mexico." E. W. Classey and R. B. D. Publ., London.

Holmes, R. T., Schultz, J. C., and Nothnagle, P. J. (1979). Bird predation on forest insects: An exclosure experiment. *Science* **206**, 462–463.

Hunter, A. F. (1991). Traits that distinguish outbreaking and nonoutbreaking Macrolepidoptera feeding on northern hardwood trees. *Oikos* **60**, 275–282.

Hutchinson, G. E. (1948). Circular causal systems in ecology. *Ann. N.Y. Acad. Sci.* **50**, 221–246.

Johnson, W. T., and Lyon, H. H. (1988). "Insects that Feed on Trees and Shrubs." Cornell Univ. Press (Comstock), Ithaca, NY.

Kaplan, R. H., and Cooper, W. S. (1984). The evolution of developmental plasticity in reproductive characteristics: An application of "adaptive coin flipping" principle. *Am. Nat.* **123**, 393–410.

Keen, F. P. (1952). Insect enemies of western forests. *Misc. Publ.—U.S., Dep. Agric.* **273.**

Klomp, H. (1966). The dynamics of a field population of the pine looper *Bupalus piniarius* L. (Lepidoptera: Geometridae). *Adv. Ecol. Res.* **3**, 207–305.

Larsson, S., Björkman, C., and Kidd, N. A. C. (1993). Outbreaks in diprionid sawflies: Why some species and not others? *In* "Sawfly Life History Adaptations to Woody Plants" (M. R. Wagner and K. F. Raffa, eds.), pp. 453–483. Academic Press, San Diego, CA.

Lees, A. D. (1983). The endocrine control of polymorphism in aphids. *In* "Endocrinology of Insects" (R. G. H. Downer and H. Laufer, eds.), pp. 369–377. Alan R. Liss, New York.

Leonard, D. E. (1970). Intrinsic factors causing qualitative changes in populations of *Porthetria dispar* (Lepidoptera: Lymantriidae). *Can. Entomol.* **102**, 239–249.

Lyons, L. A. (1962). The effect of aggregation on egg and larval survival in *Neodiprion swainei* Midd. (Hymenoptera: Diprionidae). *Can. Entomol.* **94**, 49–58.

Martineau, R. (1984). "Insects Harmful to Forest Trees." Multiscience Publications Limited with Canadian Government Publishing Centre, Supplies and Services, Ottawa.

Mason, R. R., and Wickman, B. E. (1988). The douglas-fir tussock moth in the interior Pacific Northwest. *In* "Dynamics of Forest Insect Populations: Patterns, Causes, and Implications" (A. A. Berryman, ed.), pp. 179–210. Plenum, New York.

Matthée, J. J. (1951). The structure and physiology of the egg of *Locustana pardalina* (Walk.) *Sci. Bull. Dep. Agric. For. Union S. Afr.* **316.**

May, R. M. (1974). "Stability and Complexity in Model Ecosystems." Princeton Univ. Press, Princeton, NJ.

Metcalf, C. L., and Flint, W. P. (1962). "Destructive and Useful Insects." McGraw-Hill, New York.

Miller, C. A. (1966). The black-headed budworm in eastern Canada. *Can. Entomol.* **98**, 592–613.

Morris, R. F. (1967). Influence of parental food quality on the survival of *Hyphantria cunea*. *Can. Entomol.* **99**, 24–33.

Mousseau, T. A., and Dingle, H. (1991a). Maternal effects in insect life histories. *Annu. Rev. Entomol.* **36**, 511–534.

Mousseau, T. A., and Dingle, H. (1991b). Maternal effects in insects: Examples, constraints and geographic variation. *In* "The Unity of Evolutionary Biology; Fourth International Congress of Systematic and Evolutionary Biology," pp. 745–761. Dioscorides Press Portland, OR.

Nothnagle, P. J., and Schultz, J. C. (1987). What is a forest pest? *In* "Insect Outbreaks" (P. Barbosa and J. C. Schultz, eds.), pp. 59–80. Academic Press, San Diego, CA.

Peterson, A. (1962). "Larvae of Insects. Part I. Lepidoptera and Plant-Infesting Hymenoptera." Lithographed by Edwards Brothers, Inc. Ann Arbor, MI.

Price, P. W., Cobb, N., Craig, T. P., Fernandes, G. W., Itami, J. K., Mopper, S., and Preszler, R. W. (1990). Insect herbivore population dynamics on trees and shrubs: New approaches relevant to latent and eruptive species and life table development. *In* "Insect–Plant Interactions" (E. A. Bernays, ed.), Vol. 2, pp. 1–38. CRC Press, Boca Raton, FL.

Rose, A. H., and Lindquist, D. H. (1973). Insects of eastern pine. *Can. For. Serv., Publ.* **1313.**

Rose, A. H., and Lindquist, D. H. (1977). Insects of eastern spruces, fir, and hemlock. *For. Tech. Rep. (Can. For. Serv.)* **23.**

Rose, A. H., and Lindquist, D. H. (1980). Insects of eastern larch, cedar, and juniper. *For. Tech. Rep. (Can. For. Serv.)* **28.**

Rose, A. H., and Lindquist, D. H. (1982). Insects of eastern hardwoods. *For. Tech. Rep. (Can. For. Serv.)* **29.**

Rossiter, M. C. (1987). Use of secondary host by non-outbreak populations of the gypsy moth. *Ecology,* **68**, 857–868.

Rossiter, M. C. (1991a). Environmentally-based maternal effects: A hidden force in insect population dynamics? *Oecologia* **87**, 288–294.

Rossiter, M. C. (1991b). Maternal effects generate variation in life history: Consequences of egg weight plasticity in the gypsy moth. *Funct. Ecol.* **5**, 386–393.

Rossiter, M. C. (1992). The impact of resource variation on population quality in herbivorous insects: A critical aspect of population dynamics. *In* "Effects of Resource Distribution on Animal–Plant Interactions" (M. D. Hunter, T. Ohgushi, and P. W. Price, eds.), pp. 13–42. Academic Press, San Diego, CA.

Rossiter, M. C. (1994). Maternal effects hypothesis of herbivore outbreak. *BioScience* **44**, 752–763.

Rossiter, M. C., Schultz. J. C., and Baldwin, I. T. (1988). Relationships among defoliation, red oak phenolics, and gypsy moth growth and reproduction. *Ecology* **69**, 267–277.

Rossiter, M. C., Yendol, W. G., and Dubois, N. R. (1990). Resistance to *Bacillus thuringiensis* in gypsy moth (Lepidoptera: Lymantriidae): Genetic and environmental causes. *J. Econ. Entomol.* **83**, 2211–2218.

Rossiter, M. C., Cox-Foster, D. L., and Briggs, M. A. (1993). Initiation of maternal effects in *Lymantria dispar:* Genetic and ecological components of egg provisioning. *J. Evol. Biol.* **6**, 577–589.

Stoszek, K. J., and Mika, P. G. (1978). Outbreaks, sites, and stands. *USDA For. Serv. Tech. Bull.* **1585**, 56–58.

Tenow, O. (1972). The outbreaks of *Oporinia autumnata* Bkh. and *Operophtera* spp. (Lep. Goemetridae) in the Scandinavian mountain chain and northern Finland 1862–1968. *Zool. Bidr. Uppsala, Suppl.* **2**, 1–107.

Turchin, P. (1990). Rarity of density dependence or population regulation with lags? *Nature (London)* **344**, 660–663.

Turchin, P., and Taylor, A. D. (1992). Complex dynamics in ecological time series. *Ecology* **73**, 289–305.

U.S. Department of Agriculture (1979). "A Guide to Common Insect and Diseases of Forest Trees in the Northeastern United States," USDA For. Serv. Publ. NA-FR-4. USDA, Broomall, PA.

Uvarov, B. P. (1961). Quantity and quality in insect populations. *Proc. Roy. Entomol. Soc. London, Ser. C* **25**, 52–59.

Uvarov, B. P. (1966). "Grasshoppers and Locusts," Vol. 1. Cambridge Univ. Press, London.

van der Geest, L. P. S., and Evenhuis, H. H. (1991). "Tortricid Pests: Their Biology, Natural Enemies and Control." Elsevier, Amsterdam.

Wellington, W. G. (1957). Individual differences as a factor in population dynamics: The development of a problem. *Can. J. Zool.* **35**, 293–323.

Wellington, W. G. (1964). Qualitative changes in populations in unstable environments. *Can. Entomol.* **96**, 436–451.

Wellington, W. G. (1965). Some maternal influences on progeny quality in the western tent caterpillar, *Malacosoma pluviale* (Dyar). *Can. Entomol.* **97**, 1–14.

Wellington, W. G. (1977). Returning the insect to insect ecology: Some consequences for pest management. *Environ. Entomol.* **6**, 1–8.

Wellington, W. G., Cameron, P. J., Thompson, W. A., Vertinsky, and Lansberg, A. S. (1975). A stochastic model for assessing the effects of external and internal heterogeneity on an insect population. *Res. Popul. Ecol.* **17**, 1–28.

White, T. C. R. (1976). Weather, food, and plagues of locusts. *Oecologia* **22**, 119–143.

PART IV

CASE STUDIES

Long-Term Population Dynamics of a Seed-Feeding Insect in a Landscape Perspective

Christer Solbreck

I. Introduction

Ecological processes take place on a wide range of spatial and temporal scales, and it is evident that our choices of scales will influence our perception about their dynamics (Krebs, 1991; Levin, 1992; May, 1994; Pimm, 1991; Taylor, 1991; Wiens, 1989). There is, for example, a risk that population studies are performed at scales that are too limited in time and space to convey the important aspects of their dynamics. These considerations are highly relevant for population studies of phytophagous insects.

It is of particular importance to consider extended spatial scales in insects because they are usually winged, and hence often have considerable power of movement by flight. Many species move on a landscape (or larger) scale and respond to the distribution of their plant resources at these spatial scales (Hansson *et al.*, 1992). Theoretical developments in metapopulation dynamics and allied fields, as well as practical considerations, for example, in conservation biology (Gilpin and Hanski, 1991; Hanski, 1994; Hanski and Thomas, 1994; Harrison, 1994), have spurred the interest in spatial dynamics of populations. However, theoretical developments have not been matched by a corresponding development in the exploration of spatial structure and dynamics of field populations. Do we find examples of metapopulations in the classical sense, or are populations more often of mainland–island or source and sink character (Harrison, 1991, 1994; Pulliam, 1988)? To what extent are events in local patch populations correlated (Gilpin, 1990; Harrison and Quinn, 1989) and what is the role of unusual events (Solbreck, 1991)?

The temporal aspect is also very important in insect population studies. Although insects are generally short-lived, populations often fluctuate over decades (Baltensweiler, 1993; Danell and Ericson, 1990; Myers, 1988; Royama, 1984, 1992; Swetnam and Lynch, 1993; Turchin, 1990). Interest in the role of temporal scales has been fertilized by theoretical and methodological developments (e.g., Pimm, 1991; Pimm and Redfearn, 1988; Royama, 1992; Turchin, 1990) as well as by practical problems concerning population change in fragmented landscapes and a new climate (Kareiva *et al.,* 1993; Solbreck, 1993). To explore these questions there is no alternative but to study real field populations in an extended spatial as well as temporal scale (Krebs, 1991; Taylor, 1991; Wiens, 1989). Without extended temporal and spatial perspectives we run the risk of studying nonsignificant events. Studies should encompass periods of both high and low densities. Population bottlenecks and unusual events are of particular importance if we want to understand mechanisms determining extinction rates. Brief studies of "ordinary dynamics" of populations are not likely to yield much insight into these issues; it is the extremes that are of interest.

In this chapter I describe the structure and dynamics of a simple natural population system consisting of a seed-feeding insect (*Lygaeus equestris* L., Heteroptera) and its seed resource [*Vincetoxicum hirundinaria* (Medicus), Asclepiadaceae]. The system is frequently perturbed by weather conditions, which have strong effects on both insects and plants. This system is also characterized by high patchiness in plant distribution and considerable migratory ability of the insect. The temporal scale is close to two decades and the spatial scale is approximately 10 km^2.

Long-term studies cannot be carried out to answer just one specific question. There are important reasons for this. One is the heavy investment of time in such endeavors. Another is the considerable turnover rate of ecological theory. Most original questions are likely to loose their freshness after two decades. Hence, long-term studies have to be performed against a background ensemble of continually modified ideas (cf. Taylor, 1991).

The present study has several objectives, some of which surfaced during the course of study. Hence, in addition to the problems of spatial structure and long-term change, I address the classical questions of the role of top-down or bottom-up regulation, the action of density-dependent factors, and the degree of resource tracking. I also consider weather influences in some detail. I show that weather effects can be manifold and that they may interact in complex ways with other factors. A further reason for presenting this study is that it represents a taxonomic category (Heteroptera) and a feeding guild (seed feeders) that are not overly exposed in the insect population dynamics literature.

II. Biology and Habitats

Lygaeus equestris utilizes two kinds of habitats during its yearly cycle, namely, breeding and hibernation sites. Breeding takes place in stands of the host plant *V. hirundinaria*. This long-lived perennial herb grows on rocky outcrops and at wood margins and has a very patchy distribution on the Swedish mainland.

Lygaeus equestris is both a pre- and postdispersal seed predator. Adults as well as larvae feed on green ovulae in flowers, developing and mature seeds in pods, or seeds that have been dispersed. The insect has been observed to feed on a variety of plants (Solbreck and Kugelberg, 1972). Many of these are poisonous, and the insect is aposematically colored (Sillén-Tullberg, 1985; Sillén-Tullberg, *et al.,* 1982). Despite the potential for feeding on different plant species, *L. equestris* is "functionally monophagous" in the study area. It almost exclusively occurs in patches of *V. hirundinaria* during the breeding season and larvae only occur in these places. Experimental studies have shown that *L. equestris* prefers *V. hirundinaria* seeds and that it performs best on these seeds (Kugelberg, 1973a,b, 1974). The insect is sun-loving and prefers dry, sun-exposed habitats with bare rock and soil patches, and it is frequently seen basking in the sun on cool sunny days (Solbreck, 1976). Egg batches are deposited in the soil, hence the insect needs bare soil surfaces.

The hibernation period usually starts in late August or in September and ends in late April to early May. During hibernation the insects congregate in crevices on sun-exposed rock walls and in similar places. Some of these sites are close to host plant patches, whereas others are more isolated (Solbreck and Sillén-Tullberg, 1990). There is one flight period (autumn migration) when the insects move to hibernation sites and a corresponding period in spring (spring migration) when they fly back to host plant patches. Many individuals move hundreds of meters to (at least) some kilometers, as evidenced by mark–recapture studies (Solbreck and Sillén-Tullberg, 1990). In addition to these main flight periods, the insects may also fly during the summer. These flights take place in response to food shortage (Solbreck, 1985).

Mating activities commence at the hibernation sites just before the spring migration and continue until the end of the egg-laying period. The oviposition period is rather long, but there is usually a peak in egg-laying activity around mid-June coinciding with the main flowering period of *V. hirundinaria*. The insect is normally univoltine, but under extreme conditions (very warm and sunny summers) a partial second generation may appear (Solbreck and Sillén-Tullberg, 1981). New-generation adults appear over an extended period, usually with a peak during middle to late August. Before the autumn migration there is a feeding period when the bugs accumulate large fat reserves for the winter (Solbreck, 1972).

The biotic interactions of *L. equestris* in Sweden are simple indeed. There seem to be no natural enemies at all (in contrast to the situation in southern Europe, where parasitoids may have significant effects; Solbreck *et al.,* 1989). The larva of a tephritid fly (*Euphranta connexa*) lives in the seed pods, but there is no evidence of *E. connexa* competing with *L. equestris* (Sillén-Tullberg and

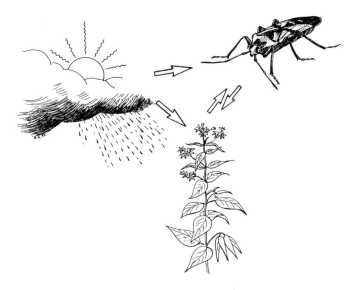

Figure 1. Main components and interactions in the population system of *L. equestris.*

Solbreck, 1990). Hence, the biotic interactions of *L. equestris* seem to be limited to those with its host plant.

Abiotic conditions, notably weather, are important for *L. equestris.* As mentioned earlier, *L. equestris* is a sun-loving insect and its population dynamics are heavily affected by weather conditions both directly and indirectly via the host plant (Sillén-Tullberg and Solbreck, 1990). The simple system of which *L. equestris* is a part can thus be summarized as a triangle of interactions between weather, host plant, and insect (Fig. 1).

III. Methods

The studies were performed at Tullgarn, about 50 km SSW of Stockholm, Sweden. This area along the coast of the Baltic Sea, measuring about 12 km², is divided into three subareas (T1, T2, and F), each with several host plant patches (see inset map in Fig. 5; see also Solbreck, 1991; Solbreck and Sillén-Tullberg, 1990). Insects move extensively between patches within subareas, but usually much less between subareas (Solbreck and Sillén-Tullberg, 1990). Unless otherwise stated, the following analysis is based on data from subarea T1, which has an uninterrupted data series from 1977 to 1994. This area of about 6 km² has 26 small and scattered host plant patches (in 1993) with a median size of 6 m² (range 0.25–187 m²).

Population sizes of *L. equestris* were estimated twice a year by mark–recapture in every host plant patch. Parent populations (*P*) were estimated in June when flowering had started in all host plant patches. The number of their adult offspring (*F*) were estimated in late August–early September (correcting for the percentage of insects still in larval stages). Using experimental data on maximum egg production per female (514), a *k*-value for each year (*t*) was calculated: $k1 = \log P(t) + \log(514/2) - \log F(t)$. This value includes egg shortfall plus egg and larval mortality. A second k-value expresses combined mortalities during autumn migration, hibernation, and spring migration: $k2 = \log F(t) - \log P(t + 1)$. In 1989, when a second generation appeared, adults of the first offspring generation were sampled in July. Food resources were estimated by counting the pods in all patches each year. Weather data were obtained from publications from The Swedish Meteorological and Hydrological Institute, Norrköping. For further details of methods used see Sillén-Tullberg and Solbreck (1990) and Solbreck and Sillén-Tullberg (1990).

IV. Dynamics of Habitats and Seed Resources

Vincetoxicum hirundinaria is a very long-lived perennial plant, and the plant patches change in size very slowly. During 18 years of study, no plant patches have disappeared and only two very small (< 0.5 m²) ones have been established. However, the amount of food resources, measured as number of seed pods, is highly variable, with differences between high and low years being more than two orders of magnitude. This variation, which is highly synchronous among patches, seems to be controlled mainly by weather conditions (Solbreck and Sillén-Tullberg, 1986; C. Solbreck, unpublished). In sunny summers, the plant is affected by drought, resulting in the abortion of flowers and young pods. Hence the sunnier it is, the fewer pods are produced.

Hibernation sites, usually being crevices in rock walls, are of course stable from year to year. However, hibernation sites may sometimes become unsuitable owing to increased shading caused by growing trees. This was observed in another area, but has not occurred at Tullgarn. Hence the spatial pattern of breeding and overwintering habitats remains essentially unchanged over almost two decades, whereas food abundance in breeding habitats varies considerably between years (C. Solbreck, unpublished).

V. Patterns of Abundance in *Lygaeus equestris* Populations

Plant patches are of widely different sizes and so are the abundances of their local insect populations. Two patches dominate and account for almost 50% of

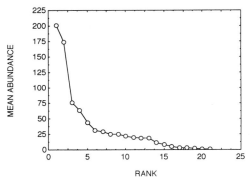

Figure 2. Host plant patches ranked according to their mean abundances of *L. equestris* during 1977–1994.

mean population densities for the study period, whereas 50% of the patches account for over 90% of all bugs (Fig. 2). However, most patches are colonized in most years and all patches have had at least some bugs in some years.

Population densities of *L. equestris* fluctuate considerably between years. These fluctuations are highly synchronous in the different patches (Fig. 3). This synchrony, the generally high percentage of patch occupancy, and the frequent interpatch movements (Solbreck and Sillén-Tullberg, 1990) justify analysis of the temporal dynamics of *L. equestris* by pooling abundances in all patches.

There are also large fluctuations in the pooled data, with long periods (5–10 years) between population peaks (Fig. 4A), and there is a significant autocorrelation with a 1-year lag in the data series [August population sizes: partial autocorrelation coefficient (lag 1) = 0.507, SE = 0.236]. Another interesting

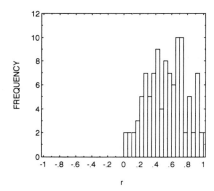

Figure 3. Distribution of correlation coefficients between *L. equestris* population densities in the 15 largest patches during 17 years. (There are 105 possible correlations between 15 data series.)

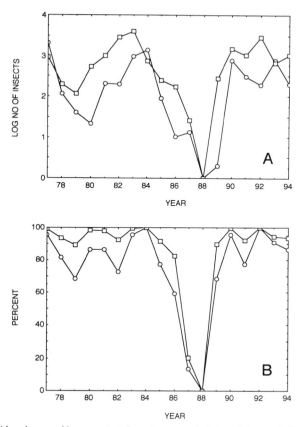

Figure 4. (A) Abundances of *L. equestris* in June (parents, circles) and August (offspring, squares), in subarea T1, 1977–1994. (1 was added to the zero values of 1988.) Notice that in both 1989 and 1990, "parent" abundance is higher than "offspring" abundance in the previous year (= same generation before overwintering). This implies that there was a considerable immigration into the area during autumn and/or spring migrations. (B) Percentage of patches (circles) and of total patch area (squares) colonized during the same period.

aspect is that a long period with a certain pattern of fluctuation was interrupted by a brief period with seemingly very different population behavior. (It is different in the sense that abundance and occupancy rates decreased dramatically, long-distance migration became important, and bivoltinism came into play. However, this may be attributed to manifestations of extreme weather conditions rather than to changes in underlying dynamics.) During the first 10 years there were moderate density fluctuations, a high percentage of patch occupancy, and an even larger occupancy with regard to total available patch area (because uncolonized patches were the smallest patches) (Fig. 4B). However, the situation changed drastically in the following 3 years as a result of unusual weather conditions. The summer of 1987 was the coolest and least sunny since measurements started in 1908, resulting in a very low autumn population, which then became extinct during the following winter (Solbreck, 1991). Not a single bug was observed during the entire summer of 1988. The following summer, however, was unusually sunny. A few immigrants arrived in early summer, but their numbers were very low and very few patches were colonized. Populations in subarea T1 did not become well established until densities in outside areas had increased to levels where high numbers of migrants were produced. This population increase was hastened by the occurrence of a second generation.

Areawide immigration was also important the next year. Population estimates of June 1989 and 1990 differ from all other years in having more bugs than during the previous late summer (Fig. 4A). This can be explained by immigration from outside areas. Populations in the two neighboring areas (T2 and F) had reached lows in 1988, but they did not become totally extinct (Fig. 5). During 1989 (both F and T2) and 1990 (only F), outside areas had considerably larger populations than T1 and could provide numerous immigrants into area T1. This was confirmed by recaptures of marked bugs from these areas (C. Solbreck, unpublished). It took 5 years (six generations) before relative densities in T1 stabilized at a level similar to the conditions prior to the 1987 crash (Fig. 5).

To summarize, whereas the system during the first 10 years of study displayed moderate fluctuations in abundance and occupancy rates and was relatively independent of immigration from outside areas, it changed character dramatically during the next few years. These changes included total extinction and dependency of long-distance migration for recolonization. Furthermore, the occurrence of two generations contributed to an extraordinary high rate of population growth following recolonization. During the last 5 years, system behavior seems to have returned to pre-1987 conditions (Figs. 4 and 5).

VI. What Factors Determine Fluctuations in *Lygaeus equestris* Populations?

Populations of *L. equestris* are affected by density-dependent factors during the summer (Fig. 6A), but not during the rest of the year (autumn migration, hibernation, and spring migration) (Fig. 6B). Density changes also seem to be more variable during summer than during winter.

Weather plays several important roles in the population dynamics of *L. equestris*. There are two direct effects, both related to sunshine hours during June/July. First, summer mortality ($k1$) is negatively correlated with sunshine hours (Fig. 7A). The sunnier it is, the lower the mortality. This agrees well with

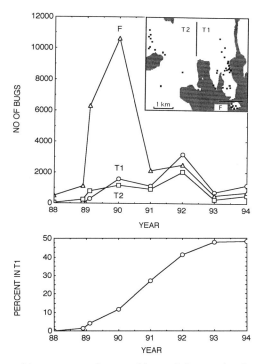

Figure 5. Abundance of *L. equestris* in three neighboring habitat archipelagoes (T1, T2, and F) following the extinction in area T1 in the winter 1987/1988, and percentage of all insects that occurred in T1. There were two generations in 1989. Squares in the inset map show locations of host plant patches. Dark areas denote water.

the sun-basking behavior of the bug (Solbreck, 1976), and with observations that *L. equestris* has problems with completing its life cycle in summers with little sunshine. Second, there is a delayed effect of sunshine hours in June–July on winter mortality (*k*2) (Fig. 7B). Evidently many bugs enter hibernation without having migrated and with poor fat reserves after cloudy summers.

Weather conditions have a further indirect effect by modifying plant seed production (Solbreck and Sillén-Tullberg, 1986). Seeds form a food resource that is involved in density regulation of *L. equestris* populations. However, because the role of food is not immediately evident, and can easily be missed, this needs an extensive explanation.

The insect feeds on seeds in pods as well as on old seeds on the ground. Hence, both present and past years' seed pod crops may be of importance. However, bug densities do not show positive correlations with pod densities of

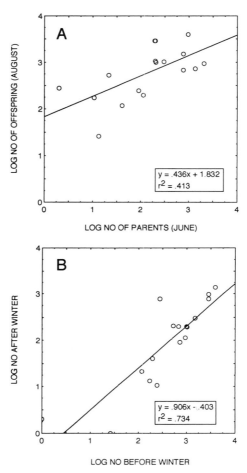

Figure 6. Density relationships of *L. equestris* in (A) summer and (B) winter habitats (including migratory periods). During the summer (A) density-dependent factors operate because the slope of the regression line is significantly smaller than 1, whereas this is not the case during the rest of the year (B). (Data from 1988 are omitted in A.)

either present or past years (Fig. 8). Surprisingly, there is instead a negative correlation between number of bugs and pods of the present year (Fig. 8A). This noncausal correlation is evidently due to sunny weather having a positive effect on bugs but a negative one on pod production. If rates of population change (k-factors) are used rather than densities, and if one looks in more detail at what

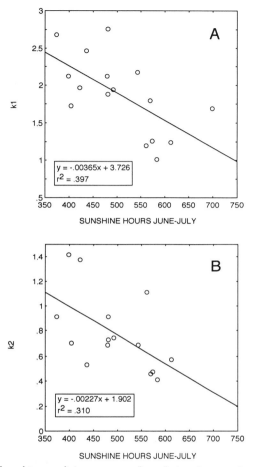

Figure 7. Effects of sunshine conditions on rates of population change in *L. equestris* (expressed as *k*-factors, see text) in (A) summer and (B) the rest of the year. Sunshine hours of June–July show significant negative correlations with both summer and "winter" *k*.

resources are actually available for the insect, this enigma of the apparent lack of food effects can be solved.

Here a digression into the microclimatic conditions of host plant patches and details of patch use by *L. equestris* is needed. The degree of sun exposure of host plant patches varies from totally exposed to totally shaded. *Lygaeus equestris* strongly prefers the sun-exposed patches and the sun-exposed parts of patches with variable exposure. When shading vegetation was removed from a large

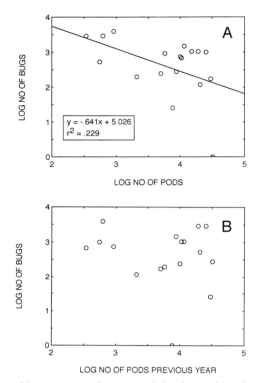

Figure 8. Abundance of *L. equestris* in relation to pod abundance of (A) the present and (B) the previous year. There is a significant *negative* correlation with pod abundance of the present year, but no significant relationship with that of the previous year.

shaded patch, insect densities increased strongly and then declined as shading vegetation recovered (Fig. 9 and Solbreck and Sillén-Tullberg, 1990). Furthermore, insects reach much higher densities in relation to food resources in the sun-exposed patches than in shaded ones (Fig. 10). It is thus evident that resources in shaded positions are not fully exploited by the insects owing to microclimatic constraints on bug activity.

On the basis of these observations, *k*-factors were analyzed in relation to pod production in a sample of totally sun-exposed patches, thus producing an index of food resources at risk. Regression analysis showed no significant correlations with the present year's food resources but a significant effect of pod production for the previous year, and particularly so if these food resources are expressed on a per capita basis (Fig. 11). In a multiple regression both sunshine conditions of the present year and per capita food resources from the previous year enter as significant variables (Table 1). The role of old seeds on the ground was confirmed

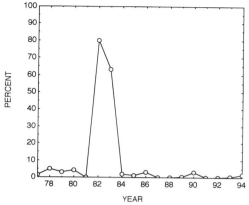

Figure 9. Relative population density of *L. equestris* (August populations as percentage of all 10 patch populations in the area of Bergshamra) in a large shaded patch that was cleared from shading vegetation in the winter 1980/1981, and where shading vegetation grew back over a few years. All populations declined in 1981, and 1982 was the first year allowing population growth following the clearing.

in a field experiment. When old seeds were added in June, parent bug survival increased, emigration decreased, and offspring production by remaining parent bugs increased (Solbreck and Sillén-Tullberg, 1990).

To summarize, food quantity and weather are the two basic factors determining population change of *L. equestris* in Sweden, weather acting both directly on the insect and indirectly through food resources. There are several converging lines of evidence for this conclusion, such as field observations of biology and

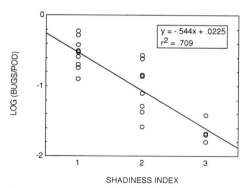

Figure 10. Maximum bug to maximum pod abundance in relation to degree of sun exposure/ shadiness in different patches at Tullgarn, 1977–1994. (1 = sun-exposed, 2 = variable, 3 = shaded patches.)

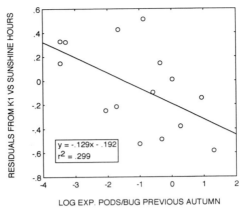

Figure 11. Residual variation in $k1$ (after removing effect of sunshine hours, see Fig. 7A) of *L. equestris* in relation to index of per capita food resources in sun-exposed habitats of the previous year. (Insect density of August the previous year is used to get an insect density measure not used to calculate $k1$).

behavior, correlations between k-factors, food, and weather, and field experiments on the role of food.

VII. General Discussion

A. Habitat Patterns and Insect Distribution

The limits of a species' distribution are set by the occurrence of suitable habitats in the landscape. For many species these limits are difficult to determine,

TABLE 1

Best Multiple Regression of $k1$ versus Weather, Food Resource, and Insect Density Variables[a]

$$k1 = 3.2537 - 0.00311 \text{ (sunshine hours June–July)} - 0.13447$$
$$\text{(sun-exposed pods/bug of the previous year)}$$

First step:	$R^2 = 44.68\%$ (sunshine hours)
Second step:	$R^2 = 62.19\%$

[a]Weather variables included temperature, precipitation, and sunshine hours of June, July, and August in all combinations. Pod densities of present and previous years were expressed as all pods, pods not attacked by *Euphranta connexa*, or pods in sun-exposed patches. Bug densities of August the previous year (to avoid using the same data as were used to calculate $k1$) were used either alone or to calculate ratios of pods per bug. Data from 1977 (no data from previous year), 1988 (zero density), and 1989 (two generations) were omitted from analysis ($N = 15$).

but for *L. equestris* they can now be fairly well defined. As for other phytophagous insects, the ultimate limit of *L. equestris* distribution during the feeding period is determined by host plant distribution. However, to define summer habitats more precisely, microclimatic conditions also have to be considered, because the bugs are concentrated in sun-exposed patches or parts of patches. Winter habitats are microclimatically favorable sites, such as southward-facing rock walls with crevices.

Both summer and winter habitats are small and patchily distributed, but they are stable with regard to size and location (Solbreck and Sillén-Tullberg, 1990; C. Solbreck, unpublished). Obviously several decades or perhaps centuries are needed to substantially change the habitat pattern even in the cultural landscape of Tullgarn. However, habitats are highly variable for the insect in two important aspects. First, habitat suitability changes seasonally forcing the insects to migrate in autumn and spring. Second, pod production varies considerably between years (Solbreck and Sillén-Tullberg, 1986, 1990; C. Solbreck, unpublished).

Flight is essential for *L. equestris* to track its variable environment (Solbreck, 1995). Migratory rates are not constant but vary in response to different environmental factors. For example, flight is possible only during warm days with weak winds (Solbreck, 1976). These weather conditions become rare at the end of the summer. Hence late-developing individuals may never have the opportunity to reach suitable hibernation sites. This is probably one of the reasons why "winter" mortality is affected by summer weather (Fig. 7B).

Flight activity during the breeding season is also affected by food conditions. Females caught in flight during the breeding period have an empty stomach and few eggs. Furthermore, experiments have shown that food shortage causes increased take-off rate (Solbreck, 1985). Thus the migratory rate increases when food is scarce, which aids the insect in tracking its variable food resources.

Populations on patchy resources, like *L. equestris,* are often assumed to be metapopulations. However, *L. equestris* does not have metapopulation structure in the classical sense of the term (Hanski and Gilpin, 1991). There are several reasons for this. For example, the insect alternates between two kinds of habitats during the year, resulting in seasonal interruptions of patch occupancy. Populations in the different plant patches change synchronously (Fig. 3), and rates of colonization and extinction are variable and intercorrelated. Finally, the high migratory potential makes all host plant patches (within T1) into essentially one common resource for the insect population. In the vocabulary of Harrison (1991), this *L. equestris* population is best described as a "patchy population," and can be added to the list of insect populations with poor fit to the classical metapopulation model (Harrison, 1991, 1994).

Although *L. equestris* does not have metapopulation structure, it nevertheless displays interesting spatial dynamics. For example, the role of migration changes with time. Over many years *L. equestris* (in T1) functioned as a "patchy population," with its dynamics well explained by internal processes. This changed to a situation where the system became dependent on immigration from outside areas during some years.

To summarize, patch synchrony is an important aspect of the dynamics within area T1, and this obviously contributed to the large-scale extinction in 1987/1988 (cf. Gilpin, 1990; Harrison and Quinn, 1989). On the other hand, the moderate asynchrony in neighboring habitat archipelagoes is another important aspect. During the population bottleneck, this created refugia that eventually provided the new immigrants.

B. Factors Affecting Population Dynamics

Studies of insect population dynamics have had a strong focus on leaf or phloem feeders or gall-makers on abundant plants. In these kinds of systems, attention has centered on the roles of food quality, natural enemies, and their interactions (e.g., Denno and McClure, 1983; Fritz, 1992; Haukioja, 1993; Karban, 1992; Larsson, 1989; Lawton and McNeill, 1979; Myers, 1988).

Seeds, as food resources for animals, differ in many ways from green plant parts. Seeds tend to be smaller and more nutritious units, they are more mobile, and they have strong temporal and spatial variation in abundance. In many ways seed feeders are more similar to predators than to other phytophages. This is why seed feeders are often called seed predators. Compared with leaf-feeding insects, seed feeders more often destroy a large fraction of their food resources, but attack rates tend to vary considerably as a result of strong fluctuations in seed abundance (Crawley, 1992; Janzen, 1971; Silvertown, 1980). It seems that the dynamics of seed–seed predator systems can often be reduced to a question of how to track food resources in space and time. For example, in the tephritid fly *Euphranta connexa,* a predispersal seed predator on *V. hirundinaria* (and a species that has parasitoids), population change can be almost totally explained by between-year variation in seed pod abundance (Solbreck and Sillén-Tullberg, 1986; C. Solbreck, unpublished).

The *L. equestris* population system is also a case of clear donor control or bottom-up regulation. This seems to be common among seed feeders, and it may be more important for other phytophages than previously thought (e.g., Dempster and Pollard, 1981; Hawkins, 1992; Ohgushi, 1992). However, food dependency was not immediately obvious in *L. equestris.* There was, for example, no positive relationship between food density and insect density in the temporal data (Fig. 8). The difficulty of detecting food dependence was due to microclimate–food interactions and to multiple weather influences, some of which had lags.

The interactions between food resources and microclimate are important for *L. equestris.* Food resources in shaded places are of little use to the insect. Furthermore, these are the sites that produce the largest quantities of seeds, and that have the least between-year variation in these resources (C. Solbreck, un-

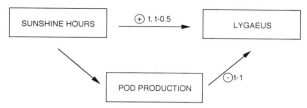

Figure 12. Summary of weather effects on *L. equestris* populations in year *t*. Plus and minus signs indicate the slopes of the regressions.

published). Such interactions between micro- or local climate and resources seem to be fairly common among insects in temperate climates (Begon, 1983; Courtney, 1986; Kingsolver, 1989; Ryrholm, 1988; Thomas, 1993), rendering large potential food resources unavailable to these insects. This microclimate–resource interaction is analogous to the situation for other kinds of plant resources where large parts may be unavailable to insects owing to, for example, poor nutritional quality (White, 1993), nonvigorous growth (Price, 1992), or inadequate plant stress (Larsson, 1989).

The influences of large-scale weather on *L. equestris* can all be traced back to the same factor, namely, summer sunshine conditions. Three effects have been found with different lags and a correlated structure as outlined in Fig. 12. The two direct effects are positively correlated although with different lags, whereas the third and indirect one is negatively correlated with the previous two and with an even longer lag.

The detailed dynamics of this kind of "weather-driven, bottom-up, food-quantity regulated" system is not immediately clear, and several aspects are in need of a theoretical analysis. The correlation structure, for example, seems to create an interesting situation. Suppose there is a sunny summer, which is positive for population growth and survival during summer and winter. Then by necessity the following summer must be "bad" in some respect. If it is sunny the insect population is likely to be limited by the poor seed crop of the previous year (which was sunny), whereas in the opposite case population growth cannot be realized because of negative direct weather influences.

The study of weather effects in population systems has received renewed attention lately, partly due to the problems envisaged in connection with the expected global warming (e.g., Ives and Gilchrist, 1993; Pollard and Rothery, 1994; Royama, 1992). There is a need to look into models that incorporate different kinds of weather effects on insect populations, taking into account various lags and correlated direct and indirect effects.

C. Long-Term Population Change

Population density of *L. equestris* fluctuates in a wavelike pattern with two peaks and two lows during 18 years (Fig. 4A). There is strong autocorrelation with a 1-year lag in these data, but the series is too short to analyze whether fluctuations have a regular period or not.

If one looks at patch occupancy rates, a somewhat different picture emerges. There seems to be a basic pattern (1977–1986 and 1991–1994) with high occupancy. However, the intervening period is different, with unusual weather conditions causing areawide extinction in one year and an extra generation in another year (Solbreck, 1991). During the "unusual period," immigration from outside areas became very important and lasting effects could be traced over several years (Fig. 5).

The period with unusual events illustrates a common and serious problem when one tries to predict extinction probabilities, namely, how to extrapolate from usual conditions to the uncommon extremes. It is also difficult to determine what ordinary conditions are. Many population systems evidently have "reddened spectra" (Pimm and Redfearn, 1988), meaning that there are trends in population development. This also seems to be the case in *L. equestris*. An analysis of weather effects on *L. equestris* suggested that short-term climatic trends, on the scale of decades, have important effects. Obviously population behavior would have been rather different had a study of equal length been performed before 1977 rather than after. For example, the frequency of years resulting in two generations would have been much larger and the production of seed pods would generally have been much lower (Solbreck, 1991). The old picture of populations fluctuating around some constant theoretical equilibrium is gradually being replaced by a view that trends over various time scales are the norm (e.g., Pimm, 1991).

D. Geographical Variation

All field population studies are unique to some degree. Any sequence of years has a particular history. Likewise, any study area, even if it is large, has a specific combination of environmental factors, which are likely to vary geographically. Thus one always has to ask how well a particular field study represents the species in general. It was shown earlier that short-term climatic trends and extreme weather conditions are important to consider for *L. equestris*. But how can the findings from Tullgarn be extrapolated geographically?

Conditions at Tullgarn seem to be typical for the insect over much of its northern range, from Denmark (Bornholm) to the northern limit in the province of Uppland in Sweden. Detailed population studies performed in an area about 50 km north of Tullgarn (Sillén-Tullberg and Solbreck, 1990; Solbreck and

Sillén-Tullberg, 1990; C. Solbreck, unpublished) provide a similar picture of population change, and observations during numerous field trips in Sweden reinforce this view. The main difference is that the host plant is more common with less patchy distribution on the islands of Gotland and Öland in the Baltic.

However, in southern Europe the situation changes. In northern Italy, a similar lygaeid species, *Tropidothorax leucopterus,* is locally common on *V. hirundinaria* and may destroy the plant completely. Hence local competition with this species seems likely (Anderson, 1991).

Farther south, trophic web interactions change drastically. For example, on Sicily, *L. equestris* has other host plants, such as *Nerium oleander,* which it shares with three or four other large lygaeid species. These species are furthermore attacked by an egg parasitoid, egg predators, and two species of tachinid flies. The tachinids often inflict very high mortality among adult bugs (Anderson, 1991; Solbreck *et al.,* 1989). Thus, although our studies seem to be relevant for the situation in northern Europe, they are of little relevance for southern Europe. The importance of competitors and enemies increases as one goes south.

VIII. Conclusions

The population dynamics of the seed-feeding bug *Lygaeus equestris* in Sweden takes place in a landscape with habitat patches determined by food resources, microclimate, and seasonal change. The insect has considerable migratory capacity and is usually able to track spatial variation. However, during a population bottleneck, with numerous empty patches, migratory rates were insufficient for rapid recolonization.

The system is weather-driven with immediate and delayed direct effects, as well as indirect effects through food resource changes. There is bottom-up regulation, but frequent weather disturbance, lagged responses, and correlations between different weather effects result in poor temporal tracking of food resources. Populations fluctuate considerably between years, with periods of several years between population peaks. Some trends in population densities are related to short-term climatic trends. The findings can be extended to large parts of northern Europe, but not to the Mediterranean region, where competitors and enemies seem to be important.

Acknowledgments

I thank Bob Denno, Tony Joern, and Stig Larsson for valuable comments on the manuscript and Rune Axelsson for drawing Fig. 1. This work has been supported by grants from The Swedish Natural Science Research Council.

References

Anderson, D. B. (1991). Seed bugs in trophic webs; Interactions with resources, competitors and enemies. Ph.D. Thesis, Swedish University of Agricultural Sciences, Uppsala.

Baltensweiler, W. (1993). Why the larch bud-moth cycle collapsed in the subalpine larch-cembrian pine forest in the year 1990 for the first time since 1850. *Oecologia* **94**, 62 –66.

Begon, M. (1983). Grasshopper populations and weather: The effects of insolation on *Chorthippus brunneus*. *Ecol. Entomol.* **8**, 361–370.

Courtney, S. P. (1986). The ecology of pierid butterflies: Dynamics and interaction. *Adv. Ecol. Res.* **15**, 51–131.

Crawley, M. (1992). Seed predators and plant population dynamics. *In* "Seeds: The Ecology of Regeneration in Plant Communities" (M. Fenner, ed.), pp. 157–191. CAB International, Wallingford.

Danell, K., and Ericson, L. (1990). Dynamic relations between the antler moth and meadow vegetation in northern Sweden. *Ecology,* **71**, 1068–1077.

Dempster, J. P., and Pollard, E. (1981). Fluctuations in resource availability and insect populations. *Oecologia* **50**, 412–416.

Denno, R. F., and McClure, M. S. (1983). "Variable Plants and Herbivores in Natural and Managed Systems" Academic Press, New York.

Fritz, R. S. (1992). Community structure and species interactions of phytophagous insects on resistant and susceptible host plants. *In* "Plant Resistance to Herbivores and Pathogens: Ecology, Evolution, and Genetics" (R. S. Fritz and E. L. Simms, eds.), pp. 240–277. Univ. of Chicago Press, Chicago.

Gilpin, M. E. (1990). Extinction of finite metapopulations in correlated environments. *In* "Living in a Patchy Environment" (B. Shorrocks and I. R. Swingland, eds.), pp. 177–186. Oxford Univ. Press, Oxford.

Gilpin, M. E., and Hanski, I. (1991). "Metapopulation Dynamics: Empirical and Theoretical Investigations." Academic Press, London.

Hanski, I. (1994). Patch-occupancy dynamics in fragmented landscapes. *Trends Ecol. Evol.* **9**, 131–135.

Hanski, I., and Gilpin, M. E. (1991). Metapopulation dynamics: Brief history and conceptual domain. *In* "Metapopulation Dynamics: Empirical and Theoretical Investigations" (M. E. Gilpin and I. Hanski, eds.), pp. 3–16. Academic Press, London.

Hanski, I., and Thomas, C. D. (1994). Metapopulation dynamics and conservation: A spatially explicit model applied to butterflies. *Biol. Conserv.* **68**, 167–180.

Hansson, L., Söderström, L., and Solbreck, C. (1992). The ecology of dispersal in relation to conservation. *In* "Ecological Principles of Nature Conservation. Applications in Temperate and Boreal Environments." (L. Hansson, ed.), pp. 162–200. Elsevier, London.

Harrison, S. (1991). Local extinction in a metapopulation context: An empirical evaluation. *Biol. J. Linn. Soc.* **42**, 73–88.

Harrison, S. (1994). Metapopulations and conservation. *In* "Large-Scale Ecology and Conservation Biology" (P. J. Edwards, R. M. May, and N. R. Webb, eds.), pp. 111–128. Blackwell, Oxford.

Harrison, S., and Quinn, J. F. (1989). Correlated environments and the persistence of metapopulations. *Oikos*, **56**, 293–298.

Haukioja, E. (1993). Effects of food and predation on population dynamics. *In* "Caterpillars: Ecological and Evolutionary Constraints on Foraging" (N. E. Stamp and T. M. Casey, eds.), pp. 425–447. Chapman & Hall, New York.

Hawkins, B. A. (1992). Parasitoid–host food webs and donor control. *Oikos*, **65**, 159–162.

Ives, A. R., and Gilchrist, G. (1993). Climate change and ecological interactions. *In* "Biotic Interactions and Global Change" (P. M. Kareiva, J. G. Kingsolver, and R. B. Huey, eds.), pp. 120–146. Sinauer Assoc., Sunderland, MA.

Janzen, D. (1971). Seed predation by animals. *Annu. Rev. Ecol. Syst.* **2**, 465–492.

Karban, R. (1992). Plant variation: Its effects on populations of herbivorous insects. *In* Plant Resistance to Herbivores and Pathogens: Ecology, Evolution, and Genetics" (R. S. Fritz and E. L. Simms, eds.), pp. 195–215. Univ. of Chicago Press, Chicago.

Kareiva, P. M., Kingsolver, J. G., and Huey, R. B., eds. (1993). "Biotic Interactions and Global Change." Sinauer Assoc., Sunderland, MA.

Kingsolver, J. G. (1989). Weather and the population dynamics of insects: Integrating physiological and population ecology. *Physiol. Zool.* **62**, 314–334.

Krebs, C. J. (1991). The experimental paradigm and long-term population studies. *Ibis* **133**, Suppl. 1, 3–8.

Kugelberg, O. (1973a). Larval development of *Lygaeus equestris* (Heteroptera, Lygaeidae) on different natural foods. *Entomol. Exp. Appl.* **16**, 165–177.

Kugelberg, O. (1973b). Laboratory studies on the effect of different natural foods on the reproductive biology of *Lygaeus equestris* (L.) (Het. Lygaeidae). *Entomol. Scand.* **4**, 181–190.

Kugelberg, O. (1974). Laboratory studies on the feeding preference and feeding behaviour in *Lygaeus equestris* (L.) (Het. Lygaeidae). *Entomol. Scand.* **5**, 49–55.

Larsson, S. (1989). Stressful times for the plant stress—Insect performance hypothesis. *Oikos* **56**, 277–283.

Lawton, J. H., and McNeill, S. (1979). Between the devil and the deep blue sea: On the problem of being a herbivore. *In* "Population Dynamics" (R. M. Anderson, B. D. Turner, and L. R. Taylor, eds.), pp. 223–244. Blackwell, Oxford.

Levin, S. A. (1992). The problem of pattern and scale in ecology. *Ecology* **73**, 1943–1967.

May, R. M. (1994). The effects of spatial scale on ecological questions and answers. *In* "Large-Scale Ecology and Conservation Biology" (P. J. Edwards, R. M. May, and N. R. Webb, eds.), pp. 1–17. Blackwell, Oxford.

Myers, J. H. (1988). Can a general hypothesis explain population cycles of forest Lepidoptera? *Adv. Ecol. Res.* **18**, 179–242.

Ohgushi, T. (1992). Resource limitation on insect herbivore populations. *In* "Effects of Resource Distribution on Animal Plant Interactions" (M. D. Hunter, T. Ohgushi, and P. Price, eds.) pp. 199–241. Academic Press, San Diego, CA.

Pimm, S. L. (1991). "The Balance of Nature? Ecological Issues in the Conservation of Species and Communities." Univ. of Chicago Press, Chicago and London.

Pimm, S. L., and Redfearn, A. (1988). The variability of population densities. *Nature (London)* **344**, 613–614.

Pollard, E., and Rothery, P. (1994). A simple stochastic model of resource-limited insect populations. *Oikos,* **69,** 287–294.

Price, P. W. (1992). Plant resources as the mechanistic basis for insect herbivore population dynamics. *In* "Effects of Resource Distribution on Animal Plant Interactions" (M. D. Hunter, T. Ohgushi, and P. Price, eds.) pp. 139–173. Academic Press, San Diego, CA.

Pulliam, H. R. (1988). Sources, sinks, and population regulation. *Am. Nat.* **132,** 652–661.

Royama, T. (1984). Population dynamics of the spruce budworm *Choristoneura fumiferana. Ecol. Monogr.* **51,** 473–493.

Royama, T. (1992). "Analytical Population Dynamics." Chapman & Hall, London.

Ryrholm, N. (1988). An extralimital population in a warm climatic outpost: The case of the moth *Idaea dilutaria* in Scandinavia. *Int. J. Biometeorol.* **32,** 205–216.

Sillén-Tullberg, B. (1985). Higher survival of an aposematic than of a cryptic form of a distasteful bug. *Oecologia* **67,** 411–415.

Sillén-Tullberg, B., and Solbreck, C. (1990). Population dynamics of a seed feeding bug, *Lygaeus equestris.* 2. Temporal dynamics. *Oikos,* **58,** 210–218.

Sillén-Tullberg, B., Wiklund, C., and Järvi, T. (1982). Aposematic coloration in adults and larvae of *Lygaeus equestris* and its bearing on Müllerian mimicry: An experimental study on predation on living bugs by the great tit *Parus major. Oikos* **39,** 131–136.

Silvertown, J. W. (1980). The evolutionary ecology of mast seeding in trees. *Biol. J. Linn. Soc.* **14,** 235–250.

Solbreck, C. (1972). Sexual cycle and changes in feeding activity and fat body size in relation to migration in *Lygaeus equestris* (L.) (Het., Lygaeidae). *Entomol. Scand.* **3,** 267–274.

Solbreck, C. (1976). Flight patterns of *Lygaeus equestris* (Heteroptera) in spring and autumn with special reference to the influence of weather. *Oikos,* **27,** 134–143.

Solbreck, C. (1985). Insect migration strategies and population dynamics. *Contrib. Mar. Sci., Suppl.* **27,** 641–662.

Solbreck, C. (1991). Unusual weather and insect population dynamics: *Lygaeus equestris* during an extinction and recovery period. *Oikos* **60,** 343–350.

Solbreck, C. (1993). Predicting insect faunal dynamics in a changing climate—A northern European perspective. *In* "Impacts of Climatic Change on Natural Ecosystems, with Emphasis on Boreal and Arctic/Alpine Areas" (J. J. Holten, G. Paulsen, and W. C. Oechel, eds.), pp. 176–185. Norwegian Institute for Nature Research, Trondheim.

Solbreck, C. (1995). Variable fortunes in a patchy landscape—The habitat templet of an insect migrant. *Res. Popul. Ecol.* (in print).

Solbreck, C., and Kugelberg, O. (1972). Field observations on the seasonal occurrence of *Lygaeus equestris* (L.) (Het., Lygaeidae) with special reference to food plant phenology. *Entomol. Scand.* **3,** 189–210.

Solbreck, C., and Sillén-Tullberg, B. (1981). Control of diapause in a "monovoltine" insect, *Lygaeus equestris* (Heteroptera). *Oikos* **36,** 68–74.

Solbreck, C., and Sillén-Tullberg, B. (1986). Seed production and seed predation in a patchy and time-varying environment. Dynamics of a milkweed–tephritid fly system. *Oecologia* **71**, 51–58.

Solbreck, C., and Sillén-Tullberg, B. (1990). Population dynamics of a seed feeding bug, *Lygaeus equestris*. 1. Habitat patch structure and spatial dynamics. *Oikos* **58**, 199–209.

Solbreck, C., Olsson, R., Anderson, D. B., and Förare, J. (1989). Size, life history and responses to food shortage in two geographical strains of a seed bug, *Lygaeus equestris*. *Oikos* **55**, 387–396.

Swetnam, T. W., and Lynch, A. M. (1993). Multicentury, regional-scale patterns of western spruce budworm outbreaks. *Ecol. Monogr.* **63**, 399–424.

Taylor, L. R. (1991). Proper studies and the art of the soluble. *Ibis* **133**, Suppl. 1, 9–23.

Thomas, J. A. (1993). Holocene climate changes and warm man-made refugia may explain why a sixth of British butterflies possess unnatural early-successional habitats. *Ecography* **16**, 278–284.

Turchin, P. (1990). Rarity of density dependence or population regulation with lags? *Nature (London)* **344**, 660–663.

White, T. C. R. (1993). "The Inadequate Environment: Nitrogen and the Abundance of Animals." Springer-Verlag, Berlin.

Wiens, J. A. (1989). Spatial scaling in ecology. *Funct. Ecol.* **3**, 385–397.

Chapter 15

Adaptive Behavior Produces Stability in Herbivorous Lady Beetle Populations

Takayuki Ohgushi

I. Introduction

A central issue in ecology is to understand the processes that govern the dynamics of populations. Though intensive studies on insect populations have undoubtedly made a major contribution to the contemporary theories in population ecology, underlying mechanisms of population regulation remain poorly understood in insect herbivores (Barbosa and Schultz, 1987; Cappuccino, 1992).

For a more exact understanding of population dynamics in insect herbivores, we need to know much more about bottom-up influences on a trophic level (Hunter et al., 1992). In contrast to the emphasis on the role of natural enemies on population regulation (Varley et al., 1973; Southwood and Comins, 1976; Hassell, 1985), the traditional argument has paid little attention to possible effects of plant resources on the dynamics of insect herbivores. However, studies concerning herbivorous insects have illustrated temporal and/or spatial resource tracking at the population level, thereby emphasizing the bottom-up influences of host plants (Mattson, 1980; Dempster and Pollard, 1981; Ohgushi and Sawada, 1985a; Solbreck and Sillén-Tullberg, 1986).

The importance of host plant characteristics as regulatory agents for population stability supports the view that resource-use traits of insects may play a significant role in determining survival and/or reproductive processes, allowing populations to respond to spatial and temporal heterogeneity of plant resources. Population ecologists, however, have long focused on the demographic consequences of environmental or biological changes, with little attention to behavioral or physiological mechanisms that may underlie changes in population dynamics.

This chapter addresses the role of adaptive life-history traits in resource use, providing a mechanistic basis for population regulation in herbivorous insects. To illustrate this mechanistic approach, I shall demonstrate how behavioral and

POPULATION DYNAMICS

physiological responses of adult females of the herbivorous lady beetle *Epilachna niponica* to host plant conditions are important in generating the fundamental patterns of population dynamics, especially population stabilization.

II. A Search for Causal Mechanisms

Although population studies have long searched for density-dependent processes to demonstrate population regulation (Dempster, 1983; Stiling, 1988; Hassell *et al.*, 1989), such phenomenological analyses at the population level will not reveal the underlying factors generating population stability. The mechanistic approach to understanding population regulation requires more attention to individual differences within a population, because important causes of population regulation may be overlooked easily using data on mean populations per generation (Hassell, 1986, 1987). Theoretical studies have emphasized individual differences in behavior and physiology as an essential component of population stability and equilibrium density (Hassell and May, 1985; Łomnicki, 1988; Nisbet *et al.*, 1989).

III. Preference–Performance Linkage: A Key to Understanding Population Dynamics in Insect Herbivores

There is growing circumstantial evidence to support the view that resource-use tactics of adult insects are critical in the population dynamics of herbivorous insects (Price, 1990, 1994; Price *et al.*, 1990). For example, density-dependent processes in many insect herbivores have often been identified in the adult stage (Dempster, 1983; Stiling, 1988), suggesting the important role of adult behavior relevant to reproduction in determining population stability. Furthermore, a number of studies testing Root's (1973) resource concentration hypothesis have suggested that the searching behavior of adult insects for favorable resources is essential in determining subsequent population densities in different vegetation textures (Kareiva, 1983; Stanton, 1983). In particular, oviposition behavior has been considered to generate the fundamental patterns of population dynamics in insect herbivores, by affecting offspring performance in terms of survival and reproduction (Preszler and Price, 1988; Craig *et al.*, 1989; Price, 1990, 1994; Price *et al.*, 1990; Ohgushi, 1992). For many herbivorous insects, the searching abilities of larvae are poor relative to those of adults; oviposition behavior is therefore of paramount importance in the process of selecting suitable host plants or plant parts for their offspring. Since natural selection favors individuals that have higher lifetime reproductive success, we can expect a strong evolutionary link between female oviposition preference and offspring performance. Other studies on the host plant or habitat selection have revealed a positive correlation between oviposition preference and relative performance of offspring (see Thompson, 1988, for a review). Some herbivorous insects, however, show poor preference–performance relationships (Karban and Courtney, 1987; Auerbach and Simberloff, 1989; Valladares and Lawton, 1991), because of impacts of natural enemies independent of plant quality, and relative shortage of suitable plants or plant parts (Thompson, 1988).

It should be noted that evaluation of the relationship between oviposition preference and offspring performance is still useful to understand population dynamics in those species that do not exhibit apparent adult preference (Price *et al.*, 1990). This is because we can remove oviposition preference from the process of resource choice in our analyses, and thus concentrate on resource-use tactics in the immature stage, such as habitat selection or foraging behavior of larvae (Schultz, 1983; Hunter, 1990). Furthermore, we should turn our attention to the effects of natural enemies and host plant characteristics on survival processes in immature insects, independent of oviposition preference. In other words, evaluation of preference–performance linkage will clarify the relative importance of these mortality agents affecting offspring performance.

Because the traditional life table approach has focused mainly on mortality agents relevant to egg and larval performance, effects of host plant selection by adult females on insect population dynamics have long been ignored. Here, I emphasize female oviposition preference, not because resource-use tactics are the most important agents determining population stability in all herbivorous insects, but because we know little about their population consequences in insect herbivores. The mechanistic approach to fully understanding the population dynamics of herbivorous insects therefore needs an evolutionary perspective based on insect–plant interactions, which will be achieved through the reduction of population ecology to individual ecology. This approach focuses on the behavioral and physiological mechanisms related to the selection of host plants or plant parts, which may underlie the population dynamics of herbivorous insects. To illustrate this mechanistic approach, I show how populations of the herbivorous lady beetle *Epilachna niponica* are regulated by oviposition tactics, based on a 5-year population study in several local populations in central Japan.

IV. Case Study: The Thistle Lady Beetle

The lady beetle *Epilachna niponica* is a univoltine species and a specialist herbivore of thistle plants (Fig. 1). Overwintering adult females emerge in early May and lay eggs in clusters on thistle leaves. After passing through four instars, new adults emerge from early July to early September, feeding on thistle leaves

Figure 1. *Epilachna niponica* on the thistle plant *Cirsium kagamontanum*. Illustration by Akiko Fukui.

through autumn. Then they enter hibernation by early November. The host plant *Cirsium kagamontanum* is a perennial herb, patchily distributed along the stream side. It grows rapidly from sprouting in late April to late June, reaching 1.5–1.8 m in height by late August, and begins flowering in mid-August. Old leaves begin to wither after summer. Here, I will show the results from two study sites, A and F. Site A (60 × 30 m) was situated at 220 m elevation on an accumulation of sandy deposits. Floods caused by heavy rainfall have often submerged and washed away the ground flora along the watercourse. On the other hand, site F (90 × 15 m) was situated at 350 m elevation, about 10 km upstream from site A. The more hardened soil deposits at this site mean that most grasses and shrubs can successfully escape serious flood damage.

The two populations were censused at 1- to 3-day intervals from early May to early November in each year from 1976 to 1980. All thistle plants in the study sites were examined; the numbers of eggs, fourth-instar larvae, pupae, and adult beetles were recorded separately for each plant. Adult beetles were individually marked with four small dots of lacquer paint on their elytra. A total of 5969 and 3507 beetles were marked at sites A and F, respectively. The numbers and daily survivals of overwintering and new adults were estimated by the Jolly–Seber stochastic model based on the mark and recapture data (Jolly, 1965; Seber, 1973). The estimate of numbers recruited and daily survival for adult beetle populations was highly reliable, because of an exceptionally high marking and recapture ratio (Ohgushi and Sawada, 1981).

A. Population Stability at Low-Density Level

We first look at annual changes in host plant abundance (Fig. 2a). Resource abundance varied independently at the two sites. Shoot numbers at site A consis-

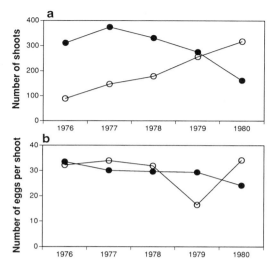

Figure 2. Annual changes in host plant and lady beetle populations over a 5-year study period at site A (open circle) and site F (solid circle). (a) Shoot numbers of thistle plants; (b) egg density (number of eggs per shoot) of the lady beetle.

tently increased up to 1980, whereas at site F they remained rather constant and then dropped in 1980 due to two large floods in the previous autumn. In spite of these two very different patterns of change in host abundance, the beetle populations coped well with the variation in food resource, exhibiting a marked temporal resource tracking (Fig. 2b). To illustrate the stability of the two populations, I calculated the standard deviation of log-transformed egg densities, which is an appropriate index for temporal variability of a population (Pimm, 1991). The index of site A and site F was 0.135 and 0.051, respectively, both of which indicated exceptionally stable populations compared to herbivorous insects previously studied (Connell and Sousa, 1983; Redfearn and Pimm, 1988; Hanski, 1990). Indeed, annual variation in egg population was mostly determined by variation in resource abundance. Host abundance explains 66 and 98% of variation in egg populations at site A and site F, respectively (Ohgushi and Sawada, 1985a).

A large-scale flood in late June in 1979 at site A washed away all the reproductive females, resulting in a considerable reduction in egg density. In spite of the large population decline, the egg population in 1980 quickly returned to the previous level of density (Fig. 2b). Thus, the "natural experiment" evidently demonstrated that the beetle population has an effective regulatory mechanism responsible for the population stability.

Note that average egg density at site A [29.8 ± 3.34 (mean ±SE) eggs per shoot] was almost identical to that at site F (29.4 ± 1.49) for the 5-year study period. This suggests that the two different populations have the same regulatory mechanism for setting a population at a certain level in relation to resource abundance. This "equilibrium" density was, however, far below the "ceiling" at which host plant defoliation occurs (Dempster and Pollard, 1981). Indeed, leaf herbivory still remained low in late June when fourth-instar larvae reached peak abundance, approximately 30% of the total leaf area at site A and less than 20% at site F (Ohgushi, 1992).

Annual variability of population densities among life stages revealed a large decline from reproductive adult to egg stage at both sites (Ohgushi, 1992). This indicates that the population density was highly stabilized during the reproductive process. In contrast, destabilization of population density tended to proceed from egg to new adult stage.

B. Density-Dependent Processes

Although there was no density-dependent survival from egg to overwintered adult stage (with the exception of larval survival at site A), the reproductive process involved two density-dependent relationships. First, lifetime fecundity negatively correlated with adult density (site A: $r = -.90$, $F = 12.49$, $P = .0385$; site F: $r = -.98$, $F = 61.00$, $P = .0044$). Second, female survival in a 10-day period sharply decreased with adult density (site A: $r = -.98$, $F = 59.27$, $P = .0046$; site F: $r = -.97$, $F = 46.65$, $P = .0064$). These negative density-dependent reproductive processes suggest that population stabilization was attained during the reproductive season. However, this analysis tells little about when and how the density-dependent processes operated within a generation. In the following, I discuss the timing of density-dependent processes operating during the reproductive season and explain how population stabilization was achieved.

C. How Population Stabilization Proceeds throughout the Reproductive Season

I examined the dependence of female survival and oviposition rate (number of eggs laid per female) on cumulative egg density in each 10-day period in May and June. In May, a significant density-dependent reduction in female survival was detected, but this was not the case for oviposition rate. However, density-dependent female survival was no longer detected in June, although oviposition rate became a significantly negative function of egg density.

I compared annual variability of cumulative egg density, expressed by the standard deviation of the log-transformed egg densities for 1976–1980, throughout the reproductive season (Fig. 3a). Annual variability in egg density sharply

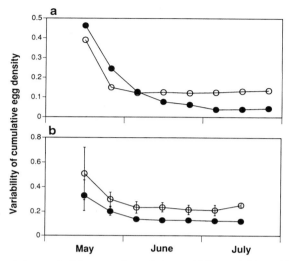

Figure 3. Seasonal changes in (a) temporal population variability and (b) spatial population variability at site A (open circle) and site F (solid circle). Variability is expressed by the standard deviation of the log-transformed egg densities. Temporal population variability was calculated for 1976 to 1980. Spatial variability was calculated at the scale of individual thistle plants; mean spatial variability (±SE) over the 5 years is shown.

declined from mid-May to early June, implying that population stabilization rapidly advanced early in the reproductive season. Also, further stabilization was evident at site F over the rest of the season.

Likewise, spatial stabilization of egg density at the scale of individual plants proceeded early in the reproductive season (Fig. 3b). Variability of cumulative egg densities among plants sharply declined from mid-May to early June. As will be shown later, the spatial stabilization in egg density is chiefly brought about by movement of females searching for suitable oviposition sites. The seasonal increase in density stabilization among plants also suggests that females tended to avoid oviposition on plants bearing more eggs.

Consequently, density-dependent female loss, including death and emigration outside the plots, coupled with subsequent density-dependent reduction in oviposition rate, undoubtedly contributed to the spatial and temporal stabilization in egg densities.

D. Oviposition Traits as Mechanistic Bases
for Population Stability

The next problem is to search for the underlying mechanisms generating the remarkable population stability that is maintained at such a low density. To do this, we need to look at properties at the individual level: life-history traits for resource use in the reproductive season when the population stabilization was actually attained.

1. Movement of Females Searching for Suitable Oviposition Sites

Since adult females determine egg distribution among thistle plants, oviposition behavior while searching for suitable oviposition sites is likely to play an important role in spatial population stabilization. In this context, I focus on movement activity in ovipositing females in terms of choice of an oviposition site.

Female mobility, expressed by the mean squared displacement per day of marked adult females, increased with adult density (site A: $r = .80$, $F = 5.38$, $P = .1029$; site F: $r = .96$, $F = 34.98$, $P = .0097$). Because females exhaust their energy supply by frequent movement, or disperse outside the sites in a density-dependent manner, female survival is likely to be lower in years of high egg density. In addition, among-plant movement reached a peak from mid-May to mid-June, and decreased consistently thereafter, suggesting that choice of a suitable oviposition site frequently occurred early in the reproductive season. Note that the enhanced movement of females early in the season was well synchronized with the spatial stabilization of egg densities among different plants (see Fig. 3b), indicating that spatial density stabilization is likely to be brought about by the frequent movement of females among thistle plants. Indeed, female mobility was highly correlated with cumulative egg density in early June, when mobility reached a peak (site A: $r = .90$, $F = 13.26$, $P = .0357$; site F: $r = .98$, $F = 65.16$, $P = .0040$). Hence, female movement during oviposition was more pronounced when egg density was high early in the season, which may partly cause the observed density-dependent reduction in oviposition rate.

2. Egg Resorption in Response to Resource Deterioration

Next, I conducted a field experiment to detect effects of host plant deterioration on egg-laying behavior. Eight plants of similar size and leaf numbers were individually covered with a nylon cage with a metal frame large enough to allow further foliage growth. One pair of adults was introduced into each cage for oviposition, which was followed until death of the female. Five cages (cages A to E) were retained untreated as controls; three cages were designed as follows. In cages F and G, each female was transferred to a new cage with an undamaged thistle 2–3 weeks after the female ended oviposition. In cage H, two nymphs of an earwig, *Anechura harmandi*, the principal egg predator, were initially introduced to remove beetle eggs laid during the experiment. Although leaf damage was around 50% consumption of the total leaf area, the females stopped laying eggs around mid-June in every cage except cage H (Fig. 4). Females in cages F and G resumed oviposition in a short time when transferred to an undamaged

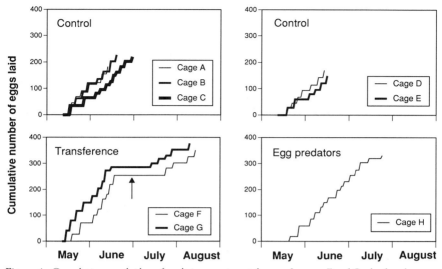

Figure 4. Cumulative eggs laid per female in experimental cages. In cages F and G, the females were transferred to another cage with an undamaged thistle at the date indicated by the vertical arrow. Cage H had two egg predators. See text for more detail. Data from Ohgushi and Sawada (1985a).

thistle, and continued oviposition until early August. In cage H, leaf damage remained less than 20% throughout the experiment because of egg removal by the earwigs. Oviposition thus continued up to late July. As a result, lifetime fecundity of the females in control cages was 215.8 ± 28.9 eggs (mean ±SE), significantly lower than that in cages free from host deterioration (366.0 ± 10.0 eggs) (Mann–Whitney *U*-test, *U* = 15, *P* = .0253).

Since females of *Epilachna* beetles resorb developing eggs in the ovary under starved conditions (Kurihara, 1975), I examined whether *E. niponica* females resorb eggs in response to host deterioration, using six caged plants with different adult numbers (Ohgushi and Sawada, 1985a). The dissected females showed that egg resorption evidently occurred in cages with high beetle density when most thistle leaves were highly exploited. On the other hand, no females resorbed eggs in cages with one pair of beetles, where 80% of the total leaf area remained intact.

The cage experiments highlighted a physiological response of ovipositing females to host deterioration: egg resorption. The process of resorption is reversible; when the host plant becomes favorable, the resorption immediately ceases and the ovaries again become productive. Also, dissection of females sampled from the natural populations confirmed that females having resorbed eggs increased rapidly after mid-June when leaf damage was much lower than that in the experiments (Ohgushi, 1992). This implies that egg resorption could contribute to the density-dependent reduction in oviposition rate in June.

Consequently, oviposition tactics of females produce a high level of stability of the lady beetle populations. Active interplant movement of females for a suitable oviposition site early in the reproductive season results in spatial stabilization in egg density. The increased movement may also reduce female survival and enhance dispersal in a density-dependent manner. Later in the season, females often resorb eggs in response to host deterioration and stop egg-laying. The combined effect of these tactics strongly reduces the reproductive rate in a density-dependent way, which in turn causes a temporally stable population maintained at a relatively low density.

E. Top-Down Influences of Natural Enemies

Since past arguments on insect population dynamics have emphasized the role of natural enemies as a regulatory agent for population stability, I summarize here the top-down influence of natural enemies in a general picture of population dynamics of the lady beetle. Although eggs and larvae were frequently subjected to large arthropod predation, principally by nymphs of the earwig *Anechura harmandi* (Ohgushi and Sawada, 1985b; Ohgushi, 1988), the role of natural enemies as a regulatory agent is unlikely. The significantly higher immature mortality caused by predation at site F resulted in a significantly lower density of new adults, compared to site A with less predation (Ohgushi and Sawada, 1985b). In spite of the large difference in adult density between the two sites, egg densities of the two populations were almost identical (see Fig. 2b). In other words, the large difference in new adult density was mostly compensated for by the regulatory process in the reproductive season. Besides, any impact of natural enemies was not found in the oviposition process, by which the population stabilization was achieved. Thus, natural enemies do not change the general pattern of population dynamics of the lady beetle.

It is concluded that the lady beetle populations have a strong bottom-up regulation, through behavioral and physiological responses of ovipositing females to spatial and seasonal deterioration of the host plant.

F. Oviposition Tactics Improve a Female's
Reproductive Success

The next problem is whether these oviposition traits really improve a female's lifetime reproductive success. To determine this, I calculated offspring lifetime fitness to evaluate the relative contribution of these traits to the lifetime reproductive success of a female. Lifetime fitness of offspring is expressed as the

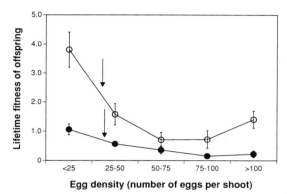

Figure 5. Lifetime fitness of offspring born on thistle plants with different egg densities. Each point represents mean ±SE for 1976 to 1980. Vertical arrows show the average population densities at site A (open circle) and site F (solid circle). The lifetime fitness of offspring grown on the *i*-th plant (F_i) is defined as follows: $F_i = E_i \times L_i \times A_i \times R_i$, where E_i = egg survival to hatching, L_i = larval survival from egg hatching to adult emergence, A_i = survival of adult females from emergence to the reproductive season in the following year, and R_i = lifetime fecundity of females estimated by reproductive life span.

expected total number of eggs produced in the following generation in terms of the expected reproductive contribution of one egg at the moment of birth.

1. Choice of Oviposition Site with Higher Offspring Fitness

The increasing stabilization in egg densities among plants over the reproductive season suggests that females tended to avoid oviposition on plants bearing more eggs. Is the egg-laying behavior an adaptation to improve the lifetime reproductive success of a female? I compared lifetime fitness of offspring grown on individual plants with different egg densities (Fig. 5). Offspring fitness declined with egg density, implying that oviposition on plants with low egg densities increases the female's reproductive success. It should also be noted that the declining tendency was much more pronounced at low egg densities, corresponding well to the equilibrium densities observed at site A and site F.

Decreasing lifetime fitness as a function of egg density involved density-dependent larval and adult mortality up to the reproductive age. Although arthropod predation caused large larval losses, it did not operate in a spatially density-dependent manner (Ohgushi, 1988). Since damage by beetle feeding reduces leaf quality in terms of amino acid and water content (Ohgushi, 1986), deterioration in leaf quality, dependent on larval density, is likely to cause the spatially density-dependent mortality. Also, thistles with high egg densities tended to produce small-sized adults (Ohgushi, 1987), which had a higher overwintering mortality.

2. Choice of Oviposition Time with Higher Offspring Fitness

Since females that resorb eggs stop oviposition, this physiological trait will not increase lifetime fitness unless the resorbed females are able to oviposit at a future time. Future oviposition should be considered as within-season and between-season oviposition. The former means a resumption of oviposition in the same season. This is more likely to occur because egg resorption is a reversible process. The latter concerns oviposition in the next reproductive season.

As already shown in the cage experiment, females often resorb eggs when host plants are highly exploited. Other disturbances to the host plant may also cause females to resorb eggs. For example, egg resorption occurred at site F in 1979 when a large June flood washed away some thistle plants and buried others with soil (Ohgushi, 1992). Following this disturbance, females stopped laying eggs for a half-month period. When the damaged thistles recovered and reflushed with new leaves in mid-July, females resumed oviposition. This suggests that females refrained from reproduction during the period unfavorable to larval survivorship due to food shortage, and became reproductive again when off-spring fitness was no longer reduced by the habitat disturbance.

Let us consider, then, the possibility of second-year reproduction by females that resorb eggs. Since almost all reproductive females at site A died by late July, I examined this possibility in population F. Of 56 females alive in mid-August, nearly 40% survived until the next spring (Ohgushi, 1992). Also, some adults were observed ovipositing in the second reproductive season, indicating that long-lived females have a relatively high probability of oviposition in the second reproductive season. Lifetime fitness of offspring at site F was remarkably depressed after mid-July (Ohgushi, 1991). In particular, no adults successfully emerged in the late cohorts, which were born in August. Females will have higher reproductive success if they stop egg-laying late in the season and postpone oviposition until the second year. In this context, female survival was enhanced following egg resorption, suggesting that there is a reproductive trade-off (T. Ohgushi, unpublished). It is thus probable that egg resorption is an adaptive response to improve lifetime reproductive success of long-lived females.

In summary, analyses based on lifetime fitness of offspring suggested that the behavioral and physiological responses of ovipositing females are adaptive to improve their reproductive success. Because of spatially density-dependent reduction in offspring fitness, avoidance of egg-laying on plants with high egg densities increases a female's reproductive success. Egg resorption may be adaptive when the females benefit by postponing reproduction. This is the case when habitat disturbance such as a large flood occurs in the reproductive season. Offspring fitness becomes extremely low owing to shortage of food available to newly hatched larvae. In addition, long-lived females refrain from laying eggs late in the season when offspring fitness is exceptionally low. Egg resorption may enhance subsequent survival of the resorbing female and her reproductive contribution to the following year. In either case, egg resorption by a female is an adaptive response that avoids unfavorable times for her offspring and postpones egg-laying to a future time with higher offspring fitness.

V. Future Directions

The population study of the lady beetle clearly demonstrated how the oviposition tactics of adult females are important determinants of the beetle's population dynamics.

Dempster (1983) and Stiling (1988) pointed out that in many herbivorous insects, density-dependent processes have been detected in the adult stage, and include factors such as adult dispersal and fecundity reduction. As shown in the lady beetle, these agents may involve behavioral or physiological responses of ovipositing females to host plant conditions. Because conventional life table analysis has not been designed to reveal the exact role of behavior and/or physiology of adult insects, we know little about how oviposition preference affects population dynamics of herbivorous insects (Preszler and Price, 1988; Price, 1990, 1994; Price *et al.*, 1990; Ohgushi, 1992). Other studies have, however, suggested the role of density-dependent oviposition as a regulatory agent generating population stability in herbivorous insects (Craig *et al.*, 1989; Romstöck-Völkl, 1990).

In the context of interactions among three trophic levels, both natural enemies and host plant have been addressed as selective forces for resource-use tactics by herbivorous insects (Price *et al.* 1980; Price, 1987). In the case of the lady beetle, seasonal pattern of changes in predatory pressure and host plant deterioration largely determined lifetime fitness of offspring on different plants or in different cohorts (Ohgushi, 1991). Thus, both enemies and host plants may provide selective forces on oviposition preference, as expressed by differences in among-plant movement during the search for oviposition sites and by egg resorption. However, the traditional argument of insect population dynamics has focused on the direct effects, or killing power, of these biological agents. The mechanistic approach I addressed here also requires paying much attention to indirect effects of natural enemies and host plant on population dynamics, as selective forces for resource-use tactics in herbivorous insects. This approach will undoubtedly provide new insights into the evolutionary implications of insect population dynamics.

Acknowledgments

I thank Peter Price and Naomi Cappuccino for reviewing this paper. Financial support was provided by a Japan Ministry of Education, Science and Culture Grant-in-Aid for General Scientific Research (#04640616 and #06640807) and for Scientific Research on Priority Areas (#319). I also thank Akiko Fukui for her illustration of *Epilachna niponica*.

References

Auerbach, M., and Simberloff, D. (1989). Oviposition site preference and larval mortality in a leaf-mining moth. *Ecol. Entomol.* **14,** 131–140.

Barbosa, P., and Schultz, J. C., eds. (1987). "Insect Outbreaks." Academic Press, San Diego, CA.

Cappuccino, N. (1992). The nature of population stability in *Eurosta solidaginis*, a nonoutbreaking herbivore of goldenrod. *Ecology* **73,** 1792–1801.

Connell, J. H., and Sousa, W. P. (1983). On the evidence needed to judge ecological stability or persistence. *Am. Nat.* **121,** 789–824.

Craig, T. P., Itami, J. K., and Price, P. W. (1989). A strong relationship between oviposition preference and larval performance in a shoot-galling sawfly. *Ecology* **70,** 1691–1699.

Dempster, J. P. (1983). The natural control of populations of butterflies and moths. *Biol. Rev. Cambridge Philos. Soc.* **58,** 461–481.

Dempster, J. P., and Pollard, E. (1981). Fluctuations in resource availability and insect populations. *Oecologia* **50,** 412–416.

Hanski, I. (1990). Density dependence, regulation and variability in animal populations. *Philos. Trans. R. Soc. London, Ser. B* **330,** 141–150.

Hassell, M. P. (1985). Insect natural enemies as regulating factors. *J. Anim. Ecol.* **54,** 323–334.

Hassell, M. P. (1986). Detecting density dependence. *TREE* **1,** 90–93.

Hassell, M. P. (1987). Detecting regulation in patchily distributed animal populations. *J. Anim. Ecol.* **56,** 705–713.

Hassell, M. P., and May, R. M. (1985). From individual behaviour to population dynamics. *In* "Behavioural Ecology" (R. M. Sibly and R. H. Smith, eds.), pp. 3–32. Blackwell, Oxford.

Hassell, M. P., Latto, J., and May, R. M. (1989). Seeing the wood for the trees: Detecting density dependence from existing life-table studies. *J. Anim. Ecol.* **58,** 883–892.

Hunter, M. D. (1990). Differential susceptibility to variable plant phenology and its role in competition between two insect herbivores on oak. *Ecol. Entomol.* **15,** 401–408.

Hunter, M. D., Ohgushi, T., and Price, P. W., eds. (1992). "Effects of Resource Distribution on Animal–Plant Interactions." Academic Press, San Diego, CA.

Jolly, G. M. (1965). Explicit estimates from capture–recapture data with both death and immigration—Stochastic model. *Biometrika* **52**, 225–247.

Karban, R., and Courtney, S. (1987). Intraspecific host plant choice: Lack of consequences for *Streptanthus tortuosus* (Cruciferae) and *Euchloe hyantis* (Lepidoptera: Pieridae). *Oikos* **48**, 243–248.

Kareiva, P. (1983). Influence of vegetation texture on herbivore populations: Resource concentration and herbivore movement. *In* "Variable Plants and Herbivores in Natural and Managed Systems" (R. F. Denno and M. S. McClure, eds.), pp. 259–289. Academic Press, New York.

Kurihara, M. (1975). Anatomical and histological studies on the germinal vesicle in degenerating oocyte of starved females of the lady beetle, *Epilachna vigintioctomaculata* Motschulsky (Coleoptera, Coccinellidae). *Kontyû* **43**, 91–105.

Lomnicki, A. (1988). "Population Ecology of Individuals." Princeton Univ. Press, Princeton, NJ.

Mattson, W. J. (1980). Cone resources and the ecology of the red pine cone beetle, *Conophthorus resinosae* (Coleoptera: Scolytidae). *Ann. Entomol. Soc. Am.* **73**, 390–396.

Nisbet, R. M., Gurney, W. S. C., Murdoch, W. W., and McCauley, E. (1989). Structured population models: A tool for linking effects at individual and population level. *Biol. J. Linn. Soc.* **37**, 79–99.

Ohgushi, T. (1986). Population dynamics of an herbivorous lady beetle, *Henosepilachna niponica*, in a seasonal environment. *J. Anim. Ecol.* **55**, 861–879.

Ohgushi, T. (1987). Factors affecting body size variation within a population of an herbivorous lady beetle, *Henosepilachna niponica* (Lewis). *Res. Popul. Ecol.* **29**, 147–154.

Ohgushi, T. (1988). Temporal and spatial relationships between an herbivorous lady beetle *Epilachna niponica* and its predator, the earwig *Anechura harmandi*. *Res. Popul. Ecol.* **30**, 57–68.

Ohgushi, T. (1991). Lifetime fitness and evolution of reproductive pattern in the herbivorous lady beetle. *Ecology* **72**, 2110–2122.

Ohgushi, T. (1992). Resource limitation on insect herbivore populations. *In* "Effects of Resource Distribution on Animal–Plant Interactions" (M. D. Hunter, T. Ohgushi, and P. W. Price, eds.), pp. 199–241. Academic Press, San Diego, CA.

Ohgushi, T., and Sawada, H. (1981). The dynamics of natural populations of a phytophagous lady beetle, *Henosepilachna pustulosa* under different habitat conditions. I. Comparison of adult population parameters among local populations in relation to habitat stability. *Res. Popul. Ecol.* **23**, 94–115.

Ohgushi, T., and Sawada, H. (1985a). Population equilibrium with respect to available food resource and its behavioural basis in an herbivorous lady beetle, *Henosepilachna niponica*. *J. Anim. Ecol.* **54**, 781–796.

Ohgushi, T., and Sawada, H. (1985b). Arthropod predation limits the population density of an herbivorous lady beetle, *Henosepilachna niponica* (Lewis). *Res. Popul. Ecol.* **27**, 351–359.

Pimm, S. L. (1991). "The Balance of Nature?: Ecological Issues in the Conservation of Species and Communities." Univ. of Chicago Press, Chicago.

Preszler, R. W., and Price, P. W. (1988). Host quality and sawfly populations: A new approach to life table analysis. *Ecology* **69,** 2012–2020.

Price, P. W. (1987). The role of natural enemies in insect populations. *In* "Insect Outbreaks" (P. Barbosa and J. C. Schultz, eds.), pp. 287–312. Academic Press, San Diego, CA.

Price, P. W. (1990). Evaluating the role of natural enemies in latent and eruptive species: New approaches in life table construction. *In* "Population Dynamics of Forest Insects" (A. D. Watt, S. R. Leather, M. D. Hunter, and N. A. C. Kidd, eds.), pp. 221–232. Intercept, Andover, UK.

Price, P. W. (1994). Phylogenetic constraints, adaptive syndromes, and emergent properties: From individuals to population dynamics. *Res. Popul. Ecol.* **36,** 3–14.

Price, P. W., Bouton, C. E., Gross, P., McPheron, B. A., Thompson, J. N., and Weis, A. E. (1980). Interactions among three trophic levels: Influence of plants on interactions between insect herbivores and natural enemies. *Annu. Rev. Ecol. Syst.* **11,** 41–65.

Price, P. W., Cobb, N., Craig, T. P., Fernandes, G. W., Itami, J. K., Mopper, S., and Preszler, R. W. (1990). Insect herbivore population dynamics on trees and shrubs: New approaches relevant to latent and eruptive species and life table development. *In* "Insect–Plant Interactions" (E. A. Bernays, ed.), pp. 1–38. CRC Press, Boca Raton, FL.

Redfearn, A., and Pimm, S. L. (1988). Population variability and polyphagy in herbivorous insect communities. *Ecol. Monogr.* **58,** 39–55.

Romstöck-Völkl, M. (1990). Population dynamics of *Tephritis conura* Loew (Diptera: Tephritidae): Determinants of density from three trophic levels. *J. Anim. Ecol.* **59,** 251–268.

Root, R. B. (1973). Organization of a plant–arthropod association in simple and diverse habitats: The fauna of collards (*Brassica oleracea*). *Ecol. Monogr.* **43,** 95–124.

Schultz, J. C. (1983). Habitat selection and foraging tactics of caterpillars in heterogeneous trees. *In* "Variable Plants and Herbivores in Natural and Managed Systems" (R. F. Denno and M. S. McClure, eds.), pp. 61–90. Academic Press, New York.

Seber, G. A. F. (1973). "The Estimation of Animal Abundance and Related Parameters." Griffin, London.

Solbreck, C., and Sillén-Tullberg, B. (1986). Seed production and seed predation in a patchy and time-varying environment. Dynamics of a milkweed–tephritid fly system. *Oecologia* **71,** 51–58.

Southwood, T. R. E., and Comins, H. N. (1976). A synoptic population model. *J. Anim. Ecol.* **45,** 949–965.

Stanton, M. L. (1983). Spatial patterns in the plant community and their effects upon insect search. *In* "Herbivorous Insects" (S. Ahmad, ed.), pp. 125–157. Academic Press, New York.

Stiling, P. (1988). Density-dependent processes and key factors in insect populations. *J. Anim. Ecol.* **57,** 581–593.

Thompson, J. N. (1988). Evolutionary ecology of the relationship between oviposition preference and performance of offspring in phytophagous insects. *Entomol. Exp. Appl.* **47**, 3–14.

Valladares, G., and Lawton, J. H. (1991). Host-plant selection in the holly leaf-miner: Does mother know best? *J. Anim. Ecol.* **60**, 227–240.

Varley, G. C., Gradwell, G. R., and Hassell, M. P. (1973). "Insect Population Ecology: An Analytical Approach." Blackwell, Oxford.

Chapter 16

Working toward Theory on Galling Sawfly Population Dynamics

Peter W. Price, Timothy P. Craig, and Heikki Roininen

I. Introduction

If theory is considered to be factually and empirically based mechanistic explanation of pattern in nature, then there is little in the field of population dynamics. Even the detection of patterns in dynamics among species has been poorly developed, and certainly mechanisms driving these patterns have been inadequately explored. Though the cornerstone of a science should be theory, its development has not been a preoccupation among many ecologists, neither in population dynamics nor in other areas.

What we like to call "ecological theory" is actually hypothesis generation, and many of the generators seem to be largely uninterested in testing their hypotheses in natural systems. Hence, we have a gap between ecological hypotheticians and ecological empiricists, with neither group devoted principally to the central issue of their science: the development of empirically based factual theory (Kareiva, 1989; Tilman, 1989; Price, 1991a).

This caricature of the state of ecology and population dynamics will be dismissed by many, no doubt. But where are the broad, repeatable patterns being studied in nature relevant to population dynamics to find the mechanistic explanation for these patterns (Tilman, 1989)? Where can we find hypotheses on population dynamics in nature tested so thoroughly and convincingly that they can be regarded as factually based theory rather than hypothesis?

Following the lead of Darwin, who began with empirical fact and worked up through pattern detection to factually based theory, we have emphasized the need for emulating his approach in the field of population dynamics (Price, 1991a). Roughgarden et al. (1989, p. 3) appreciated, as we do, Darwin's admonitions to "let theory guide your observations" and "all observation must be for or against some view if it is to be of any service." But Darwin used empirical facts on which to found his theory: in our view a sound strategy for understanding broad patterns in nature.

A new synthesis in population dynamics is emerging, we think, with the search for broad patterns in nature. Broad patterns in the life-history characteristics of outbreak species is one example (e.g., Nothnagle and Schultz, 1987; Wallner, 1987; Barbosa *et al.*, 1989; Price *et al.*, 1990; Hunter, 1991). Another search for pattern involves the linkage, or its lack, between a female's ovipositional preference and larval performance and the evolutionary basis for strong to absent positive associations (e.g., Singer, 1986; Thompson, 1988; Courtney and Kibota, 1990; Price *et al.*, 1990). A third approach has been to find pattern in the kinds of species that attack most frequently vigorous and young plants or plant modules, and the dynamical consequences of such utilization patterns (Price *et al.*, 1990; Price, 1991b). All of these are initiatives taken in the last 10 years.

The new synthesis will also undoubtedly encompass the fields of behavior, ecology, and evolution and the underlying physiological mechanisms involving interactions among trophic levels. The evolution of life histories, the phylogenetic constraints on adaptation, the chemical ecology of food plant heterogeneity, host selection and utilization, female behavior and behavioral variation in populations, and the behavior of carnivores will all become essential elements in unraveling the bottom-up and top-down forces in trophic systems determining the population dynamics of insect herbivores (Price, 1992; Karr *et al.*, 1992; Hunter and Price, 1992).

Our attitudes expressed here have been influenced by studies on gall-forming sawflies over the past 15 years that we and our colleagues have undertaken. We believe that we have developed the basis for a theory on the population dynamics of some of these gallers, a claim we are making for the first time. Therefore, the remainder of this chapter is devoted to recounting the processes employed in the development of hypotheses, their repeated testing by observations and experiments, the broad geographical perspective of the studies, and ultimately the development of factually based theory explaining a broad pattern in nature. Though we can claim a broad pattern in one genus of galling sawflies, this is still a very narrow perspective relative to the field of population dynamics at large.

II. Studies on a Stem-Galling Sawfly

Our initial research was concentrated around Flagstaff, Arizona, along temporary streams draining the San Francisco Peaks, and at rare permanent springs in the area.

A. Life History

We described the life history of *Euura lasiolepis* (Hymenoptera: Tenthredinidae) on its only host plant, *Salix lasiolepis* (Salicaceae) for the first time

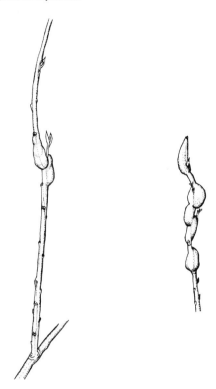

Figure 1. Galls of *Euura lasiolepis* on arroyo willow, *Salix lasiolepis*. The season depicted is early spring when new shoots are beginning to develop, and the sawfly is about to pupate. Illustration by Mary Bayless.

(Price and Craig, 1984). Adults emerge in the spring and mate, and females oviposit into rapidly growing young willow shoots with their sawlike ovipositor. Gall formation depends on the adult female's injection of stimulating substances, a process that is not understood. Galls develop through the summer and larvae feed over a prolonged period, overwintering in the gall and pupating in the spring (Fig. 1).

The important point to note is that emergence of females is synchronized with rapid willow growth in the spring, the female initiates one gall at a time, and larvae establish and feed all their immature lives exactly where the female oviposits. Therefore, there is the potential for a strong preference–performance linkage between ovipositional choices and larval success in establishment at a feeding site, and successful growth and survival.

B. Preference–Performance Linkage

We have recorded repeatedly in natural populations and in experiments that females attack most the longer classes of shoots in a population of shoots within and between willow clones (Price and Clancy, 1986a; Craig *et al.*, 1986, 1989; Preszler and Price, 1988). Three separate experiments using watering treatments all confirmed that females prefer the longest shoots available, and these are found on plants receiving the most water (Price and Clancy, 1986a; Craig *et al.*, 1989; Preszler and Price, 1988). In nature the longest shoots are found on plants that are young, especially juvenile growth, and plants with high water supply, such as at a permanent spring (Craig *et al.*, 1988a; Price *et al.*, 1990).

There is also a very clear pattern of higher survival of larvae on longer shoots, found repeatedly in natural populations and experiments (Price and Clancy, 1986a; Craig *et al.*, 1989; Preszler and Price, 1988). In fact, the preference–performance linkage in this species is probably the strongest discovered to date, given a literature with an apparent perplexingly poor relationship between female ovipositional choices and larval performance (cf. Thompson, 1988; Craig *et al.*, 1989; Courtney and Kibota, 1990; but see Damman and Feeny, 1988). In this *Euura* species, mother does know best!

As willow ramets age they produce shoots of decreasing length, becoming less and less susceptible to sawfly attack. Thus, young plants in a juvenile state, basal juvenile shoots on an old rootstock, or regrowth after damage provide the resources that sustain populations in nature. When disturbance is minimal and juvenile plants are uncommon, long juvenile shoots in a population of willows provide a generally low carrying capacity for the *Euura* population. The high preference–performance relationship linked to long shoots results in resources in short supply. The current conditions around Flagstaff, of fire suppression, low populations of vertebrate herbivores, and meltwater from the San Francisco Peaks diverted to Flagstaff's water supply, lead to a shortage of resources for *Euura*. The repeated observations on attack and survival being highest on the longest shoots in a population gave rise to the Plant Vigor Hypothesis (Price, 1991b), discussed in Section IV,B.

C. Chemical Ecology

Attraction of females to long shoots is not accounted for by protein concentration in shoots or total phenolics concentrations (Waring and Price, 1988). However, long shoots have higher concentrations of certain phenolic glycosides than do short shoots, especially tremulacin and salicortin (Price *et al.*, 1989). Recently, Heikki Roininen and Ritta Julkunen-Tiitto (personal communication) showed that females were stimulated strongly to oviposit when exposed to green paper discs dipped in tremulacin. Salicortin provided a much weaker stimulus.

Hence, we have proximate and ultimate mechanistic explanations for why females attack long shoots. Long shoots have high levels of tremulacin, with a stimulating effect on oviposition behavior. The ultimate cause for this physiological response is the higher survival of progeny in the longer shoots. The proximate mechanistic reason for higher survival in long shoots has yet to be explored.

Another important chemical interaction mediates competition for oviposition sites (Craig *et al.*, 1988b). Damage of the plant at the oviposition site results in a scar with phenolics at the surface. As these oxidize the scars brown, with the probable oxidation of phenolics to quinones. The scar becomes most repellent to subsequent females about two hours after the original oviposition. Such use of a plant product, equivalent to territorial chemical defense at an oviposition site by many insects (Price, 1984), has not been described before, but it acts as a mediator of competition among females, providing rapid negative feedback at high population density and diminishing larval competition.

D. Behavior

Four aspects of female behavior are central to the understanding of population dynamics in this species. First, female preference is well developed and we understand the proximate and ultimate mechanisms involved, as explained earlier. Second, females avoid used oviposition sites, resulting in more even dispersion of eggs, reduced larval competition, and rapid negative feedback in dense populations. Third, females adjust primary sex ratios in an adaptive manner in response to resource quality (Craig *et al.*, 1992). Fourth, sawflies are remarkably philopatric, with very low probability of leaving the natal clone (Stein *et al.*, 1994).

Facultative sex ratio variation is practiced by these haplodiploid sawflies, with a higher percentage of females in higher-quality plants. Craig *et al.* (1992) found a mean of 60% females produced among progeny on high-water-treatment plants and 36% females among progeny on low-water-treatment plants. As a result, population growth rate can be much higher on plants growing well and rapidly than on relatively water-stressed plants.

E. Facilitation and Resource Regulation

An unusual feature of these galling sawflies is that they modify resources to the advantage of individuals in the same generation and in subsequent generations. Short-term facilitation occurs when the galling process immediately accelerates shoot growth distal to the gall, making the shoot more attractive to subsequent females in the same generation, within the next few days (Craig *et al.*, 1990a).

Resource regulation (which could also be called long-term facilitation) occurs when galls, after emergence of adult sawflies, undergo necrosis and tissue death, with frequent shoot death of heavily galled shoots. In the spring after shoot death, more juvenile shoots sprout from dormant buds proximal to the damage, usually replacing what would have been less vigorous, more distal shoots with proximal shoots invigorated by natural pruning as a result of galling activity. Hence sawflies, especially on favorable plants, maintain or regulate shoot resources in a favorable condition over several generations (Craig *et al.*, 1986). In essence, the negative feedback of physiological aging on ramet growth is countered by the positive feedback of pruning.

F. Third Trophic Level

Carnivores in the system killing *Euura* larvae have a generally weak impact, and do not help to explain population variation from clone to clone, or from year to year. Small chalcidoid parasitoids, like *Pteromalus* spp., are constrained by gall size, which is largest on clones in the wettest sites where populations are highest (Price and Clancy, 1986b; Price, 1988). They cannot reach larvae in large galls for oviposition. The larger ichneumonid *Lathrostizus euurae* is limited by the rate of gall toughening (Craig *et al.*, 1990b). In experiments lasting for three sawfly generations, Woodman (1990) found no significant differences in *Euura* populations among treatments with uncaged clone segments, caged segments without parasitoids present, and segments with the full complement of parasitoids present. Ants have little impact on *Euura* populations, although strong but very local effects from ant predation have been documented for other galling genera of sawflies in the Flagstaff area (Woodman and Price, 1992). Mountain chickadees can have heavy but spasmodic impact, but they do not change the general pattern of dynamics (cf. Price, 1992).

G. Stability in Natural Populations

1. Natural Populations

The interactions discussed so far result in remarkably stable population dynamics of *Euura lasiolepis* in the Flagstaff area. Clearly, the overriding influences on population dynamics are the bottom-up effects of resource supply and the effects of water availability to plants on this supply. Low winter precipitation results in a reduced number of shoots per stem initiated, reduced shoot growth, reduced synchrony of sawflies with rapidly growing shoots, and reduced survival of early larvae in galls (Price and Clancy, 1986a; Price, 1992). High winter precipitation has the opposite effects.

The stability of the populations depends on a pattern of winter precipitation in which years with low precipitation are interspersed with years of moderate to

TABLE 1
Correlation Matrix among Eleven Generations of *Euura lasiolepis* (1983 to 1993)
on Fifteen Clones of *Salix lasiolepis* in the Flagstaff, Arizona Area[a]

Year	1984	1985	1986	1987	1988	1989	1990	1991	1992	1993
1983	0.91	0.99	0.89	0.75	0.55	0.66	0.85	0.91	0.87	0.99
1984		0.90	0.93	0.94	0.80	0.89	0.61	0.70	0.64	0.91
1985			0.89	0.77	0.56	0.66	0.84	0.91	0.95	0.99
1986				0.89	0.76	0.79	0.60	0.71	0.63	0.91
1987					0.93	0.96	0.40	0.50	0.44	0.78
1988						0.94	0.21	0.30	0.25	0.57
1989							0.30	0.38	0.33	0.67
1990								0.98	0.99	0.82
1991									0.98	0.89
1992										0.84

[a]Values in the table are correlation coefficients squared, r^2, providing an estimate of the proportion of the variance accounted for by the linear correlation between two years. Critical values for r^2 are 0.264, $P < 0.05$, and 0.411, $P < 0.01$. The dashed line encompasses values where 50% or less of the variance is accounted for, during a series of drought years, 1987–1989.

high precipitation. Populations track resource supply rapidly from year to year, and do not go locally extinct. Over the landscape, willow clones in wet sites, and the mixture of younger and older clones, provide a mosaic of resource qualities for sawflies, but each clone tends to support a predictable sawfly density relative to others. As a result, correlations are high between populations per clone from one year to the next, and even from one decade to the next (Table 1). Only during a series of years with below normal precipitation (1987–1989) did correlations fall to 50% or less of the variance accounted for. Reduced r^2 values resulted from strong declines in gall numbers on the clones in wet sites. In spite of this unusual series of years, populations in 11 years in wet and drier sites varied by less than $7\times$ in the former and $23\times$ in the latter (Fig. 2). This is remarkable compared to populations of outbreak forest insects that typically change over 3–5 orders of magnitude in a similar period of time (Price *et al.*, 1990).

Where resources are particularly favorable for sawflies, competition for oviposition sites provides rapid negative feedback limiting population growth (Craig *et al.*, 1990a).

2. A Perturbation Experiment

Early in our studies, an alternative to resource supply as a major limitation on population size per clone was the possibility that clones with low densities simply lacked an adequate supply of immigrating females. Therefore, populations of females were increased on 15 clones in 1984 by bagging four large

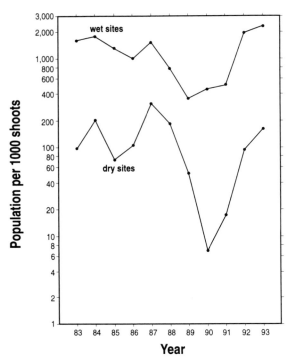

Figure 2. Populations of galls per 1000 shoots on 15 clones of arroyo willow from generations initiated in 1983 to 1993. Most clones (12), labeled as "dry sites," occur along the temporary Schultz Creek, a drainage from the San Francisco Peaks fed by snowmelt water in the spring season. The "wet sites" are represented by three clones, two at Coyote Spring and one receiving extra water from roof drainage.

branches per clone and leaving four adjacent branches per clone as controls. Galls were introduced into each bag and males and females emerged naturally into the bags. About 25 females per bag emerged.

The number of galls produced per female in the bags showed no significant trend across all clones, even though natural populations of galls varied from 9 to 2070 galls per 1000 shoots (Fig. 3). However, emergence of adults per 100 galls was very low where populations were naturally low, and relatively high where populations were naturally high in the wetter sites (Fig. 3), and a significant correlation existed between populations in the field and survival in galls in the treatment.

Clearly, there is a strong resource quality influence on population size per clone, and supplemented adult populations returned to background levels within a generation.

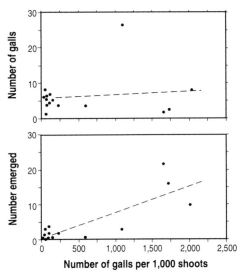

Figure 3. Data from the perturbation experiment in 1984. The X axis is the number of galls per 1000 shoots on the 15 natural willow clones in the 1984 generation. The Y axis is the number of galls formed per female in the bagged treatments (top) and the number of adults emerged per 100 galls (below). Top: $y = 115.82 + 0.04x$, $n = 15$, $r^2 = 0.13$, N.S. Bottom: $y = 0.30 + 0.01x$, $n = 15$, $r^2 = 0.68$, $P < 0.01$.

III. The Search for Generality

All species of *Euura* sawflies attack woody plants, and all have similar life histories (Price and Roininen, 1993). Some gall stems, and others gall buds, petioles, leaf midribs, and intermediate sites. The genus is Holarctic in distribution and restricted to willow plant hosts (*Salix:* Salicaceae). Some hosts are shrubs and others are trees. In the search for generality and pattern across *Euura* species we have adopted a comparative approach within a group of phylogenetically related species. These studies suggest that related species are constrained in their ecology by critical aspects of their common phylogenetic background: The Phylogenetic Constraints Hypothesis (Price, 1994a,b) discussed in Section IV,A.

A. Patterns of Preference and Performance

We have studied the ecology of *Euura* species in representative locations over the geographic range of the genus: from Arizona to Alaska in the United States, in Finland, and in Japan. In all the species we have studied there is a

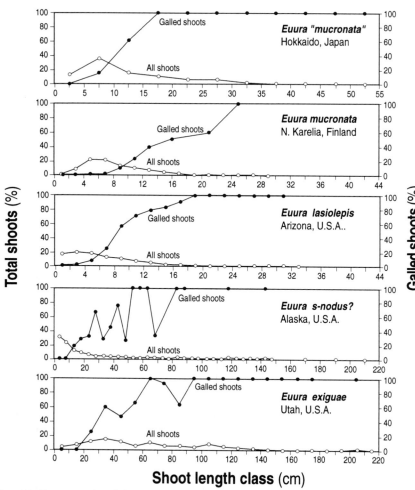

Figure 4. Five examples of the distribution of shoot length classes on willows, expressed as the percentage of total shoots in each shoot length class and the percentage of shoots galled in each shoot length class. Host willow species are, in order, *Salix sachalensis* ($n = 206$ shoots, 173 galls), *S. cinerea* ($n = 224$ shoots, 52 galls), *S. lasiolepis* ($n = 1230$ shoots, 344 galls), *S. interior* ($n = 276$ shoots, 53 galls), and *S. exiguae* ($n = 276$ shoots, 257 galls).

strong preference for attack of long shoots and strong preference–performance linkage. As with *E. lasiolepis*, although longer shoots are relatively rare, they are heavily attacked (Fig. 4). As would be expected if females respond to specific phenolic glycoside concentrations in long shoots, individual nodes on long shoots are more attacked per node than those on short shoots.

We have now studied eight *Euura* species in enough detail to detect clear evidence of the preference–performance pattern. These species are *E. amerinae* in Finland (Roininen *et al.*, 1993a, 1995), *E. atra* in Finland (H. Roininen and P. W. Price, unpublished), *E. exiguae* in Utah, U.S.A. (Price, 1989), *E. lasiolepis* in Arizona, U.S.A. (see references in Section II, B), *E. mucronata* in Finland (Price *et al.*, 1987a,b; Roininen *et al.*, 1988), *E. "mucronata"* in Japan (P. W. Price and T. Ohgushi, unpublished), a new species of midrib galler in Arizona (Woods *et al.*, 1995), and *E. s-nodus?* in Alaska (P. W. Price and H. Roininen, unpublished). The present taxonomy of the genus is not well developed (Smith, 1979), and just as the former species, *E. atra*, contains several sibling species (Roininen *et al.*, 1993b), so we feel that the bud gallers, usually named *E. mucronata*, represent a species complex (Price and Roininen, 1993). Hence, we argue that *E. "mucronata"* in Japan is a species distinct from its namesake in Finland.

Additional *Euura* species show a positive response of galling as shoot length increases, although more detailed studies have not been completed. These species are *Euura* n. sp. stem galler in Illinois, *Euura* petiole galler in Arizona, and *E. lappo* bud galler in Finland, with additional *E. mucronata* bud gallers on four *Salix* host species in Finland and Germany (Price and Roininen, 1993).

We have evidence consistent with the data for *E. lasiolepis* that at least 10 additional species show similar plant–herbivore interactions. For this genus, the pattern is broad.

B. Patterns in Shrub and Tree Attack, Common and Rare Species

There are many more *Euura* species on shrubby willows than on trees in the genus *Salix*. This is clearly because shrubs produce new juvenile shoots from the rootstock frequently, providing more predictable and prolonged resources than on trees, which undergo ontogenetic and physiological aging from a relatively early age (cf. Kearsley and Whitham, 1989; Roininen *et al.*, 1993a). Trees pass rapidly through a phase that is susceptible to stem-galling sawflies, making the population dynamics of gallers predictable but unstable. For example, a young stand of *Salix pentandra* was colonized by *Euura amerinae* in 1983, and the population flushed and crashed to extinction within 8 years, probably under the influence of ontogenetic tree aging (Roininen *et al.*, 1993a). *Euura atra* can persist in treelike willows, such as *Salix alba*, only if trees are heavily pruned in parks or along roadsides (H. Roininen and P. W. Price, unpublished).

On the basis of these relationships we can infer that the pattern of willow species growth and water supply have strongly predictable consequences on the *Euura* species that attack them. In predictably moist conditions, a strongly shrubby willow may provide abundant resources for galling sawflies, which may be very common. Such is the case for *Salix cinerea* in Finland and the bud galler,

E. mucronata, which has high and predictable populations in most sites where the willow grows (cf. Price *et al.,* 1987a,b). At the edge of a willow species' range, limited by water supply, as in *Salix lasiolepis* in Arizona, we can expect a much more patchy distribution of *Euura* species, and a shift toward an uncommon insect in a landscape. *Euura* species attacking trees may actually be rare in a landscape, because of uncommon disturbance resulting in young stands of willow trees and the ephemeral susceptibility of these trees. Commonness and rarity in the *Euura* sawflies is dictated strongly from the bottom up, through resource supply, and is highly predictable.

C. The Role of Carnivores

Throughout all our studies on *Euura* species we have never found an example where their natural enemies help to account for attack and survival patterns, or population dynamics. Generally, carnivores play a relatively passive role in population dynamics, attacking where possible, but not changing the gross dynamics dictated by resource supply (see Harrison and Cappuccino, Chapter 7, this volume). We have studied the role of natural enemies carefully on five species of *Euura* and in all cases their effects are weak. The species studied are *E. amerinae* (Roininen *et al.,* 1993a, 1995), *E. exiguae* (Price, 1989), *E. lasiolepis* (Price and Clancy, 1986b; Price, 1988; Craig *et al.,* 1990b), *E. mucronata* (Price *et al.,* 1987a,b), and *Euura* n. sp., a midrib galler in Arizona (Woods *et al.,* 1995).

D. Resource Regulation

Any galling sawfly that causes death of shoot meristems, directly or indirectly, acts as a plant pruner, affecting plant architecture and resource supply to subsequent generations of gallers (cf. Roininen and Tahvanainen, 1991). The shoot and bud gallers are involved in such mortality. The effects of resource regulation have been documented clearly for *E. lasiolepis* (Craig *et al.,* 1986) and for *E. mucronata* (Roininen *et al.,* 1988). How general resource regulation is by the genus *Euura* remains to be seen, and its role in population dynamics is not clearly understood. Such regulation is most effective where populations per clone are already high. Therefore, persistence of populations in the most favorable sites and during prolonged stressful conditions may be facilitated by resource regulation, accentuating the quality of small refugia during severe drought stress.

E. Chemical Ecology

To date, only one other *Euura* species has been tested for the females' responses to chemical cues during oviposition. For *E. amerinae,* Kolehmainen *et al.* (1994) found that the main phenolic glycoside in the *S. pentandra* host species, 2'-*O*-acetylsalicortin, elicited strong ovipositional behavior. This response was thought to provide a probable mechanistic explanation for the ability of females to select long shoots preferentially during oviposition. As in *S. lasiolepis,* longer shoots had higher concentrations of phenolics than shorter shoots, and females of *E. amerinae* oviposited more in the longer shoots, just as in *E. lasiolepis.* Probably, other *Euura* species will be found to respond to major phenolic glycosides in their host plants.

IV. Theory on *Euura* Population Dynamics

A. Phylogenetic Constraints Hypothesis

We have argued that in order to understand the population dynamics of a species it is essential to incorporate evolutionary, behavioral, and ecological factors into a mechanistic, empirically based theory (Price *et al.,* 1990). The macroevolutionary background of morphology, life history, and phylogeny dictate the way in which an insect herbivore interacts with the plant host. The *Phylogenetic Constraints Hypothesis* states that macroevolutionary traits in a lineage set limits on the range of life-history patterns and behaviors that can evolve (Price *et al.,* 1990; Price, 1994a,b). For example, the ovipositor of tenthredinid sawflies, shaped like a saw, requires oviposition into soft plant tissue, usually when a module is young. This necessitates the evolution of a life history with females emerging in synchrony with rapid plant growth, for example, in a north temperate spring. In turn, this sets the timing of larval feeding relative to leaf age. In the galling sawflies there is the additional constraint that a gall must be initiated in young undifferentiated plant tissue, close to a meristem or into a meristem. The more active the meristem, the higher is the probability of rapid gall development, selecting for close association between rapid plant growth and sawfly oviposition. The phylogenetic constraints cause an *adaptive syndrome* of characters to evolve that minimize the effects of the constraints and maximize effective utilization of host plant parts. Much of the ecology of the species therefore depends on its evolutionary background and the population dynamics is an almost inevitable consequence of the evolved plant–herbivore interaction. Therefore, we have called ecological aspects such as dispersion, commonness or rarity, and population dynamics *emergent properties* (Price *et al.,* 1990).

The flow of influence from phylogenetic constraints, to the adaptive syndrome, to emergent properties constitutes the Phylogenetic Constraints Hypothesis. We have shown repeatedly how a sawlike ovipositor and the galling habit have constrained the lineage of the *Euura* genus to the utilization of rapidly growing plant modules. We have also shown that the adaptive syndrome includes a tight preference–performance linkage that maximizes survival of eggs and larvae. Given these evolved traits, the emergent properties are predictable once we know the growth habit and ecology of the host plant and predictability of soil moisture. The uniformly weak effects of carnivores in these systems make the bottom-up forces of paramount importance, and leave us with a relatively simple ecology to explain.

We maintain that we have tested the Phylogenetic Constraints Hypothesis relevant to *Euura* sawflies repeatedly, and all results are consistent with the hypothesis. Our tests had the potential to falsify the hypothesis, with either field-based data from natural populations or numerous experiments. However, results were remarkably consistent across the many series of *Euura* studied, and the broad geographical distribution to the genus. No *Euura* species are known from the tropics or desert regions, so we have studied the genus in the most relevant vegetation types available.

Therefore, we claim to have developed a theory on the population dynamics of *Euura* sawflies. The theory is empirically and factually based, and explains mechanistically the pattern of plant–herbivore interactions and population dynamics observed in the genus. Aspects of the theory provide both proximate and ultimate mechanistic explanations, as with female preference for rapidly growing modules, yet some questions remain, such as why larvae survive best in galls on long shoots. The theory is highly predictive because any new species studied should fit the pattern described in relation to host plant growth form and demography, soil moisture for plant growth, and gall location.

B. Plant Vigor Hypothesis

Another hypothesis that we have tested repeatedly is the *Plant Vigor Hypothesis*. This states that "plant modules which are growing vigorously, or have grown vigorously to become relatively large in a population of modules, are favorable to certain kinds of herbivores" (Price, 1991b, p. 245). Our studies on *Euura* species are remarkably uniform and consistent with this hypothesis and account for the distribution of gallers on plants of different age, and on large modules in a population on a single plant. For *Euura* we also have a mechanistic explanation of why long modules are utilized and the population consequences of module variation in space and time. Therefore, we feel that the Plant Vigor Hypothesis can be incorporated into the theory on *Euura* population dynamics.

V. Toward More General Theory

A broader general theory can be developed by studying other genera of galling sawflies, as we are doing (cf. Price and Roininen, 1993; Price *et al.*,

1994), and related free-feeding sawflies, with research in progress. We have also argued that many endophytic insects, such as shoot borers, have constraints similar to those of gallers on their population dynamics (Price *et al.*, 1990; Price, 1992). Such arguments have been expanded to pest species with eruptive population dynamics, with a different set of phylogenetic constraints, adaptive syndromes, and emergent properties (Price *et al.*, 1990; Price, 1992, 1994a,b).

As the hypothetical framework expands, so the demands for testing hypotheses increase, which rapidly become time-consuming and costly. Commonly, the relevant questions have not been addressed before, requiring new experiments and field observations. As research scope broadens there is a remarkable reluctance among funding agencies to foot the bill. However, if we are to build more general theory on population dynamics, these are the kinds of challenges that must be overcome.

Acknowledgments

Research by P. W. P. and T. P. C. was supported by National Science Foundation Grants DEB-7816152, DEB-8021754, BSR-83144594, BSR-8705302, BSR-8715090, BSR-9020317, and DEB-9318188. H. R. was supported by the Finnish Academy. We are grateful for reviews of this paper by Naomi Cappuccino, Timothy Carr, and Jens Roland.

References

Barbosa, P., Krischik, V., and Lance, D. (1989). Life-history traits of forest-inhabiting flightless Lepidoptera. *Am. Midl. Nat.* **122**, 262–274.

Courtney, S. P., and Kibota, T. T. (1990). Mother doesn't know best: Selection of hosts by ovipositing insects. *In* "Insect–Plant Interactions" (E. A. Bernays, ed.), pp. 161–188. CRC Press, Boca Raton, FL.

Craig, T. P., Price, P. W., and Itami, J. K. (1986). Resource regulation by a stem-galling sawfly on the arroyo willow. *Ecology* **67**, 419–425.

Craig, T. P., Price, P. W., Clancy, K. M., Waring, G. M., and Sacchi, C. F. (1988a). Forces preventing coevolution in the three-trophic-level system: "Willow, a gall-forming herbivore, and parasitoid. *In* "Chemical Mediation of Coevolution" (K. Spencer, ed.), pp. 57–80. Academic Press, San Diego, CA.

Craig, T. P., Itami, J. K., and Price, P. W. (1988b). Plant wound compounds from oviposition scars used as oviposition deterrents by a stem-galling sawfly. *J. Insect Behav.* **1**, 343–356.

Craig, T. P., Itami, J. K., and Price, P. W. (1989). A strong relationship between oviposition preference and larval performance in a shoot-galling sawfly. *Ecology* **70**, 1691–1699.

Craig, T. P., Itami, J. K., and Price, P. W. (1990a). Intraspecific competition and facilitation by a shoot-galling sawfly. *J. Anim. Ecol.* **59**, 147–159.

Craig, T. P., Itami, J. K., and Price, P. W. (1990b). The window of vulnerability of a shoot-galling sawfly to attack by a parasitoid. *Ecology* **71**, 1471–1482.

Craig, T. P., Price, P. W., and Itami, J. K. (1992). Facultative sex ratio shifts by a herbivorous insect in response to variation in host plant quality. *Oecologia* **92**, 153–161.

Damman, H., and Feeny, P. (1988). Mechanisms and consequences of selective oviposition by the zebra swallowtail butterfly. *Anim. Behav.* **36**, 563–573.

Hunter, A. F. (1991). Traits that distinguish outbreaking and nonoutbreaking Macrolepidoptera feeding on northern hardwood trees. *Oikos* **60**, 275–282.

Hunter, M. D., and Price, P. W. (1992). Playing chutes and ladders: Heterogeneity and the relative roles of bottom-up and top-down forces in natural communities. *Ecology* **73**, 724–732.

Kareiva, P. (1989). Renewing the dialogue between theory and experiments in population ecology. *In* "Perspectives in Ecological Theory" (J. Roughgarden, R. M. May, and S. A. Levin, eds.), pp. 68–88. Princeton Univ. Press, Princeton, NJ.

Karr, J. R., Dionne, M., and Schlosser, I. J. (1992). Bottom-up versus top-down regulation of vertebrate populations: Lessons from birds and fish. *In* "Effects of Resource Distribution on Animal–Plant Interactions" (M. D. Hunter, T. Ohgushi, and P. W. Price, eds.), pp. 243–286. Academic Press, San Diego, CA.

Kearsley, M. J. C., and Whitham, T. G. (1989). Developmental changes in resistance to herbivory: Implications for individuals and populations. *Ecology* **70**, 422–434.

Kolehmainen, J., Roininen, H., Julkunen-Tiitto, R., and Tahvanainen, J. (1994). Importance of phenolic glycosides in host selection of the shoot galling sawfly, *Euura amerinae,* on *Salix pentandra. J. Chem. Ecol.* (in press).

Nothnagle, P. J., and Schultz, J. C. (1987). What is a forest pest? *In* "Insect Outbreaks" (P. Barbosa and J. C. Schultz, eds.), pp. 59–80. Academic Press, San Diego, CA.

Preszler, R. W., and Price, P. W. (1988). Host quality and sawfly populations: A new approach to life table analysis. *Ecology* **69**, 2012–2020.

Price, P. W. (1984). "Insect Ecology," 2nd ed. Wiley, New York.

Price, P. W. (1988). Inversely density-dependent parasitism: The role of plant refuges for hosts. *J. Anim. Ecol.* **57**, 89–96.

Price, P. W. (1989). Clonal development of coyote willow, *Salix exigua* (Salicaceae), and attack by the shoot-galling sawfly, *Euura exiguae* (Hymenoptera: Tenthredinidae). *Environ. Entomol.* **18**, 61–68.

Price, P. W. (1991a). Darwinian methodology and the theory of insect herbivore population dynamics. *Ann. Entomol. Soc. Am.* **84**, 465–473.

Price, P. W. (1991b). The plant vigor hypothesis and herbivore attack. *Oikos* **62**, 244–251.

Price, P. W. (1992). Plant resources as a mechanistic basis for insect herbivore population dynamics. *In* "Effects of Resource Distribution on Animal–Plant Interactions" (M. D. Hunter, T. Ohgushi, and P. W. Price, eds.), pp. 139–173. Academic Press, San Diego, CA.

Price, P. W. (1994a). Phylogenetic constraints, adaptive syndromes, and emergent properties: From individuals to population dynamics. *Res. Popul. Ecol.* **36**, 1–12.

Price, P. W. (1994b). Patterns in the population dynamics of insect herbivores. *In* "Individuals, Populations and Patterns in Ecology" (S. R. Leather, A. D. Watt, N. J. Mills, and K. E. A. Walters, eds.), pp. 109–117. Intercept, Andover, UK.

Price, P. W., and Clancy, K. M. (1986a). Multiple effects of precipitation on *Salix lasiolepis* and populations of the stem-galling sawfly, *Euura lasiolepis. Ecol. Res.* **1**, 1–14.

Price, P. W., and Clancy, K. M. (1986b). Interactions among three trophic levels: Gall size and parasitoid attack. *Ecology* **67**, 1593–1600.

Price, P. W., and Craig, T. P. (1984). Life history, phenology, and survivorship of a stem-galling sawfly, *Euura lasiolepis* (Hymenoptera: Tenthredinidae), on the arroyo willow, *Salix lasiolepis,* in northern Arizona. *Ann. Entomol. Soc. Am.* **77**, 712–719.

Price, P. W., and Roininen, H. (1993). The adaptive radiation of gall induction. *In* "Sawfly Life History Adaptations to Woody Plants" (M. R. Wagner and K. F. Raffa, eds.), pp. 229–257. Academic Press, San Diego, CA.

Price, P. W., Roininen, H., and Tahvanainen, J. (1987a). Plant age and attack by the bud galler, *Euura mucronata. Oecologia* **73**, 334–337.

Price, P. W., Roininen, H., and Tahvanainen, J. (1987b). Why does the bud-galling sawfly, *Euura mucronata,* attack long shoots? *Oecologia* **74**, 1–6.

Price, P. W., Waring, G. L., Julkunen-Tiitto, R., Tahvanainen, J., Mooney, H. A., and Craig, T. P. (1989). The carbon-nutrient balance hypothesis in within-species phytochemical variation of *Salix lasiolepis. J. Chem. Ecol.* **15**, 1117–1131.

Price, P. W., Cobb, N., Craig, T. P., Fernandes, G. W., Itami, J. K., Mopper, S., and Preszler, R. W. (1990). Insect herbivore population dynamics on trees and shrubs: New approaches relevant to latent and eruptive species and life table development. *In* "Insect–Plant Interactions" (E. A. Bernays, ed.), Vol. 2, pp. 1–38. CRC Press, Boca Raton, FL.

Price, P. W., Clancy, K. M., and Roininen, H. (1994). Comparative population dynamics of the galling sawflies. *In* "The Ecology and Evolution of Gall-forming Insects" (P. W. Price, W. J. Mattson, and Y. Baranchikoff, eds.), pp. 1–11. *USDA For. Serv. N. Centr. For. Exp. Stn. Gen. Tech. Rep,* **NC-174.**

Roininen, H., and Tahvanainen, J. (1991). Impact of the bud galler, *Euura mucronata,* on its resources and on the architecture of its willow host. *Salix cinerea. In* "The Ecology and Evolution of the Host Plant Relationships Among Willow-Feeding Sawflies." University of Joensuu, Joensuu, Finland.

Roininen, H., Price, P. W., and Tahvanainen, J. (1988). Field test of resource regulation by the bud-galling sawfly, *Euura mucronata,* on *Salix cinerea. Holarctic Ecol.* **11**, 136–139.

Roininen, H., Price, P. W., and Tahvanainen, J. (1993a). Colonization and extinction in a population of the shoot-galling sawfly, *Euura amerinae. Oikos* **68**, 448–454.

Roininen, H., Vuorinen, J., Tahvanainen, J., and Julkunen-Tiitto, R. (1993b). Host preference and allozyme differentiation in a shoot-galling sawfly, *Euura atra. Evolution (Lawrence, Kans.)* **47**, 300–308.

Roininen, H., Price, P. W., and Tahvanainen, J. (1995). Bottom-up and top-down influences in the trophic system of a willow, a galling sawfly, parasitoids and inquilines. *Oikos* (in press).

Roughgarden, J., May, R. M., and Levin, S. A. (1989). Introduction. *In* "Perspectives in

Ecological Theory" (J. Roughgarden, R. M. May, and S. A. Levin, eds.), pp. 3–10. Princeton Univ. Press, Princeton, NJ.

Singer, M. C. (1986). The definition and measurement of oviposition preference in plant-feeding insects. *In* "Insect–Plant Interactions" (J. R. Miller and T. A. Miller, eds.), pp. 65–94. Springer, New York.

Smith, D. R. (1979). Suborder Symphyta. *In* "Catalog of Hymenoptera in America North of Mexico" (K. V. Krambein, P. D. Hurd, D. R. Smith, and B. D. Brooks, eds.), pp. 3–137. Smithsonian Institution Press, Washington, DC.

Stein, S. J., Price, P. W., Craig, T. P., and Itami, J. K. (1994). Dispersal of a galling sawfly: Implications for studies on insect population dynamics. *J. Anim. Ecol.* **63,** 666–676.

Thompson, J. N. (1988). Evolutionary ecology of the relationship between oviposition preference and performance of offspring in phytophagous insects. *Entomol. Exp. Appl.* **47,** 3–14.

Tilman, D. (1989). Discussion: Population dynamics and species interactions. *In* "Perspectives in Ecological Theory" (J. Roughgarden, R. M. May, and S. A. Levin, eds.), pp. 89–100. Princeton Univ. Press, Princeton, NJ.

Wallner, W. E. (1987). Factors affecting insect population dynamics: Differences between outbreak and non-outbreak species. *Annu. Rev. Entomol.* **32,** 317–340.

Waring, G. L., and Price, P. W. (1988). Consequences of host plant chemical and physical variability to an associated herbivore. *Ecol. Res.* **3,** 205–216.

Woodman, R. L. (1990). Enemy impact and herbivore community structure: Tests using parasitoid assemblages, predatory ants, and galling sawflies on arroyo willow. Ph.D. Dissertation, Northern Arizona University, Flagstaff.

Woodman, R. L., and Price, P. W. (1992). Differential larval predation by ants can influence willow sawfly community structure. *Ecology* **73,** 1028–1037.

Woods, J. O., Carr, T. G., Price, P. W., and Stevens, L. (1995). Growth of coyote willow and the attack and survival of a midrib galling sawfly, *Euura* sp. *Oecologia* (in press).

Chapter 17

Host Suitability, Predation, and Bark Beetle Population Dynamics

John D. Reeve, Matthew P. Ayres, and Peter L. Lorio, Jr.

I. Introduction

Many species of bark beetles (Coleoptera: Scolytidae) undergo dramatic population fluctuations that lead to extensive damage of coniferous forests. Considerable effort has been devoted to determining the causes of bark beetle outbreaks. The most common explanation is that outbreaks occur when there is an abundance of trees with low resistance to attacking adults, and which are suitable for beetle reproduction (Rudinsky, 1962; Stark, 1965; Berryman, 1973, 1976; Raffa and Berryman, 1983; Berryman and Ferrell, 1988; Christiansen and Bakke, 1988; Raffa, 1988). Outbreaks are often thought to occur as a result of stress on the trees that limits the effectiveness of the oleoresin system as a defense against attacking adult beetles (Berryman, 1972). Factors that have been implicated as agents of stress include drought, competition between trees, disease or defoliation by other insects, storms, and aging. The decline of outbreak populations to endemic levels is often attributed to an increase in the resistance of potential host trees, or simply to the depletion of trees suitable for beetle reproduction.

In this chapter, we examine the hypothesis that outbreaks in the southern pine beetle (SPB), *Dendroctonus frontalis* Zimmermann (Coleoptera: Scolytidae), are generated by environmental factors that affect host-tree suitability, especially water availability. There is a substantial literature linking climatic patterns to SPB outbreaks (Wyman, 1924; Craighead, 1925; Beal, 1927, 1933; St. George, 1930; King, 1972; Kroll and Reeves, 1978; Kalkstein, 1981; Michaels, 1984). Nonetheless, Turchin *et al.* (1991) found no relationship between the per capita rate of increase of SPB and any of three climatic variables (water deficits, winter temperatures, and summer temperatures). Here, we evaluate a physiological model for southern pine that predicts *decreased* suitability for SPB under moderate water stress, but increased suitability under severe water stress

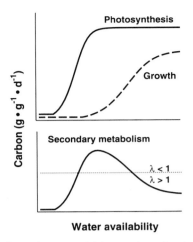

Figure 1. Hypothesized effects of water availability on the carbon budgets of pine. Moderate water deficits limit growth but not photosynthesis, thus as carbon availability for secondary metabolism increases, oleoresin production rises, and suitability for bark beetles is reduced. At very low water availability, trees constrict stomata to limit transpiration, causing photosynthesis and secondary metabolism to plummet, and suitability for bark beetles to increase. Thus attacking bark beetles are predicted to have high reproductive success ($\lambda > 1$) under conditions of high water availability or extreme drought stress, but low reproductive success ($\lambda < 1$) under moderate water stress.

(Fig. 1). Although this model is derived from simple physiological principles, it leads to more complex predictions than have usually been considered in judging the role of climatic effects on bark beetle population dynamics; it has the potential to reconcile some of the conflicting conclusions about the role of climatic patterns in driving the population dynamics of SPB and other bark beetle species.

In contrast with host resistance, natural enemies generally have been assigned a minor role in controlling bark beetle dynamics (Rudinsky, 1962; Christiansen and Bakke, 1988; Raffa, 1988), although they have sometimes been credited with accelerating the decline of outbreaks (Berryman, 1973; Berryman and Ferrell, 1988; Grégoire, 1988). However, a number of studies using exclusion cages around host trees have shown that natural enemies can inflict substantial mortality on immature bark beetles (Linit and Stephen, 1983; Miller, 1984, 1986; Riley and Goyer, 1986; Weslien, 1992). Furthermore, a 30-year-time-series analysis suggests that SPB outbreaks are driven by some population process acting in a delayed density-dependent manner (Turchin *et al.*, 1991), as might be expected if specialist natural enemies are influencing population dynamics. Among the potential natural enemies, clerid beetles (Coleoptera: Cleridae) are likely candidates in many systems for influencing bark beetle population dynamics (Thatcher and Pickard, 1966; Berryman, 1967; Moore, 1972; Mills, 1985; Weslien and Regnander, 1992; Weslien, 1994). Typically, clerids are specialist predators of bark beetles that prey on both larval and adult life stages. *Thanasimus dubius* is the clerid species most abundant in the SPB system. It has long been recognized as a potentially important source of mortality for SPB (Fiske, 1908; Thatcher and Pickard, 1966; Moore, 1972), and its abundance relative to SPB has been used to predict future SPB activity (Billings, 1988). Here we report preliminary results from a research program designed to (1) quantify the potential effects of *T. dubius* on SPB populations, (2) compare the numeric effects of *T. dubius* relative to the effects of host suitability, and (3) evaluate the potential role of *T. dubius* in influencing SPB population dynamics.

II. Moisture Stress and Host Suitability

A. Models of Host Suitability Applied to Bark Beetle Populations

Several general models have been proposed to explain population variation in herbivorous insects based on the physiological condition of host plants. However, the various hypotheses can lead to different predictions, and the data are equivocal. The plant stress hypothesis (White, 1974) suggests that insect performance is favored in stressed plants (e.g., water limited) because of increased nitrogen availability in consumed tissue. This hypothesis is often linked to arguments that stressed plants synthesize less defensive chemicals (Rhoades, 1979), and that climatic effects can allow the release of endemic insect populations to epidemic levels (Greenbank, 1956; Mattson and Haack, 1987a,b). Larsson (1989) has discussed difficulties in rationalizing the plant stress hypothesis with many experimental results, although he concludes that performance of cambium feeders (e.g., bark beetles) is often enhanced on stressed hosts. The plant vigor hypothesis (Price, 1991), which may be viewed as in conflict with the plant stress hypothesis, is supported by observations that many herbivorous insects feed preferentially on vigorously growing plants. Waring and Cobb (1992) concluded from their review of over 450 studies that stress has strong effects on herbivore population dynamics—positive, negative, or nonlinear. Many studies lack the details of induced plant stress (e.g., level, timing, and duration of water deficit), which are essential for understanding the nature of plant as well as herbivore responses. Further, it seems to be generally unrecognized that physiological changes associated with ontogeny of plants, apart from stress effects, may alter host suitability for herbivorous insects (Kozlowski, 1969).

For the case of bark beetles, plant stress has long been a favored explanation for outbreaks. Wyman (1924), Craighead (1925), and St. George (1930) all implicated climatic effects, especially drought, in weakening host trees and permitting increases in SPB populations. These early formulations of the stress hypothesis, like modern versions, assumed that tree "vigor" is positively correlated with the ability to resist attack by herbivores, perhaps because fast-growing trees were thought to have larger carbon budgets with which to produce defensive chemicals (Waring and Pitman, 1980; Waring, 1983; Raffa, 1988). Drought stress of trees remains one of the most commonly cited explanations for bark beetle outbreaks. In a volume devoted to the dynamics of forest insect populations (Berryman, 1988), four of five chapters dealing with injurious bark beetle species specifically invoke drought stress as a key factor in the release of beetle outbreaks (*Dendroctonus micans,* Grégoire 1988; *Dendroctonus ponderosae,* Raffa 1988; *Ips typographus,* Christiansen and Bakke 1988; *Scolytus ventralis,* Berryman and Ferrell 1988).

In spite of a long history in the literature, the linkage between drought stress and SPB population dynamics is only tenuously supported by the evidence. There have been few experimental tests, probably owing to the logistical problems of working with large trees. Manipulations of loblolly pine have shown that severe water deficits can enhance the success of attacking SPB (Lorio and Hodges, 1977), as predicted by the stress hypothesis, but these treatments may have exceeded normal climatic variation. Time-series analyses failed to detect any sign of climatic effects on SPB populations (Turchin *et al.,* 1991). Pine stands at highest risk for SPB infestations frequently appear to be those in which trees grow fastest, not slowest (Lorio *et al.,* 1982; Lorio and Sommers, 1986). In fact, the earliest scientific accounts suggest that SPB may prefer vigorously growing, "healthy" trees (Hopkins, 1892, 1921).

B. An Alternative Model for the Effects of Water Availability on Pines

We have been testing an alternative physiological model for the effects of water availability on the suitability of southern pines for bark beetles (Lorio *et al.,* 1990). Our hypotheses are derived from principles of plant growth–differentiation balance (Loomis, 1932, 1953), and a conceptual model of loblolly pine suitability for SPB (Lorio *et al.,* 1990). Growth–differentiation balance allows two sets of predictions relevant to bark beetles: one dealing with static carbon budgets (Fig. 1) and another incorporating phenological patterns of tree growth (Lorio, 1986, 1988). Our static model predicts a curvilinear response of secondary metabolism to water availability, implying that drought may increase or decrease suitability for bark beetles depending on the initial water status of the trees and the severity of the drought (Fig. 1). Moderate water stress is predicted to limit growth more than photosynthesis, resulting in an increase in the proportion and total amount of carbon allocated to secondary metabolism. In contrast, severe water deficits are predicted to limit carbon assimilation, thereby restricting secondary metabolism and increasing tree suitability for bark beetles. Phenological considerations can be overlain on this static model. Pine trees in the southeastern United States grow very rapidly in the spring and early summer, but by midsummer tree growth typically becomes limited by soil water deficits. This change in environmental conditions corresponds to an ontogenetic transition in the cambium from the production of earlywood to latewood. The latewood is highly invested with vertical resin ducts that function in the synthesis and transport of oleoresin (a mixture of monoterpenes and resin acids that impedes attacking bark beetles, but also plays a role in secondary attraction of SPB). Annual variation in climatic conditions influences the timing of seasonal water deficits and, within the limits of endogenous controls, the ontogenetic transition from earlywood to latewood. Measurements of resin flow from standardized wounds to the face of the cambium indicate that spring and early summer is a time of relatively low resin flow in loblolly pine. The transition to latewood formation is accompanied by an increase in resin flow, presumably because there are more vertical resin ducts, and because growth limitations imposed by water deficits leave more carbon for secondary metabolism. The predictable increase in SPB infestations during spring and early summer (Billings, 1979) may occur because tree physiology tends to be growth-dominated at that time (Zahner, 1968; Lorio and Sommers, 1986). Climatic patterns that protract the period of earlywood production in southern pine (e.g., high precipitation in midsummer) are predicted to favor population growth in SPB by protracting the time when trees are suitable for beetle colonization.

C. Experimental Results from 1990

In our first attempt at testing these predictions, Dunn and Lorio (1993) manipulated the water balance of 11-year-old loblolly pine using a combination of rain-exclusion shelters and irrigation. Compared to irrigated trees, trees treated with rain-exclusion shelters had lower xylem water potential, lower cambial growth, lower photosynthetic rates, and lower resin flow. Irrigated trees suffered 149–218% more SPB attacks than sheltered trees, but those SPB that attacked irrigated trees produced 32–52% fewer eggs per pair of attacking adults. This experiment did not provide an entirely satisfactory test of our growth–differentiation balance model because the summer in which it was conducted (1990) happened to be unusually hot and dry, and rain shelters that were designed to produce moderate deficits, limiting growth but not photosynthesis, actually produced such severe drought stress that photosynthetic rates were 30–60% lower than those of irrigated trees. Consequently, we were unable to evaluate whether the reduced resin flow in sheltered trees resulted from an increase, decrease, or no change in the proportional commitment of carbon to secondary metabolism. Nonetheless, resin flow was very high in all trees, and attacking SPB averaged only 2.9–7.9 eggs per attacking pair, which indicates that even severe drought stress was insufficient to render the trees into high-quality hosts for attacking SPB.

D. Experimental Results from 1991

In 1991, we initiated a similar study with eight 22-year-old loblolly pine at each of three treatment levels: rain shelters to exclude precipitation, natural precipitation, and natural precipitation plus irrigation. Rain shelters were complete on 10 April and irrigation began on 5 June. SPB attack was induced on half of the trees on 22 July. The attacked trees were cut 10 days later. Bolts were removed from each tree at 1 and 3 m above ground, and two 500-cm^2 areas on opposite sides of each bolt were dissected to determine SPB performance.

Treatments were successful in producing a gradient of soil water availability. By 22 July, estimated soil water storage (Zahner and Stage, 1966) ranged from 2 to 12 to 17 cm in sheltered, control, and irrigated treatments, respectively. A calculated soil water deficit of >25 cm had accumulated in the sheltered treatment by the time of attack. Predawn water potential did not differ among treatments initially, but by mid-June, the sheltered trees began to show lowered water potentials, indicating development of moderate water deficits.

Water regime influenced tree height growth and cambial growth, but not photosynthesis (Fig. 2). Height growth was reduced at very modest water deficits (control versus irrigated), whereas cambial growth was reduced only at somewhat greater water deficits (sheltered versus control). The resin yield of sheltered trees rose dramatically in late June to nearly twice that of control and irrigated trees (Fig. 3). Because carbon assimilation was not affected by the water treatments (Fig. 2), sheltered trees were apparently committing a larger proportion of their carbon budgets to secondary metabolism and a lower proportion to growth (i.e., decreased growth : differentiation). There was a strong negative relationship between beetle attack success and tree resin production (Fig. 4). In the upper trunk, SPB attack densities did not differ across treatments (18–22 attacking pairs per 500 cm^2). However, beetles attacking sheltered trees produced less gallery than those attacking control or irrigated trees (mean ± SE = 145 ± 3 versus 216 ± 25 versus 348 ± 48 cm per 500 cm^2, respectively), and had much lower reproductive success (4.6 ± 1.4 versus 7.1 ± 1.3 versus 12.5 ± 4.0 eggs per adult pair, respectively). In the lower trunk, SPB reproductive success was even more severely reduced in sheltered trees relative to irrigated trees (1.7 ± 0.4 versus 3.1 ± 1.0 versus 11.3 ± 4.3 eggs per adult pair, respectively). Presumably beetles encountering high resin flow were forced to spend more time

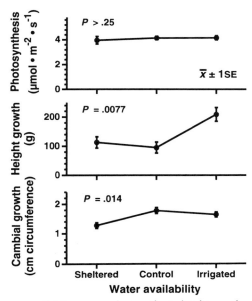

Figure 2. Effects of water availability on net photosynthesis, height growth, and cambial growth in 22-year-old loblolly pine. *P* values show results of tests for linear trend. Photosynthesis was measured at midmorning on nine dates during the summer. Height growth includes the second through the fourth flushes, with the first flush used as a covariate. Cambial growth is cumulative circumference growth at breast height (1.37 m) during the treatment year, with circumference growth in the previous year used as a covariate.

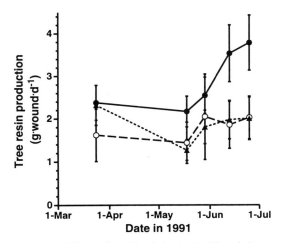

Figure 3. Resin production in 1991 as affected by sheltering (solid circles), control (open circles), and irrigation (solid triangles).

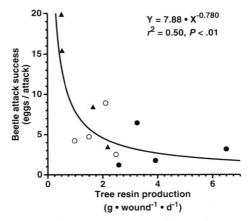

Figure 4. Beetle attack success (eggs per attacking adult) as a function of resin production in individual trees exposed to different water regimes: sheltering (solid circles), control (open circles), and irrigation (solid triangles). Attack success was determined by dissecting four 500-cm^2 sections of phloem from each tree just after oviposition was completed.

removing resin from their attack galleries and had less time for excavating oviposition galleries. SPB attacking irrigated trees left 2.7 to 6.6 times more progeny than their counterparts attacking sheltered trees.

These results seem counterintuitive if it is assumed that water deficits and growth limitations are inherently bad for trees and good for bark beetles. However, results strongly support hypotheses derived from plant growth–differentiation balance (Loomis, 1932, 1953). Trees do not partition carbon according to fixed ratios. Environmental effects can dramatically alter commitments of carbon to growth versus secondary metabolism. In the case of water deficits that limit growth but not photosynthesis, the accumulating carbon supports cellular differentiation (e.g., secondary wall thickening, the formation of vertical resin ducts) and can lead to increased secondary metabolism (e.g., oleoresin synthesis), resulting in a more resistant tree. Conversely, well-watered trees tend to sustain growth further into the summer, which limits carbon for cellular differentiation and secondary metabolism, and produces a less resistant tree.

E. Future Studies

These results suggest new and largely untested predictions about the role of climatic patterns in bark beetle population dynamics. For example, years with high precipitation, especially if it prolongs the period of earlywood growth, are expected to favor SPB population increases. Similarly, productive sites with relatively high water availability are expected to provide better habitat for SPB. This latter prediction is consistent with stand risk ratings for SPB in the Kisatchie National Forest of Louisiana (Lorio *et al.*, 1982; Lorio and Sommers, 1986). Future research will include the development of models that input precipitation patterns and soil attributes to predict patterns of secondary metabolism in loblolly pine and rates of population increase in SPB. One of the key unknowns is the level of water deficits at which tree secondary metabolism begins to drop and bark beetle success increases (Fig. 1).

III. Predation and SPB Dynamics

A. Background on Interactions between *Thanasimus dubius* and SPB

Once attack of the host tree by SPB is under way, a large assemblage of natural enemies prey upon or parasitize various stages of SPB, both inside and outside the tree. One of the first enemies to appear is adult *Thanasimus dubius* (F.) (Coleoptera: Cleridae), which are attracted by SPB pheromones and volatiles emitted by the damaged host tree (Vité and Williamson, 1970; Dixon and Payne, 1979, 1980). Adult *T. dubius* catch and consume the adult SPB arriving on the host tree. At the same time, *T. dubius* mate and oviposit on the bark surface (Thatcher and Pickard, 1966). After hatching, *T. dubius* larvae enter the phloem, where they feed on the larval progeny of those SPB that successfully entered the tree.

A number of studies suggest that *T. dubius* is the most important natural enemy of SPB, although its role in SPB dynamics still remains uncertain. Moore (1972) estimated that 24% of SPB larvae were killed by insect predators and parasitoids, and attributed half of this mortality to *T. dubius* larvae. Linit and Stephen (1983) estimated that 26% of the mortality of SPB larvae was due to natural enemies, especially *T. dubius*. These results must underestimate the total impacts of *T. dubius* on SPB because they do not include adult predation. As early as 1908, Fiske noted that on trees where *T. dubius* was numerous, the fragmentary remains of adult SPB could be found in quantity within crevices in the bark. Thatcher and Pickard (1966) released SPB and *T. dubius* adults into a room containing a freshly cut pine log, and found that *T. dubius* reduced by about 50% the number of SPB successfully attacking the log. Studies of *T. dubius* in laboratory arenas provide further indications that *T. dubius* can capture and consume large numbers of SPB (Turnbow *et al.*, 1978; Nebeker and Mizell, 1980; Turnbow and Franklin, 1980; Frazier *et al.*, 1981).

Clearly, *T. dubius* can inflict considerable mortality on SPB, but additional information is required to define its role in SPB population dynamics. For example, we need more detailed estimates of the mortality inflicted on SPB by

T. dubius (both adults and larvae) across a natural range of predator and prey densities. We also need information on the numerical response of *T. dubius* to fluctuations in SPB density. In addition, theory has emphasized the importance of both predator and prey movement in population dynamics (Kareiva and Odell, 1987; Murdoch and Stewart-Oaten, 1989; Hassell *et al.*, 1991; Ives, 1992a,b; Murdoch *et al.*, 1992), but at present we have information only for SPB (Turchin and Thoeny, 1993). Finally, developments in age-structured population models suggest that the relative durations of the predator and prey life cycles can have important effects on dynamics (Nisbet and Gurney, 1982; Hastings, 1984; Nunney, 1985; Murdoch *et al.*, 1987; Godfray and Hassell, 1989), but until now we had little information on the development of *T. dubius* under field conditions.

B. Functional Response of *Thanasimus dubius*

To estimate the mortality inflicted by adult *T. dubius* on SPB, we conducted laboratory experiments in which SPB were exposed to different predator densities. The experimental arenas were bolts (80 cm long) cut from mature loblolly pines, with 50 cm enclosed by a spherical cage. Predator densities were set to 0, 10, 20, and 40 adult *T. dubius* per bolt, which encompasses the range of predator densities we have observed in the field (0 to 1.25 predators per dm^2 of bark surface area). The predator treatments were crossed with prey densities of 100, 200, and 400 SPB per bolt, to create a range of attack densities representative of what might occur over one day in nature. At the end of the experiment, we counted the number of SPB elytra in the cage to estimate the number of SPB eaten by *T. dubius* prior to entering the bolt. Then bolts were dissected to count the number of SPB that had successfully entered the phloem. Each treatment combination was replicated three times.

Predation by adult *T. dubius* substantially reduced the proportion of SPB successfully attacking the bolt (Fig. 5). As predator density increased from 0 to 40 adults per bolt, the proportion of SPB entering the bolt fell from about 60 to 20%. Moreover, there appeared to be an impact on attack success even at the lowest predator density (10 adults per bolt). Adult *T. dubius* consumed 17–53% of the attacking SPB, with the proportion eaten increasing with predator density (Fig. 5). However, the proportion eaten declined with increasing prey density, implying some saturation of the predator's functional response at high SPB densities. These results suggest that predation by adult *T. dubius* in the field could cause substantial mortality of SPB and slow colonization of individual host trees, and reduce the total number of trees that are attacked.

C. Numerical Response of *Thanasimus dubius*

To examine numerical responses of *T. dubius*, we initiated a long-term sampling program to monitor adult densities of SPB and *T. dubius* in the

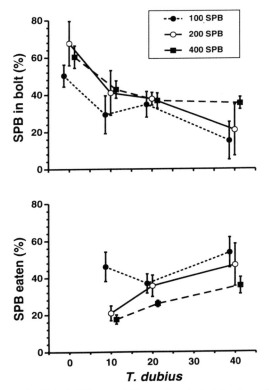

Figure 5. Proportion of SPB (±1 SE) successfully entering the bolt by the end of the experiment, and the proportion eaten by *T. dubius* (±1 SE) as a function of SPB and *T. dubius* density. *Thanasimus dubius* density had a significant negative effect on the proportion entering the bolt ($P < 0.001$), whereas SPB density had a significant positive effect ($P < 0.015$). Conversely, *T. dubius* density had a significant positive effect on the proportion eaten ($P < 0.001$), whereas SPB density had a significant negative effect ($P < 0.001$).

Kisatchie National Forest in central Louisiana. Since 1989, we have maintained three to nine baited multiple funnel traps (Lindgren, 1983) deployed in transects across each of four Ranger Districts. The traps are baited with frontalin (the aggregation pheromone of SPB) and turpentine; this combination of chemicals is highly attractive to both SPB and *T. dubius* (Vité and Williamson, 1970; Payne *et al.*, 1978; Dixon and Payne, 1979). The bulk of insect captures typically occur during the first 6 months of each year. Consequently, we estimated the relative abundance of SPB and *T. dubius* as individuals caught per trap per day for each district, during the first 6 months of each year. The four districts showed similar temporal patterns of SPB and *T. dubius* abundance, so we averaged the trap catches across them to yield a mean value for each year of the survey. Graphical

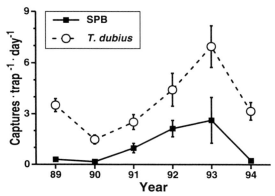

Figure 6. Trap catch per day of SPB (solid squares) and *T. dubius* (open circles) within the Kisatchie National Forest, for the period 1989–1994.

analysis suggests a numerical response by *T. dubius* to changes in the density of SPB (Fig. 6). As SPB populations changed from endemic levels in 1989–1990 to a mild outbreak in 1993, followed by a collapse in 1994, captures of *T. dubius* paralleled these changes in prey density (Fig. 6). This pattern of numerical response is consistent with the hypothesis that *T. dubius* predation could impact SPB population dynamics, suppressing or at least moderating outbreaks. Although *T. dubius* abundance varied over time, adults were always present, even when SPB densities were very low (Fig. 6). We now know that some *T. dubius* enter a prolonged diapause (see the following), so some of the *T. dubius* trapped during low SPB density could have been individuals emerging from trees attacked by SPB one or more years previously.

D. Life History of *Thanasimus dubius*

Laboratory studies of *T. dubius* have indicated a development time from egg to adult of about 110–240 days, depending on temperature (Nebeker and Purser, 1980; Lawson and Morgan, 1992). However, recent results indicate that in nature many *T. dubius* individuals undergo an extended period of development inside the host tree. Figure 7 shows the emergence of *T. dubius* adults from a tree attacked by SPB in October 1992. Thirty-two *T. dubius* adults emerged in the spring of 1993, which is about the time that would be expected based on previous laboratory studies. However, another 57 *T. dubius* emerged during the following autumn and winter, and another 15 emerged this past autumn, approximately 2 years after the time of SPB attack and *T. dubius* oviposition. Apparently, this prolonged development is not unusual. Bark samples taken from trees long va-

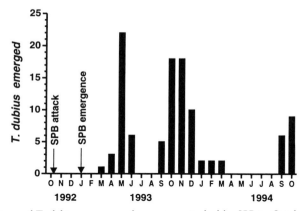

Figure 7. Pattern of *T. dubius* emergence from a tree attacked by SPB in October 1992. A large emergence trap, covering a 150-cm length of the tree, was used to collect the emerging adults. Also shown is the time at which SPB emergence occurred.

cated by SPB frequently contain large numbers of immature *T. dubius* that are apparently in a state of diapause. At a site in East Texas, 49% of 307 pupation chambers still contained live *T. dubius* immatures almost 2 years after SPB attack, and at a site in central Louisiana of similar age, 36% of 181 chambers still contained living *T. dubius*. By comparison, we estimate that SPB may complete about six generations per year in central Louisiana (based on climatic records and the temperature sensitivity of developmental processes; Gagne *et al.*, 1982; Wagner *et al.*, 1984; M. P. Ayres, unpublished analyses).

These revisions in our understanding of the *T. dubius* life cycle have two important implications for the population dynamics of *T. dubius* and SPB. First, the protracted life history of *T. dubius* makes it a viable candidate for the delayed density dependence thought responsible for generating population cycles in SPB, which operates with a 1-year time lag (Turchin *et al.*, 1991). If the generation time of *T. dubius* were only 6 months in the field, as we had previously thought, then we would have rejected *T. dubius* as the force driving SPB population cycles. Second, theoretical studies suggest that a long development time in the predator, relative to that of the prey, has a strong destabilizing effect on predator–prey dynamics (Nunney, 1985; Murdoch *et al.*, 1987; Godfray and Hassell, 1989). On these grounds we would expect a *T. dubius*–SPB system to exhibit long-term oscillations in predator and prey abundance, unless some rapidly acting factor compensates for the destabilizing effect of this delay in *T. dubius* development (Nunney, 1985).

IV. Conclusions

SPB population dynamics, like those of many other bark beetles, are characterized by extreme fluctuations (up to four orders of magnitude within 5 years; Turchin *et al.*, 1991). Historically, explanations of bark beetle population dynamics have recognized the potential role of natural enemies, but have usually emphasized variation in host suitability. Our ongoing work with SPB is aimed at (1) elucidating and testing physiologically explicit models of environmental effects on tree suitability for bark beetles, (2) assessing the numerical impacts of natural enemies on SPB population dynamics, and (3) evaluating the relative contributions of host suitability and natural enemies in producing observed population dynamics.

SPB population fluctuations show pattern in both time and space. Time-series analyses indicate a cyclical pattern to SPB outbreaks (Turchin *et al.*, 1991). The periodicity of these outbreaks is difficult to explain with any climatic mechanisms (e.g., effects of precipitation patterns on suitability of host trees) unless there is some periodicity to the climatic patterns; none has yet been discovered (Turchin *et al.*, 1991), although our revised understanding of moisture effects on tree physiology suggests that linear models may be an inadequate test (Fig. 1). It seems more likely that the cyclic tendencies of SPB populations are due to some biological agent(s) that act in a delayed density-dependent fashion (e.g., natural enemies or competitors). SPB population fluctuations are also characterized by large-scale spatial synchrony. For example, severe SPB outbreaks occurred in 1985–1986 from East Texas through Louisiana, Mississippi, and Alabama (Price *et al.*, 1992). This spatial synchrony is difficult to explain if population dynamics are primarily driven by natural enemies, because most natural enemies of SPB are other arthropods that seem unlikely to move more than a few kilometers per generation. It seems more likely that the spatial synchrony is driven by forces that act on a regional scale, such as climatic variation. Even infrequent climatic events may be adequate to synchronize population cycles across broad geographic areas (Royama, 1984).

Our research supports the hypothesis that SPB population dynamics are influenced by both density-dependent interactions with *T. dubius* and environmental effects on host trees. *Thanasimus dubius* adults at natural densities can kill up to 53% of the SPB adults attempting to colonize a tree (Fig. 5). By comparison, moderate drought stress can increase resin flow in trees and reduce SPB reproductive success by 63–85% relative to that in irrigated trees (Figs. 3 and 4). In both cases, these are probably underestimates of the potential effects on SPB population dynamics. The *T. dubius* larvae feed on SPB larvae beneath the bark, but we presently can only estimate effects of adult predation. Similarly, tree water balance and physiological status probably influence the growth and survival of SPB larvae, but we presently can only estimate effects on attacking adults. The expected effects of these two forces on SPB population dynamics are quite different. *Thanasimus dubius* has a long development time, which makes it a good candidate for producing outbreak cycles, whereas climatic effects on host-tree suitability may influence the amplitude of outbreaks and synchronize population cycles on a regional scale.

Acknowledgments

We thank F. Gerhardt, S. Harrison, P. Price, J. Roland, J. Ruel, A. Shumate, A. Swanson, and R. Wilkens for comments on the manuscript, and R. Wilkens for his assistance in preparing the figures.

References

Beal, J. A. (1927). Weather as a factor in southern pine beetle control. *J. For.* **25**, 741–742.

Beal, J. A. (1933). Temperature extremes as a factor in the ecology of the southern pine beetle. *J. For.* **31**, 329–336.

Berryman, A. A. (1967). Estimation of *Dendroctonus brevicomis* (Coleoptera: Scolytidae) mortality caused by insect predators. *Can. Entomol.* **99**, 1009–1014.

Berryman, A. A. (1972). Resistance of conifers to invasion by bark beetle–fungus associations. *Science* **22**, 598–602.

Berryman, A. A. (1973). Population dynamics of the fir engraver, *Scolytus ventralis* (Coleoptera: Scolytidae). I. Analysis of population behavior and survival from 1964 to 1971. *Can. Entomol.* **105**, 1465–1488.

Berryman, A. A. (1976). Theoretical explanation of mountain pine beetle dynamics in lodgepole pine forests. *Environ. Entomol.* **5**, 1225–1233.

Berryman, A. A. (1988). "Dynamics of Forest Insect Populations: Patterns, Causes, and Implications." Plenum, New York.

Berryman, A. A., and Ferrell, G. T. (1988). The fir engraver beetle in western states. *In* "Dynamics of Forest Insect Populations: Patterns, Causes, and Implications" (A. A. Berryman, ed.), pp. 556–578. Plenum, New York.

Billings, R. F. (1979). Detecting and aerially evaluating southern pine beetle outbreaks. *South. J. Appl. For.* **3**, 50–54.

Billings, R. F. (1988). Forecasting southern pine beetle infestation trends with pheromone traps. *In* "Integrated Control of Scolytid Bark Beetles" (T. L. Payne and H. Sarenmaa, eds.), pp. 295–306. Virginia Polytechnic Institute and State University, Blacksburg.

Christiansen, E., and Bakke, A. (1988). The spruce bark beetle of Eurasia. *In* "Dynamics of Forest Insect Populations: Patterns, Causes, and Implications" (A. A. Berryman, ed.), pp. 480–504. Plenum, New York.

Craighead, F. C. (1925). Bark beetle epidemics and rainfall deficiency. *J. Econ. Entomol.* **18**, 577–584.

Dixon, W. N., and Payne, T. L. (1979). Attraction of entomophagous and associate insects of the southern pine beetle to beetle- and host tree-produced volatiles. *Environ. Entomol.* **8**, 178–181.

Dixon, W. N., and Payne, T. L. (1980). Sequence of arrival and spatial distribution of entomophagous and associate insects on southern pine beetle-infested trees. *Tex. Agric. Exp. Stn.* [*Misc. Publ.*] **MP-1432**.

Dunn, J. P., and Lorio, P. L., Jr. (1993). Modified water regimes affect photosynthesis, xylem water potential, cambial growth, and resistance of juvenile *Pinus taeda* L. to *Dendroctonus frontalis* (Coleoptera: Scolytidae). *Environ. Entomol.* **22**, 948–957.

Fiske, W. F. (1908). Notes on insect enemies of wood boring Coleoptera. *Proc. Entomol. Soc. Wash.* **9**, 23–27.

Frazier, J. L., Nebeker, T. E., Mizell, R. F., and Calvert, W. H. (1981). Predatory behavior of the clerid beetle *Thanasimus dubius* (Coleoptera: Cleridae) on the southern pine beetle (Coleoptera: Scolytidae). *Can. Entomol.* **113**, 35–43.

Gagne, J. A., Wagner, T. L., Sharpe, P. J. H., Coulson, R. N., and Fargo, W. S. (1982). Reemergence of *Dendroctonus frontalis* (Coleoptera: Scolytidae) at constant temperatures. *Environ. Entomol.* **11**, 1216–1222.

Godfray, H. C. J., and Hassell, M. P. (1989). Discrete and continuous insect populations in tropical environments. *J. Anim. Ecol.* **58**, 153–174.

Greenbank, D. O. (1956). The role of climate and dispersal in the initiation of outbreaks of the spruce budworm in New Brunswick. *Can. J. Zool.* **34**, 453–476.

Grégoire, J. C. (1988). The greater European spruce beetle. *In* "Dynamics of Forest Insect Populations: Patterns, Causes, and Implications" (A. A. Berryman, ed.), pp. 456–478. Plenum, New York.

Hassell, M. P., May, R. M., Pacala, S., and Chesson, P. L. (1991). The persistence of host–parasitoid associations in patchy environments. I. A general criterion. *Am. Nat.* **138**, 568–583.

Hastings, A. (1984). Delays in recruitment at different trophic levels: Effects on stability. *J. Math. Biol.* **21**, 25–44.

Hopkins, A. D. (1892). Notes on a destructive forest tree scolytid. *Science* **20**, 64–65.

Hopkins, A. D. (1921). The southern pine beetle: A menace to the pine timber of the southern states. *Farmers' Bull.* **1188**.

Ives, A. (1992a). Continuous-time models of host–parasitoid interactions. *Am. Nat.* **140**, 1–29.

Ives, A. (1992b). Density-dependent and density-independent parasitoid aggregation in model host–parasitoid systems. *Am. Nat.* **140**, 912–937.

Kalkstein, L. S. (1981). An improved technique to evaluate climate–southern pine beetle relationships. *For. Sci.* **27**, 579–589.

Kareiva, P., and Odell, G. M. (1987). Swarms of predators exhibit "preytaxis" if individual predators use area-restricted search. *Am. Nat.* **130**, 233–270.

King, E. W. (1972). Rainfall and epidemics of the southern pine beetle. *Environ. Entomol.* **1**, 279–285.

Kozlowski, T. T. (1969). Tree physiology and forest pests. *J. For.* **67**, 118–123.

Kroll, J. C., and Reeves, H. C. (1978). A simple model for predicting annual numbers of southern pine beetle infestations in East Texas. *South. J. Appl. For.* **2**, 62–64.

Larsson, S. (1989). Stressful times for the plant stress–insect performance hypothesis. *Oikos* **56**, 277–283.

Lawson, S. A., and Morgan, F. D. (1992). Rearing of two predators, *Thanasimus dubius* and *Temnochila virescens*, for the biological control of *Ips grandicollis* in Australia. *Entomol. Exp. Appl.* **65**, 225–233.

Lindgren, B. S. (1983). A multiple funnel trap for scolytid beetles (Coleoptera). *Can. Entomol.* **115**, 299–302.

Linit, M. J., and Stephen, F. M. (1983). Parasite and predator component of within-tree southern pine beetle (Coleoptera, Scolytidae) mortality. *Can. Entomol.* **115**, 679–688.

Loomis, W. E. (1932). Growth–differentiation balance vs. carbohydrate–nitrogen ratio. *Proc. Am. Soc. Hortic. Sci.* **29**, 240–245.

Loomis, W. E. (1953). Growth and differentiation—An introduction and summary. *In* "Growth and Differentiation in Plants" (W. E. Loomis, ed.), pp. 1–17. Iowa State Coll. Press, Ames.

Lorio, P. L., Jr. (1986). Growth–differentiation balance: A basis for understanding southern pine beetle-tree interactions. *For. Ecol. Manage.* **14**, 259–273.

Lorio, P. L., Jr. (1988). Growth–differentiation balance relationships in pines affect their resistance to bark beetles (Coleoptera: Scolytidae). *In* "Mechanisms of Woody Plant Defenses Against Insects" (W. J. Mattson, J. Levieux, and C. Bernard-Dagan, eds.), pp. 73–92. Springer-Verlag, New York.

Lorio, P. L., Jr., and Hodges, J. D. (1977). Tree water status affects induced southern pine beetle attack and brood production. *USDA For. Serv. Res. Pap.* **SO-135.**

Lorio, P. L., Jr., and Sommers, R. A. (1986). Evidence of competition for photosynthates between growth processes and oleoresin synthesis in *Pinus taeda* L. *Tree Physiol.* **2**, 301–306.

Lorio, P. L., Jr., Mason, G. N., and Autry, G. L. (1982). Stand risk rating for the southern pine beetle: Integrating pest management with forest management. *J. For.* **80**, 212–214.

Lorio, P. L., Jr., Sommers, R. A., Blanche, C. A., Hodges, J. D., and Nebeker, T. E. (1990). Modeling pine resistance to bark beetles based on growth and differentiation balance principles. *In* "Process Modeling of Forest Growth Responses to Environmental Stress" (R. K. Dixon, R. S. Meldahl, G. A. Ruark, and W. G. Warren, eds.), pp. 402–409. Timber Press, Portland, OR.

Mattson, W. J., and Haack, R. A. (1987a). The role of drought stress in provoking outbreaks of phytophagous insects. *In* "Insect Outbreaks" (P. Barbosa and J. C. Schultz, eds.), pp. 365–407. Academic Press, New York.

Mattson, W. J., and Haack, R. A. (1987b). The role of drought in outbreaks of plant-eating insects. *BioScience* **37**, 110–118.

Michaels, P. J. (1984). Climate and the southern pine beetle in Atlantic Coastal and Piedmont regions. *For. Sci.* **30**, 143–156.

Miller, M. C. (1984). Mortality contribution of insect natural enemies to successive generations of *Ips calligraphus* (Germar) (Coleoptera, Scolytidae) in loblolly pine. *Z. Angew. Entomol.* **98**, 495–500.

Miller, M. C. (1986). Survival of within-tree *Ips calligraphus* (Col.: Scolytidae): Effects of insect associates. *Entomophaga* **31**, 305–328.

Mills, N. J. (1985). Some observations on the role of predation in the natural regulation of *Ips typographus* populations. *Z. Angew. Entomol.* **99**, 209–320.

Moore, G. E. (1972). Southern pine beetle mortality in North Carolina caused by parasites and predators. *Environ. Entomol.* **1**, 58–65.

Murdoch, W. W., and Stewart-Oaten, A. (1989). Aggregation by parasitoids and predators: Effects on equilibrium and stability. *Am. Nat.* **134**, 288–310.

Murdoch, W. W., Nisbet, R. M., Blythe, S. P., Gurney, W. S. C., and Reeve, J. D. (1987). An invulnerable age class and stability in delay-differential parasitoid–host models. *Am. Nat.* **129**, 263–282.

Murdoch, W. W., Briggs, C. J., Nisbet, R. M., Gurney, W. S. C., and Stewart-Oaten, A. (1992). Aggregation and stability in metapopulation models. *Am. Nat.* **140**, 41–58.

Nebeker, T. E., and Mizell, R. F. (1980). Behavioral considerations in quantifying the impact of *Thanasimus dubius* (F.) adults on bark beetle populations. *USDA For. Serv. Tech. Bull.* **1630**, 98–108.

Nebeker, T. E., and Purser, G. C. (1980). Relationship of temperature and prey type to development time of the bark beetle predator *Thanasimus dubius* (Coleoptera: Scolytidae). *Can. Entomol.* **112**, 179–184.

Nisbet, R. M., and Gurney, W. S. C. (1982). "Modelling Fluctuating Populations." Wiley, New York.

Nunney, L. (1985). The effects of long time delays in predator–prey systems. *Theor. Popul. Biol.* **27**, 202–221.

Payne, T. L., Coster, J. E., Richerson, J. V., Edson, L. J., and Hart, E. R. (1978). Field response of the southern pine beetle to behavioral chemicals. *Environ. Entomol.* **7**, 578–582.

Price, P. W. (1991). The plant vigor hypothesis and herbivore attack. *Oikos* **62**, 244–251.

Price, T. S., Doggett, C., Pye, J. M., and Holmes, T. P. (1992). "A History of Southern Pine Beetle Outbreaks in the Southeastern United States." Georgia Forestry Commission, Macon.

Raffa, K. F. (1988). The mountain pine beetle in western North America. *In* "Dynamics of Forest Insect Populations: Patterns, Causes, and Implications" (A. A. Berryman, ed.), pp. 506–530. Plenum, New York.

Raffa, K. F., and Berryman, A. A. (1983). The role of host plant resistance in the colonization behavior and ecology of bark beetles (Coleoptera: Scolytidae). *Ecol. Monogr.* **53**, 27–49.

Rhoades, D. F. (1979). Evolution of plant chemical defense against herbivores. *In* "Herbivores: Their Interaction with Secondary Plant Metabolites" (G. A. Rosenthal and D. H. Janzen, eds.), pp. 3–54. Academic Press, New York.

Riley, M. A., and Goyer, R. A. (1986). Impact of beneficial insects on *Ips* spp. (Coleoptera: Scolytidae) bark beetles in felled loblolly and slash pines in Louisiana. *Environ. Entomol.* **15**, 1220–1224.

Royama, T. (1984). Population dynamics of the spruce budworm, *Choristoneura fumiferana*. *Ecol. Monogr.* **54**, 429–462.

Rudinsky, J. A. (1962). Ecology of Scolytidae. *Annu. Rev. Entomol.* **7**, 327–348.

Stark, R. W. (1965). Recent trends in forest entomology. *Annu. Rev. Entomol.* **10**, 303–324.

St. George, R. A. (1930). Drought-affected and injured trees attractive to bark beetles. *J. Econ. Entomol.* **23,** 825–828.

Thatcher, R. C., and Pickard, L. S. (1966). The clerid beetle, *Thanasimus dubius,* as a predator of the southern pine beetle. *J. Econ. Entomol.* **59,** 955–957.

Turchin, P., and Thoeny, W. T. (1993). Quantifying dispersal of southern pine beetles with mark–recapture experiments and a diffusion model. *Ecol. Appl.* **3,** 187–198.

Turchin, P., Lorio, P. L., Jr., Taylor, A. D., and Billings, R. F. (1991). Why do populations of southern pine beetles (Coleoptera: Scolytidae) fluctuate? *Environ. Entomol.* **20,** 401–409.

Turnbow, R. H., and Franklin, R. T. (1980). The effects of temperature on *Thanasimus dubius* oviposition, egg development and adult prey consumption. *J. Ga. Entomol. Soc.* **15,** 456–459.

Turnbow, R. H., Franklin, R. T., and Nagel, W. P. (1978). Prey consumption and longevity of adult *Thanasimus dubius. Environ. Entomol.* **7,** 695–697.

Vité, J. P., and Williamson, D. L. (1970). *Thanasimus dubius:* Prey perception. *J. Insect Physiol.* **106,** 233–239.

Wagner, T. L., Gagne, J. A., Sharpe, P. J. H., and Coulson, R. N. (1984). A biophysical model of southern pine beetle, *Dendroctonus frontalis* Zimmermann (Coleoptera: Scolytidae), development. *Ecol. Modell.* **21,** 125–147.

Waring, G. L., and Cobb, N. S. (1992). The impact of plant stress on herbivore population dynamics. *In* "Plant–Insect Interactions" (E. A. Bernays, ed.), Vol. 4, pp. 167–226. CRC Press, Boca Raton, FL.

Waring, R. H. (1983). Estimating forest growth and efficiency in relation to tree canopy area. *Adv. Ecol. Res.* **13,** 327–354.

Waring, R. H., and Pitman, G. B. (1980). A simple model of host resistance to bark beetles. *For. Res. Notes* **65,** 1–2.

Weslien, J. (1992). The arthropod complex associated with *Ips typographus* (L.) (Coleoptera, Scolytidae): Species composition, phenology, and impact on bark beetle productivity. *Entomol. Fenn.* **3,** 205–213.

Weslien, J. (1994). Interaction within and between species at different densities of the bark beetle *Ips typographus* and its predator *Thanasimus formicarius. Entomol. Exp. Appl.* **71,** 133–143.

Weslien, J., and Regnander, J. (1992). The influence of natural enemies on brood production in *Ips typographus* (Coleoptera, Scolytidae) with special reference to egg-laying and predation by *Thanasimus formicarius* (L.) (Coleoptera, Cleridae). *Entomophaga* **37,** 333–342.

White, T. C. R. (1974). A hypothesis to explain outbreaks of looper caterpillars, with special reference to populations of *Selidosema suavis* in a plantation of *Pinus radiata* in New Zealand. *Oecologia* **16,** 279–301.

Wyman, L. (1924). Bark-beetle epidemics and rainfall deficiency. *USDA For. Serv. Bull.* **8,** 2–3.

Zahner, R. (1968). Water deficits and growth of trees. *In* "Water Deficits and Plant Growth" (T. T. Kozlowski, ed.), pp. 191–254. Academic Press, New York.

Zahner, R., and Stage, R. (1966). A procedure for calculating daily moisture stress and its utility in regressions of tree growth and weather. *Ecology* **47,** 64–74.

Chapter 18

The Dominance of Different Regulating Factors for Rangeland Grasshoppers

Gary E. Belovsky and Anthony Joern

I. Introduction

Developing a unified, conceptual framework that explains population fluctuations presents a fundamental and critical challenge to ecologists. Empirically and theoretically, this is a formidable undertaking since so many mechanisms can potentially contribute to population fluctuations, including predation, intraspecific competition, and abiotic-induced mortality. This complexity has largely stifled progress as well as interest in the problem. Yet, these issues underlie all of population ecology and cannot be left unresolved. Though our perspective on each population mechanism is not new, we believe that our integration of these processes, using grasshoppers as model organisms, offers some important resolutions to the general problem of population fluctuations.

Previously, rather than viewing these mechanisms in an integrated fashion, ecologists have viewed one or another of them to be *generally dominant* in controlling most populations in nature and this has led to considerable debate for more than 50 years. However, an approach that integrates these mechanisms does not eliminate the issue of which mechanism dominates in limiting a *particular* population. Given our emphasis on integrating mechanisms, it may seem contradictory to refer to a dominant mechanism; however, we will demonstrate that even when a number of mechanisms are operating together, only one density-dependent mechanism can exert a regulating influence at a time in a population. This does not mean that other mechanisms are unimportant in controlling population size, for they help define which mechanism assumes a dominant regulating influence. For example, holding predation and food availability constant, a population experiencing high abiotic-induced mortality may be regulated by predation, whereas another population with lower abiotic-induced mortality may be food-limited.

Within this integrated approach, it is possible for mechanisms to change in

POPULATION DYNAMICS

importance under different environmental conditions, and our approach shifts the focus away from determining which mechanism is *generally* limiting for all populations to defining the environmental and biotic conditions where different mechanisms come to dominate the regulation of a species' population. For example, what conditions lead to a population being limited by predators versus being limited by food resources? Insights gained under this perspective will help to identify which data are most important for assessing population dynamics. Finally, our integrated approach indicates that it is possible to observe alternative stable states for a population that emerge when different conditions (e.g., historical effects) arise in a homogeneous environment or as a heterogeneous environment varies over time.

To evaluate our integrated perspective, we argue that an experimental approach combined with field observations is necessary. Given the environmental variation occurring over time for each population of a species and over space by different populations of the same species, we do not see how the issue of population limitation can be clearly and easily resolved by statistically analyzing large data sets composed of population numbers obtained over time and space in nature (Royama, 1993; Strong, 1984, 1986a,b). Likewise, though experimental population studies can demonstrate how a particular mechanism operates, these results have to be combined with observational studies of field populations to demonstrate whether the experiments provide insights into nature.

In this chapter, we first review data from our studies on grasshoppers that led to our integrated perspective. Second, we develop graphical representations for the operation of different ecological mechanisms. Third, we develop a graphical model of population dynamics that integrates the proposed mechanisms. Fourth, we describe the experimental and observational methods and review representative results from our continuing long-term studies with grasshoppers that are designed to evaluate the model. Finally, we contrast the insights gained by our approach with those provided by others, placing our findings into a general ecological context.

II. Grasshoppers, a First Visit

We evaluated our integrated perspective using grasshoppers in western U.S. grasslands. Grasshoppers (Orthoptera, Acrididae) and grasslands are well suited for this. Grasshoppers are hemimetabolous so they do not dramatically change their behavior during ontogeny; grasshoppers are easy to observe and enumerate given their body size and high activity; most grasshoppers are not highly specialized herbivores, which ensures that their population dynamics do not rely on changes in the availability of a few plant species; many grasshoppers are regularly abundant, even though their population densities can vary dramatically over time and space; grasshoppers are important primary consumers; and grasshoppers are a major source of food to many predators, parasitoids, and parasites. Grasslands are structurally simple terrestrial environments compared to forests, making observations of grasshoppers relatively simple, and grasslands vary dramatically in productivity within sites among years and between sites within a year, making evaluation of environmental differences straightforward.

Although the standard notion is that grasshopper densities are related to annual weather (Andrewartha and Birch, 1954, 1984), the population mechanisms associated with weather have not been established. Furthermore, anomalous patterns emerge between weather conditions and grasshopper densities in northern and southern regions. In the south, densities tend to decrease with hot and dry conditions, whereas in the north they tend to increase (Capinera, 1987; Capinera and Thompson, 1987; Capinera and Horton, 1989; Lockwood and Lockwood, 1991). Therefore, our understanding of grasshopper population dynamics may not be as firm as commonly thought.

Background data for assessing the importance of different mechanisms that might be important to grasshopper populations in grasslands come from more than 15 years of experience gained independently by each of the authors at two different sites (Belovsky at the National Bison Range in Montana, a Palouse prairie, and Joern at Arapaho Prairie in Nebraska, a Sandhill mixed-grass prairie). Each of us initially chose a research approach based on the supposition that a single mechanism would dominate the regulation of grasshopper populations at our study sites, but each of us thought that a different mechanism would be important (for Belovsky, food, and for Joern, predation). Eight years ago, we began to discuss our separate studies, because we were coming to disparate conclusions (Belovsky in Montana was observing food limitation and Joern in Nebraska was observing predator, especially avian, limitation). Rather than arguing over who was correct, we wondered what differences between our study sites might explain our diverging conclusions or what each of us might have overlooked.

Our willingness to seek a unifying explanation was obviated because each of us had employed experimental methods that provided unambiguous results. However, we had each employed different experimental manipulations in the different systems. Belovsky in Montana eliminated all predators and varied initial grasshopper density (nymphs) and plant availability in the field to observe how these factors affected adult grasshopper density and reproduction; these values were compared with unmanipulated field populations. Joern in Nebraska eliminated avian predation and compared grasshopper population densities in the presence and absence of birds. Therefore, it was obvious that each of us needed to conduct the other's experiments at our own study sites.

Comparative results on intraspecific competition for food, predation, and abiotic factors that were obtained from common experiments with the most abundant grasshoppers at each site (*Melanoplus sanguinipes, M. femurrubrum,* and *Ageneotettix deorum* in Montana and *A. deorum* in Nebraska) are summarized in the following. Other factors (e.g., parasitoids) could be important at some times and places, but our goal was to examine a manageable set of mechanisms that we already suspected to be important, and then use these to construct a general framework for the integration of mechanisms that are simultaneously operating.

A. Predators Eliminated and the Initial Grasshopper Density and Plant Availability Varied

These experiments were conducted in small cages (0.1–1.0 m²) placed over field vegetation that maintained abiotic conditions for the grasshoppers and plants (Belovsky and Slade, 1993, 1995; Joern and Klucas, 1993). The cages were stocked with grasshopper nymphs at densities comparable to the range of values observed in the field at hatching time. Though the use of small enclosed populations is sometimes criticized as being too artificial, we believe that they provide an important experimental tool for identifying the presence of particular population mechanisms and describing how they operate, because the necessary manipulations cannot be maintained outside of enclosures. The important caveat is that enclosed populations must be contrasted with unnmanipulated field populations to ensure that the observations are pertinent.

We obtained many similar results at both sites. First, adult grasshopper density at a given plant availability stabilized at a single constant density, if the initial density of nymphs was sufficiently large (see Fig. 3d). Second, the adult density increased with increasing plant availability in Montana (Fig. 1a), where availability was increased by providing vegetation with supplemental water (Fig. 1b) and/or nitrogen (Schmitz, 1993). *Ageneotettix deorum* densities also increased with supplemental water and nitrogen (J. Moorehead, personal communication). In Nebraska, *A. deorum* densities did not change with supplemental water and/or nitrogen in most years, but decreased in abundance with supplemental water and/or nitrogen in another year (Fig. 1c). However, *M. sanguinipes* increased in abundance with supplemental water and nitrogen (Fig. 1c).

The different responses for *A. deorum* at the two sites may reflect generally lower plant quality in Nebraska. This is supported by measures of plant solubility in HCl + pepsin, which are correlated with digestibility to grasshoppers (Belovsky and Slade, 1995) and plant protein content (Heidorn and Joern, 1987): 20–30% soluble in Nebraska versus 32–41% in Montana in July. The smaller-bodied *A. deorum* needs plants of higher nutritional value than *M. sanguinipes* (Belovsky, 1986; Belovsky and Slade, 1996a), and supplemental water and nitrogen tend to increase plant biomass while diminishing its nutrient content (Belovsky and Slade, 1995; Joern, 1989b, 1990). Therefore, many of the plants in Nebraska, which are nutritionally poorer than in Montana, may become too poor

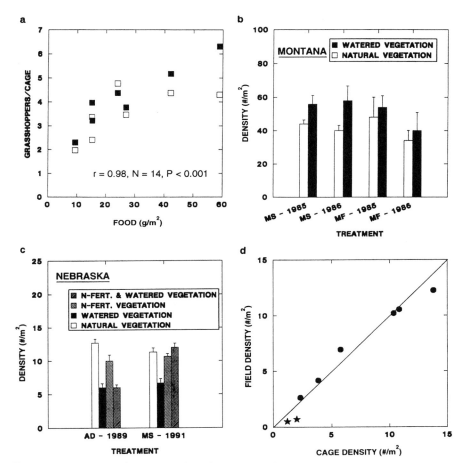

Figure 1. A summary of experimental results using caged populations of grasshoppers. (a) The relationship between grasshopper density and food abundance (product of plant biomass and its solubility in HCl + pepsin, see text) for *M. sanguinipes* (solid squares) and *M. femurrubrum* (open squares) in Montana (Belovsky and Slade, 1995). (b) The effect of supplemental water for vegetation on the survival to the adult stage for *M. sanguinipes* (MS) and *M. femurrubrum* (MF) in Montana (Belovsky and Slade, 1995). (c) The effect of supplemental water and/or nitrogen fertilizer for vegetation on survival to the adult stage for *A. deorum* (AD) and *M. sanguinipes* (MS) in Nebraska (Joern, 1989b, 1990). (d) Comparison of densities attained in cages by *M. sanguinipes* in Montana (circles) (Belovsky and Slade, 1995) and *A. deorum* in Nebraska (stars) (Joern and Klucas, 1993) with field densities. All error bars represent the standard error with sample size equal to 6.

for *A. deorum* to feed upon when they receive supplemental water and nitrogen; whereas the higher-quality plants in Montana may still be of sufficient quality after supplementation.

When constant grasshopper densities obtained in cages were compared with unmanipulated field densities, differences emerged. In Montana, field and experimental population densities were similar (Fig. 1d), indicating that field populations were food-limited (i.e., bottom-up control, *sensu* Hunter and Price, 1992). In Nebraska, the field population densities were lower than experimental densities, indicating that food was not limiting (Fig. 1c). Because the cages did not appreciably modify abiotic conditions (Belovsky and Slade, 1993), differences in density-independent mechanisms cannot explain discrepancies or concordance between cage and field densities.

B. Elimination of Avian Predators

Avian predation was examined in other experiments (100-m^2 areas covered with netting to exclude birds and not covered) (Joern, 1986, 1992; Belovsky and Slade, 1993). Different average results were obtained at the two sites, but infrequent annual variation was observed. In Nebraska, the densities of all grasshopper species tended to be lower in the presence of birds (7 of 9 years and locations), indicating that birds generally exert a limit on grasshopper densities (Fig. 2a) (top-down control, *sensu* Hunter and Price, 1992). In Montana, the densities of some large-bodied grasshopper species declined in the presence of birds, but the smaller-bodied common grasshoppers increased in abundance with birds (Fig. 2b), indicating that birds did not limit (16 of 18 years and locations), but enhanced the abundance of common species, which is expected if food (bottom-up control) limits their abundance and the larger-bodied species are superior competitors for food (Belovsky, 1986; Belovsky and Slade, 1993).

C. Our Observations and Support from Elsewhere in the Western United States

The foregoing results indicate the importance of long-term experiments, because our original identification of disparities between the sites was based on neither of us having observed the other's results at our sites (e.g., predator limitation in Montana). However, in recent years we have each observed periodic changes in the dominance of population-limiting mechanisms in particular years and locations. This has led us to ask under what conditions the grasshopper populations might shift between predator- and food-limited regulation.

To assess whether patterns exist between grasshopper populations that were predator-limited versus food-limited, we reviewed literature on grasshoppers from the western United States. We restricted our review to studies that employed field experiments (Fowler *et al.*, 1991; Evans, 1989; Brusven and Fielding, 1990; Wang and Walgenbach, 1989, 1990; Thompson *et al.*, 1989; E. Evans, personal communication; M. Ritchie, personal communication; Bock

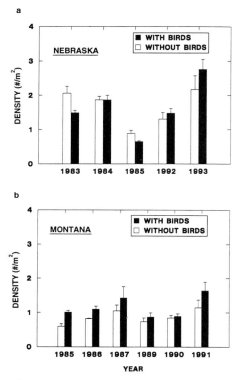

Figure 2. A summary of experimental results using avian exclosures. (a) In most years at one location in Nebraska, in the presence of birds grasshopper numbers decreased or did not change, but in 1993 numbers increased. (b) In most years at one location in Montana, in the presence of birds grasshopper numbers increased (also see Fig. 8d), but in 1992 numbers decreased (Fig. 8c). All error bars represent the standard error with sample size equal to 3.

et al., 1992), and where the data indicated food- or predator-limited populations. Because different methods were often used in these studies, we employed only qualitative comparisons (e.g., density classes, body size classes, evidence for interspecific competition, mainly C_3 or C_4 grasses). The literature review reinforced our emerging conclusions that populations could be either predator- or food-limited and predictable patterns were associated with one or the other mode of limitation. Sites exhibiting predator-limited populations tended to be dominated by small-bodied grasshoppers that have low densities and where C_4 grasses dominate. Sites exhibiting food-limited populations tend to be dominated by large-bodied grasshoppers that have high densities and compete interspecifically and where C_3 grasses dominate.

These patterns translate geographically and taxonomically, because C_4 grasses dominate in the south (Teeri and Stowe, 1976), and Gomphocerine grasshoppers, which generally are small, tend to be more abundant with C_4 grasses (Otte, 1984). Furthermore, these patterns have well-established ecological explanations. Grasshoppers tend to harvest less, and less thoroughly assimilate, C_4 grasses as compared with C_3 grasses (Caswell et al., 1973; Caswell and Reed, 1976), and Gomphocerine species compete less (Joern and Klucas, 1993; Chase and Belovsky, 1994) and have narrower feeding niches than Melanopline or Oedopodine grasshoppers (Joern, 1979a,b, 1989a; Joern and Lawlor, 1980; Otte and Joern, 1977; Chapman, 1990).

Results from our studies, as well as from the literature review, confirmed our earlier conclusions that grasshopper populations may be limited differently. We know that a similar array of mechanisms existed at our sites and presumably most western U.S. sites, but a different mechanism appears to emerge as dominant (i.e., limiting) at each site. Why might the same mechanisms operating at our sites, and presumably at other sites, differ in magnitude and consequently lead to differences in grasshopper population limitation? Furthermore, why will annual shifts in how populations are limited emerge and how frequently might these shifts be observed?

III. Graphical Depictions of Population Mechanisms Identified in Our Studies

To develop an integrated perspective, we focused on demographic responses most likely to exhibit significant density-dependent and -independent effects. The density-dependent responses are particularly important since these are necessary for populations to be regulated (Sinclair, 1989). We considered three responses: (1) survival of hatchlings to the adult stage in the absence of predators, which assesses the impacts of food competition and density-independent mortality; (2) the production of hatchlings per female for the next generation, which assesses the impacts of food competition and density-independent processes; and (3) predatory mortality. To date, data addressing the mechanisms underlying each of the responses (competition for food, avian predation, and density-independent processes) come from the studies in Montana with *M. sanguinipes*.

A. Probability of Surviving from Hatching to the Adult Stage in the Absence of Predators

An individual's probability of surviving (proportion of all individuals surviving) should be a function of initial hatchling density. This probability will be constant (density independent) over a range of low hatchling densities and then

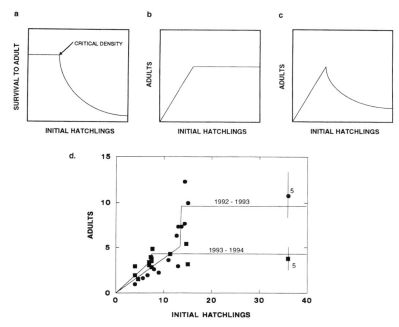

Figure 3. A mechanistic view of survival probability from hatching to the adult stage in the absence of predators. (a) Survival probability is density independent until a critical density is attained and food competition reduces survival. The relationship between survival and initial density in (a) can lead to a constant density of adults, if individuals do not compete equally (b), or to a declining density of adults, if all individuals compete equally (c). (d) Survival observed in experimental Montana populations of M. *sanguinipes* over two generations.

will decrease hyperbolically after hatchling density reaches a critical value. Density-independent survival depends on abiotic conditions and nutrition before per capita food intake begins to decline due to competition (Fig. 3a). After the critical density is attained, survival will decline hyperbolically as per capita food intake (available food/hatchlings) declines (Schoener, 1973; Lomnicki, 1988).

The decline in survival with food competition can result in two patterns: a constant density of adults or a decreasing density of adults as hatchling density continues to increase. The constant adult density reflects exploitative competition where some individuals are more capable of acquiring resources (e.g., different stages of development or genetic differences), so that the less capable individuals die (Fig. 3b) (Lomnicki, 1988). The declining adult density reflects exploitative competition where all individuals are relatively similar in their capabilities for acquiring food, so they all experience reduced survival (Fig. 3c). The Montana data indicate that survival of hatchlings to the adult stage in experimental cages leads to constant, but different annual, adult densities as hatchling density increases (Fig. 3d). The two years were very different (1992 was warm and dry and the food plants were more digestible; 1993 was very cool and wet, and the food plants were less digestible) (Belovsky and Slade, 1996b). Therefore, annual differences in food resources and abiotic conditions are important, especially because abiotic conditions strongly impact food resource quality and quantity (Belovsky and Slade, 1995).

B. Production of Hatchlings per Adult Female

Hatchlings produced per adult female should exhibit a pattern like that observed for survival: until a critical initial hatchling density is attained, per capita production should remain constant (density independent), and then should decline as per capita food intake declines with increasing initial hatchling density (density dependent) (Fig. 4a). The density-independent production is set by a female's ability to process food and convert it into young, given food quality and the abiotic environment. This is equivalent to White's (1978, 1984, 1993) claim that insect reproduction can change in a density-independent fashion as food quality changes.

The Montana data for reproduction were obtained in two ways. First, adult females reared in field cages at different initial hatchling densities were dissected to count ovariole relics that remain after each egg is produced (Joern and Klucas, 1993). These data (Fig. 4b) indicate that egg production/adult female followed the pattern hypothesized in Fig. 4a. Second, hatchling production/adult female cannot be directly measured because the hatchlings observed in the cages at the start of the next year cannot be attributed to a particular adult female. Consequently, the density of hatchlings was divided by the density of adult females in the cage during the previous year to estimate the hatchlings produced/adult female; this also led to the hypothesized pattern (Fig. 4c).

As observed for the survival function, the reproduction functions also varied between years: egg and hatchling production/adult female were lower in 1993 than 1992. These differences are consistent with the previous described annual variations in abiotic conditions and food quality. Therefore, annual differences in food quantity and quality are critical in setting reproductive output.

C. Probability That an Individual Grasshopper Is Killed by Predators

An individual's probability of being killed by predators (proportion of the population killed by predators) should be a function of initial hatchling density. This function will be a combination of the functional and numerical responses for the array of predators in the environment (Holling, 1966). The combined func-

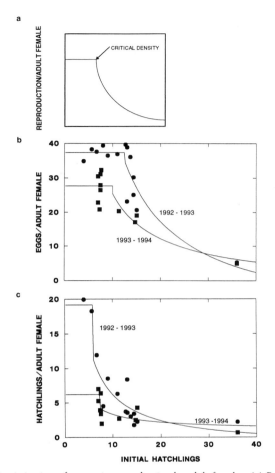

Figure 4. A mechanistic view of per capita reproduction by adult females. (a) Per capita reproduction should be density independent until a critical density is attained and food competition reduces reproduction. The relationship between per capita reproduction and initial density based on ovariole relicts (b) and hatchling production (c) for *M. sanguinipes* in experimental populations over two generations in Montana.

tions should show the probability increasing as initial prey density increases, but at some critical initial density the probability will begin to decline as initial density increases and the predators become "saturated" with prey, unable to kill any more.

The combined functional and numerical responses or the responses for individual predator species are difficult to obtain in the field. However, grasshoppers

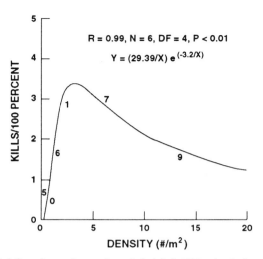

Figure 5. The probability of a grasshopper being killed (kills/100 individuals present/day) by avian, mammalian, spider, and ant predators at different grasshopper densities. Data are for *M. sanguinipes* in Montana that were tethered in the field over 6 years (Belovsky *et al.*, 1990); each year is labeled by the year's last digit (1985, 1986, 1987, 1989, 1990, 1991). The line and equation are the best-fit function to these observations.

can be tethered on nylon monofilament in the field and monitored for predator removal. In most cases, the predator can be identified (Belovsky *et al.*, 1990). This was done in different years to obtain predation probabilities at different grasshopper densities. As hypothesized, the predation function initially increases as hatchling density increases and then declines (Fig. 5). Although spiders, ants, and mammals killed grasshoppers, birds were the major predator creating this function (Belovsky *et al.*, 1990), and this response was independently verified with experimental studies of optimal foraging behavior in cowbirds (*Molothrus ater*), one of the principal avian predators at the site (Belovsky, 1990, 1993). A similar response has been observed for parasitoids that attack grasshoppers (Smith, 1965; M. Lietti de Guibert, J. B. Slade, and G. E. Belovsky, unpublished).

IV. An Integrated Model of Population Limitation

We developed a theoretical framework to integrate the population responses and their underlying mechanisms described earlier. We then determined whether food and predator limitation can both be accommodated by this framework, as well as the contribution played by variable abiotic conditions. With a theoretical framework, field experiments can be performed that manipulate the mechanisms presumed to be operating within populations and then observing whether changes in population responses (survival and reproduction) agree with expectations.

A. Basic Modeling Strategy

A model that captures the three population processes is Ricker Curve (sometimes referred to as recruitment curve) analysis, which is particularly suited for univoltine (i.e., nonoverlapping generations that are produced annually) species like our grasshoppers (Varley *et al.*, 1973). This analysis plots the density of offspring initiating one generation (N_t) against the density of offspring produced by that generation to initiate the next generation (N_{t+1}), given constant environmental conditions (Fig. 6; Ricker, 1954, 1958). The curve (Ricker Curve) represents the entire range of population responses (N_t and N_{t+1}) for the environmental conditions; however, a population at a given site and time will exhibit only a single pair of N_t and N_{t+1} values. When environmental (abiotic and biotic) conditions vary between sites or over time (a likely situation), it is not proper to empirically construct a Ricker Curve by plotting observed N_t and N_{t+1} values, because each observed pair of values resides on a different Ricker Curve. Therefore, an experiment must be used to reconstruct the Ricker Curve (i.e., range of N_t and N_{t+1} values) when environmental conditions are constant; this is what the experiments described earlier allowed us to do.

A reference line ($N_t = N_{t+1}$) can be added to the Ricker Curve plot (Fig. 6). An intersection between the line and the Ricker Curve denotes an equilibrium population (self-sustaining condition), because the density of hatchlings initiating the population at time t equals the density of hatchlings produced by them to initiate the next generation. We do not expect the population to be at equilibrium given the annual variability in environmental conditions; however, the equilibrium still has importance, because it defines a state that is attracting the population, and this state defines the mechanism that is dominant over others in terms of regulating the population. Because there are two graphical representations of the function relating adult density to the initial density of hatchlings (see the foregoing), a Ricker Curve analysis must be developed for each (Fig. 6).

B. Scenario 1: Adult Density without Predators Declines as Initial Hatchling Density Increases

In this case the Ricker Curve is unimodal (Fig. 6.I), which means that the reference line, if it intersects the Ricker Curve, can only intersect it at a single point.

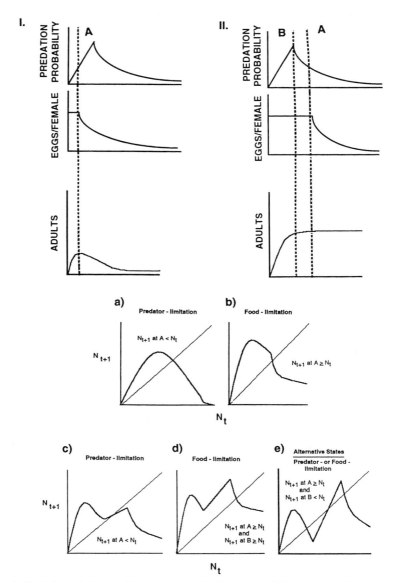

Figure 6. Possibilities for Ricker Curves (a–e) are developed given different relationships for survival, reproduction, and predation; straight lines are reference lines that denote equilibria at the intersection with the Ricker Curve. (I) Ricker Curves (a) or (b) can be observed when the critical density for the predation function (A) is greater than the critical densities for survival and reproduction (B), and either survival function (Figs. 3b and 3c) is observed. (II) Ricker Curves (c–e) can be observed when the critical density for the predation function (A) is less than the critical density for reproduction and the survival function attains a constant value (Fig. 3b).

C. Scenario 2: Adult Density without Predators Is Constant as Initial Hatchling Density Increases

In this case the Ricker Curve can be unimodal or bimodal (Fig. 6.II). To be unimodal the critical density at which reproduction becomes density dependent (point A in Fig. 6.II) must be less than the density at which the predation function peaks (point B in Fig. 6.II). A unimodal Ricker Curve can only be intersected by the reference line once. If the critical density at which reproduction becomes density dependent (point A in Fig. 6.II) is greater than the density at which the predation function peaks (point B in Fig. 6.II), then the Ricker Curve is bimodal. A bimodal Ricker Curve, if it is intersected by the reference line, can be intersected once or three times.

D. Operation of the Ricker Curve Model

If the reference line intersects the Ricker Curve once (unimodal or bimodal curve), two modes of population limitation are possible. If the intersection occurs at an N_t value greater than or equal to the critical density where density dependence is observed for either survival in the absence of predation or reproduction, whichever is smaller, then the population is food-limited (Figs. 6b and 6d: bottom-up control). If the intersection occurs at an N_t value less than the critical value where density dependence is observed for either survival in the absence of predation or reproduction, whichever is smaller, then the population is predator-limited (Figs. 6a and 6c: top-down control). Either of these outcomes is typical of the classic density-dependent perspective argued by some ecologists (e.g., Lack, 1954).

If the reference line intersects the bimodal Ricker Curve three times, possibilities for population limitation are diversified, because this denotes the existence of multiple stable equilibria (Fig. 7a). The intersection closest to the origin is a predator-limited equilibrium, the intersection farthest from the origin is a food-limited equilibrium, and the intermediate intersection is an unstable equilibrium (saddle point). The saddle point delimits regions where the population at lower N_t values moves toward the predator-limited equilibrium and the population at higher N_t values moves toward the food-limited equilibrium. Therefore, even though the population is limited by density-dependent processes (food and predation), it is possible for the population depending on the density of individuals founding it to be attracted to either the predator- or food-limited equilibrium; this is not classic density dependence (e.g., Lack, 1954). Furthermore, given the proper shape of the Ricker Curve (Fig. 7b), the population might switch chaotically over time between predator and food limitation (Hastings *et al.*, 1993).

If there is no intersection between the reference line and Ricker Curve, then the population cannot persist unless the environment varies between years, and

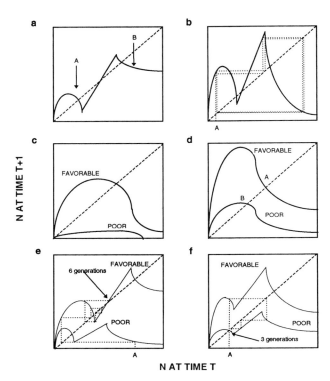

N AT TIME T

Figure 7. Various Ricker Curve outcomes are portrayed, where the dashed line is the reference line that denotes equilibria where it intersects the Ricker Curve. (a) A bimodal Ricker Curve can provide multiple stable equilibria depending on initial densities in the population: (A) represents an initial density leading to predator limitation and (B) represents food limitation. (b) A bimodal Ricker Curve can produce a population that varies between predator- and food-limited generations (dotted line trajectory), if the population is initiated with (A) individuals (c). If environmental conditions generally produce a low Ricker Curve (poor) that is not intersected by the reference line, the population must occasionally experience better environmental conditions that produce a higher Ricker Curve (favorable) that intersects the reference line, if the population is to persist. (d) If environmental conditions vary between generations, favorable conditions will tend to produce a food-limited population (A) and poor conditions will tend to produce a predator-limited population (B). (e) When environmental conditions generally are favorable and populations tend to be food-limited, an occasional poor year can produce a number of generations (dotted line trajectory) that are predator-limited in subsequent favorable years (six generations in this example). (f) The opposite can emerge when poor years are common and occasional favorable years emerge.

some years provide a favorable environment so the Ricker Curves in those years lie above the reference line and are intersected (Fig. 7c). This case does not imply density-independent population limitation, because density dependence (food or predator limitation) is occurring at all N_t values, even if the Ricker

Curve does not intersect the reference line. However, it does imply that the population is not attracted (regulated) toward an equilibrium except in occasional favorable years. This is equivalent to the classic density-independent perspective of some population ecologists (e.g., Andrewartha and Birch, 1954, 1984), but the population growth trajectories still result from density-dependent mechanisms (Sinclair, 1989).

Environmental variation between years not only can create the classic density-independent view, but can result in a different Ricker Curve for each year. The existence of different Ricker Curves in different years indicates that a population can be limited by predation and food in different years; that is, as the reference line intersects each year's Ricker Curve at a different N_t, the population can be attracted to an equilibrium that is defined by a different dominant density-dependent process (Fig. 7d). The Ricker Curve changes shape as annual variations in food quantity and quality affect survival and reproduction and abiotic-induced mortality varies.

Annual shifts in observed modes of population limitation can occur in several ways, but they are most likely for bimodal Ricker Curves, because less variation is needed to create curves that produce either predator or food limitation, or both.

1. As the Ricker Curve is lowered (reduce N_{t+1} for a given N_t), predator limitation is more likely, and as the curve is raised, food limitation is more likely (Fig. 7d). The effects of each population process on the Ricker Curve must be considered. Increased abiotic mortality and predator efficiency (greater proportion killed at a density and greater critical density at which the proportion killed begins to decline) decrease the Ricker Curve, making predator limitation more likely. Food quality and quantity have diverging effects. Increased food quality increases density-independent survival and reproduction and allows a larger population to be supported, possibly permitting an escape from the effects of predation. This raises the Ricker Curve and fosters food limitation. However, increased food quality and quantity increase the critical density at which food competition occurs, decreasing the likelihood of food competition. The net effect is that increased food quality and quantity increase the likelihood of food limitation.

2. As annual environmental conditions shift between being generally favorable (higher Ricker Curves) to being occasionally unfavorable (lower Ricker Curve), or the reverse, the occasional year can have a profound effect on the mode of limitation observed. This requires that the conditions generally observed produce bimodal Ricker Curves that are intersected three times by the reference line. For example, when favorable environmental conditions are generally observed and food limitation is expected, an occasional unfavorable year would have a population initiated with a large density of hatchlings (Fig. 7e). The unfavorable year with a large initial density would experience food competition,

even though predator limitation is expected, and produce very few hatchlings for the next year, which is once again favorable (Fig. 7e). This creates a series of years with favorable conditions that are predator-limited, even though food limitation is expected (Fig. 7e). If unfavorable conditions are generally expected and favorable conditions appear only occasionally, the opposite effect (i.e., food limitation when predator limitation is expected) can be produced (Fig. 7f). Therefore, varying environmental conditions can create historical effects so that population dynamics, though deterministic and predictable, cannot be discerned without examining variations in the Ricker Curve over time.

Although the conclusions about population limitation are apparent when the Ricker Curves are examined, they are not intuitive when the mechanisms are considered in isolation, as is typical of classic density-dependent and -independent views. Furthermore, these conclusions may be counterintuitive (e.g., food competition is fostered as food quantity and quality increase). A variety of possibilities for population dynamics does not imply lack of understanding, because this variety can emerge from a simple model. The simple model also illustrates the need to view population dynamics in a broader and more long-term perspective (Belovsky and Joern, 1996).

V. Grasshoppers Revisited: Ricker Curves

The precautions that must be taken in measuring a Ricker Curve were discussed in the preceding section, that is, the range of N_t and N_{t+1} values must be acquired from a single set of environmental conditions (abiotic and biotic). Our experimental system meets these criteria, where replicate populations (9-m^2 field enclosures) at a given site and year are initiated with a range of hatchling densities (N_t) and the production of hatchlings in the next year is measured (N_{t+1}). In this way, a Ricker Curve can be constructed for each site and year. The experiments can also be used to assess whether a predator- or food-limited equilibrium is attracting the field population by predicting its mode of limitation given its initial hatchling density and the Ricker Curve, and comparing it with the observed mode in the experiment.

A. Observed Ricker Curves

To date, we have the information to construct two Ricker Curves for *M. sanguinipes,* a single site over 2 years in Montana. In both years (generations), the Ricker Curves were bimodal and intersected the reference line at three points (Figs. 8a and 8b). This was expected because the observed functions for survival, reproduction, and predation (Figs. 3c, 5c, and 5d) met the conditions for this type of Ricker Curve (Fig. 6). Finally, the higher Ricker Curve in 1992–1993 compared to 1993–1994 (as well as greater survival and reproduction: Figs. 3c,

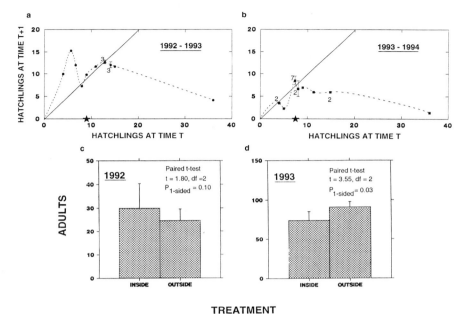

TREATMENT

Figure 8. Experimentally determined Ricker Curves (dashed lines) for M. *sanguinipes* over two generations in Montana are presented in (a) and (b). Ricker Curves were fit to data from experimental populations using a Spline-smoothing algorithm, and standard deviations are presented for populations that had the same initial densities (numbers refer to the sample size), indicating the confidence in generated curves. The reference line is a solid line. The initial density of hatchlings observed in the field in each year is marked by a star. (c) and (d) present the results from avian exclosure experiments (total number in exclosure or control area). Error bars represent standard errors (sample size = 3); statistical tests are one-sided, because birds are expected to decrease grasshopper numbers.

5c, and 5d) appears to be due to food plants having higher quality and quantity in 1992 (Belovsky and Slade, 1996b).

The Ricker Curves predicted shifting modes of population limitation between these two generations:

1. *In 1992–1993*, field initial hatchling densities produced predator limitation of adults, but food limitation of reproduction (Fig. 8a), because the densities fell between the saddle point (intermediate intersection) and the food-limited equilibrium (intersection farthest from origin). This was observed (Fig. 8c) and agrees with the idea that predator limitation is more likely as years vary from unfavorable to favorable conditions (food quality was lower in 1991 than 1992; Belovsky and Slade, 1996b).

2. *In 1993–1994*, initial hatchling density produced food limitation of adults and reproduction (Fig. 8b), because densities fell above the food-limited

equilibrium (intersection farthest from origin). This was observed (Fig. 8d) and, as pointed out earlier, a shift from predator to food limitation may be more likely as years vary from favorable to unfavorable (food quality was higher in 1993 than 1994; Belovsky and Slade, 1996b).

An interesting additional property emerges: the populations within each generation (year) were at equilibrium with the food resource (i.e., utilized all available food so that food was competed for), because reproduction was food-limited in 1992–1993 even though survival was predator-limited, and survival and reproduction were food-limited in 1993–1994. However, in neither of the generations was the population at an equilibrium ($N_t = N_{t+1}$) based on density. This poses wide-ranging implications for traditional perspectives in population ecology and food web theory that we discuss later.

B. Implications for Our Grasshopper Populations

Results from Montana indicate that the conditions that enhance the likelihood for modes of population limitation to shift between sites and years are exhibited, because the Ricker Curves vary with year and bimodal Ricker Curves, especially if they intersect the reference line at three points, can most easily be shifted between food (bottom-up control) and predator limitation (top-down control). Combining the observed Ricker Curves with the Ricker Curve scenarios developed earlier provides possible explanations why some western U.S. grasshopper populations, like those at the Montana site, tend to be bottom-up controlled and others, like those at the Nebraska site, tend to be top-down controlled. For top-down control to be expected, the Ricker Curves at the Nebraska site, which are yet unmeasured, and at other predator-limited sites must be lower than, and the density at which food competition begins must be higher than, observed at the Montana site. These conditions can arise when:

1. *Food is of lower quality in Nebraska than Montana.* This is supported by the data presented earlier on plant solubility in HCl + pepsin. In addition, Nebraska grasses tend to be coarser, tend to be represented by more C_4 species that are less digestible than C_3 species, and tend to grow in tufts, which is related to lower quality and cropping rates.

2. *Food is of greater quantity in Nebraska than in Montana.* Plant biomass measures at the two sites ($\approx$250 g-dry/m² in Nebraska versus $\approx$190 g-dry/m² in Montana) support this. However, we believe that the observed differences in food quantity cannot account for the different modes of limitation observed in Montana and Nebraska, whereas this may not be the case for the entire western United States, such as the shortgrass steppe.

3. *Abiotic-induced mortality and predator efficiency are greater in Nebraska than in Montana.* At present, we have no evidence to suggest that abiotic conditions reduce survival in Nebraska more than in Montana. However, there is

evidence that avian predators, although of comparable abundance at both sites, may be more successful at capturing grasshoppers in Nebraska (85%: Joern, 1988) than in Montana (22–28%: Belovsky *et al.*, 1990); this may be due to greater cover in Montana (fewer tuft-forming grasses).

VI. Insights for Population Ecology

Our integrated approach resembles population models developed by others for a variety of species (e.g., insects: Southwood and Comins, 1976; Berryman, 1987; insects and fish: Peterman *et al.*, 1979; mammals: Sinclair *et al.*, 1990; Haber and Walters, 1980). These earlier models were phenomenological (i.e., relying on the general shapes of population growth functions) rather than mechanistic (i.e., relying on explicit functions defining the operation of mechanisms that impact the population). Therefore, our approach has implications for an array of other organisms and general ecological issues. These are summarized in the following.

A. Study Design

There are no shortcuts for assessing population dynamics. Our results indicate the need for long-term data sets, which are based on experimental manipulations coupled with careful descriptions of populations. The need for experimental populations arises because different Ricker Curves emerge in different habitats for a given year and between years in the same habitat. This seriously handicaps the ability to statistically unravel population dynamics from large data sets composed of field population censuses made at the same site or different sites over time (Royama, 1993). It further questions the utility of existing data sets for addressing issues like density vagueness, density veiled, and density dependence (*sensu* Strong, 1984, 1986a,b; Stiling, 1988).

B. Fallacy of Searching for Single Explanations

Much of the 50-year debate over population limitation arises in part from weaknesses in the available data sets, but largely stems from the desire of ecologists to attribute limitation to a single mechanism (e.g., predation, food, and so on). Our Ricker Curve approach integrates the operation of a number of potentially limiting mechanisms and indicates that one density-dependent mechanism *must* limit a population at a given site and time, but the limiting density-dependent mechanism can change over time and between sites. Consider two populations of a species with identical density-dependent responses produced by food competition and predation. One population has lower density-independent survival and reproduction; this population may be predator-limited, whereas the other is food-limited. Even though the populations are respectively predator- and food-limited, is it correct to view density-independent effects as unimportant? We need to consider all the mechanisms together.

Our findings with the Ricker Curve model have similarities to Sinclair's (1989) distinction that density-independent mechanisms (e.g., abiotic-induced mortality) cannot regulate populations; regulation requires density-dependent processes. For example, density-independent processes can depress populations to densities where food competition may not occur. However, for density-independent processes to be limiting, they must reduce the population to such a low density that its predators cannot persist in the environment; otherwise, predation, a density-dependent process, will limit the population. Therefore, density-independent processes are very important in determining which density-dependent processes regulate (attract toward an equilibrium) the population, but the density-independent processes are not limiting in themselves. This is one of the attributes that emerges when all population mechanisms are considered to be operating simultaneously.

C. Misconception of Invariant Density Dependence and Equilibrium

Ecologists often consider the mechanism limiting a population to be constant over time, even though occasional divergences during periods of exceptional environmental conditions (e.g., drought) occur. This has been taken to the extreme by considering that populations not fluctuating greatly over time are largely limited by density-dependent mechanisms, whereas widely fluctuating populations are limited by density-independent mechanisms (Horn, 1966). As pointed out in the foregoing, density-independent mechanisms cannot regulate a population, but variation in the intensity of their operation, as the environment varies over time, can change which density-dependent mechanism limits a population. Therefore, the mode of population regulation for a population is not constant.

As discussed earlier, there are additional ways that the mode of population regulation can change over time for a population or be different for populations in similar environments. When the population can exhibit multiple stable equilibria (bimodal Ricker Curve that intersects the reference line at three points), historical factors can produce different modes of limitation given different densities of founder individuals or different chronologies of favorable and unfavorable time periods. A population with multiple stable equilibria can exhibit chaotic shifts in its mode of limitation over time, even in a constant environment.

The issue of equilibria in population ecology must also be reexamined given our model. The classical notion of population equilibrium as a constant density, where $N_t = N_{t+1}$, is unlikely to be attained in our population model. However, this equilibrium is still very important for ecologists to consider and estimate, because the equilibrium acts as an attractor as environmental conditions vary over time and defines the mode of limitation operating on the population. Therefore, the unlikely event of observing a population at equilibrium does not mean that an understanding of population dynamics based on equilibrial models is invalid.

Our model and experiments with grasshopper populations indicate the potential for a different type of equilibrium, even if the population does not attain the equilibrial densities. It is possible that food resources are fully utilized (consumptive equilibrium), even at population densities below the equilibrial density $(N_t = N_{t+1})$. This arises because the equilibrial density in a food-limited population is set at a point where per capita food intake permits an individual to replace only itself reproductively (Schoener, 1973). However, if the population is food-limited, but not at the equilibrial density, there exists a range of population densities at which individuals still consume the same total quantity of food but have more or less on a per capita basis, so that reproduction is greater or less than a replacement level. Therefore, populations do not have to be at equilibrial densities to "track" changing food resources (Roughgarden, 1979).

The distinction between an equilibrium based on density and consumption has important implications for food web theory. Observing that a predator reduces the density of prey does not imply top-down dynamics with the expectation of trophic cascades (Hairston *et al.*, 1960; Slobodkin *et al.*, 1967; Hairston, 1989; Fretwell, 1977; Oksanen, 1990, 1991; Oksanen *et al.*, 1981; Carpenter and Kitchell, 1984, 1987, 1988), if the reduced density of prey still maintains the equilibrial consumption. This was theoretically argued by Schmitz (1992).

D. Multiple Stable Equilibria

The importance of multiple stable equilibria in population dynamics has been downplayed after Connell and Sousa (1983) reviewed the evidence. Their argument that little, if any, evidence supports the existence of multiple stable equilibria was appropriate given the types of long-term data available. Nonetheless, the issue of populations with multiple stable equilibria has continued to be raised, even though data neither support nor refute their existence (e.g., Berryman, 1987; Sinclair *et al.*, 1990; Messier, 1994; Knowlton, 1992; Leonardsson, 1994). However, we believe that the Ricker Curves measured by us for a grasshopper population in two different years clearly demonstrate for the first time the existence of the necessary dynamics to produce multiple stable states.

VII. Conclusions

Our findings support the need for long-term studies of population dynamics that combine experimental manipulations and field observations to decipher how ecological mechanisms (e.g., predation, food competition, and biotic-induced mortality) operate in concert on the population over time. This is necessary because these mechanisms can have different effects over time and between habitats.

Our studies indicate that one of the key mechanisms that changes over time and between sites for herbivore populations is the influence of plant quality and quantity. Plant quality and quantity impact density-independent survival and reproduction and determine the strength of food competition. For example, given two populations of the same herbivore with identical functions relating predatory mortality to prey density and abiotic-induced mortality, the population with better food resources will more likely be food-limited, whereas a population with poorer food resources will more likely be predator-limited. Therefore, studies of population dynamics must consider food resources and how they vary over time and space (Price, 1984, 1992), especially if we are to resolve issues like population limitation and top-down versus bottom-up control.

Acknowledgments

We wish to thank J. B. Slade and Jon Chase for commenting on this manuscript. The work was funded by The Grasshopper Integrated Pest Management Program (USDA/ APHIS), the National Science Foundation (DEB-9317984), and the Utah State University Agricultural Experiment Station.

References

Andrewartha, H. G., and Birch, L. C. (1954). "The Distribution and Abundance of Animals." Univ. of Chicago Press, Chicago.
Andrewartha, H. G., and Birch, L. C. (1984). "The Ecological Web." Univ. of Chicago Press, Chicago.
Belovsky, G. E. (1986). Generalist herbivore foraging and its role in competitive interactions. *Am. Zool.* **25**, 51–69.
Belovsky, G. E. (1990). How important are nutrient constraints in optimal foraging models or are spatial/temporal factors more important? *NATO ASI Ser., Ser. G* **20**, 255–278.
Belovsky, G. E. (1993). Modeling avian foraging: Implications for assessing the ecological effects of pesticides. *In* "Wildlife Toxicology and Population Modeling: Integrated Studies of Agroecosystems" (R. J. Kendall and T. E. Lacher, Jr., eds.), pp. 131–145. CRC Press, Boca Raton, FL.
Belovsky, G. E., and Joern, A. (1996). A population model balancing food, predation and abiotic-induced mortality. In preparation.
Belovsky, G. E., and Slade, J. B. (1993). The role of vertebrate and invertebrate predators in a grasshopper community. *Oikos* **68**, 193–201.
Belovsky, G. E., and Slade, J. B. (1995). Dynamics of some Montana grasshopper

populations: Relationships among weather, food abundance and intraspecific competition. *Oecologia* **101**, 383–396.

Belovsky, G. E., and Slade, J. B. (1996a). Grasshopper foraging behavior: Selection of plant species. Submitted for publication.

Belovsky, G. E., and Slade, J. B. (1996b). Ricker Curves for grasshoppers: The potential for multiple equilibria. In preparation.

Belovsky, G. E., Slade, J. B., and Stockhoff, B. A. (1990). Susceptibility to predation for different grasshoppers: An experimental study. *Ecology* **71**, 624–634.

Berryman, A. A. (1987). The theory and classification of outbreaks. *In* "Insect Outbreaks" (P. Barbosa and J. C. Schultz, eds.), pp. 3–30. Academic Press, San Diego, CA.

Bock, C. E., Bock, J. H., and Grant, M. C. (1992). Effects of bird predation on grasshopper densities in an Arizona grassland. *Ecology* **73**, 1706–1717.

Brusven, M. A., and Fielding, D. J. (1990). Grasshopper ecology on BLM land in south central Idaho: Rangeland management alternatives. *USDA-APHIS-PPQ, Grasshopper IPM Proj. Annu. Rep.* **FY-90**, 15–18.

Capinera, J. L. (1987). Population ecology of rangeland grasshoppers. *In* "Integrated Pest Management on Rangeland: A Shortgrass Prairie Perspective" (J. L. Capinera, ed.), pp. 162–182. Westview Press, Boulder, CO.

Capinera, J. L., and Horton, D. R. (1989). Geographic variation in effects of weather on grasshopper infestation. *Environ. Entomol.* **18**, 8–14.

Capinera, J. L., and Thompson, D. C. (1987). Dynamics and structure of grasshopper assemblages in shortgrass prairie. *Can. Entomol.* **119**, 567–575.

Carpenter, S. R., and Kitchell, J. F. (1984). Plankton community structure and limnetic primary production. *Am. Nat.* **124**, 159–172.

Carpenter, S. R., and Kitchell, J. F. (1987). The temporal scale of variance in limnetic primary production. *Am. Nat.* **129**, 417–433.

Carpenter, S. R., and Kitchell, J. F. (1988). Consumer control of lake productivity. *BioScience* **38**, 764–769.

Caswell, H., and Reed, F. C. (1976). Plant–herbivore interactions: The indigestibility of C_4 bundle sheath cells by grasshoppers. *Oecologia* **26**, 151–156.

Caswell, H., Reed, F. C., Stephenson, S. N., and Werner, P. A. (1973). Photosynthetic pathways and selective herbivory: A hypothesis. *Am. Nat.* **107**, 465–478.

Chapman, R. F. (1990). Food selection. *In* "Biology of Grasshoppers" (R. F. Chapman and A. Joern, eds.), pp. 39–72. Wiley, New York.

Chase, J. M., and Belovsky, G. E. (1994). Experimental evidence for the included niche. *Am. Nat.* **143**, 514–527.

Connell, J. H., and Sousa, W. P. (1983). On the evidence needed to judge ecological stability or persistence. *Am. Nat.* **121**, 789–824.

Evans, E. W. (1989). Interspecific interactions among phytophagous insects of tallgrass prairie: An experimental test. *Ecology* **70**, 435–444.

Fowler, A. C., Knight, R. L., George, T. L., and McEwen, L. C. (1991). Effects of avian predation on grasshopper populations in North Dakota grasslands. *Ecology* **72**, 1775–1781.

Fretwell, S. D. (1977). The regulation of plant communities by the food chains exploiting them. *Perspect. Biol. Med.* **20**, 169–185.

Haber, G. C., and Walters, C. J. (1980). Dynamics of the Alaska–Yukon caribou herds and management implications. In "Proceedings of the 2nd International Reindeer/Caribou Symposium, 1979, Roros, Norway" (E. Reimers, E. Gaare, and S. Skjenneberg, eds.), pp. 645–663. Direktoratet for vilt og ferskvannsfisk, Trondheim.

Hairston, N. G., Smith, F. E., and Slobodkin, L. B. (1960). Community structure, population control, and competition. Am. Nat. **94**, 421–425.

Hairston, N. G., Jr. (1989). "Ecological Experiments: Purpose, Design and Execution." Cambridge Univ. Press, Cambridge, UK.

Hastings, A., Hom, C. L., Ellner, S., Turchin, P., and Godfray, H. C. J. (1993). Chaos in ecology: Is Mother Nature a strange attractor? Annu. Rev. Ecol. Syst. **24**, 1–33.

Heidorn, T. J., and Joern, A. (1987). Feeding preference and spatial distribution of grasshoppers (Acrididae) in response to nitrogen fertilization of Calamovilfa longifolia. Funct. Ecol. **1**, 369–375.

Holling, C. S. (1966). The functional response of predators to prey density and its role in mimicry and population regulation. Mem. Entomol. Soc. Can. **48**, 1–87.

Horn, H. S. (1966). Measurement of 'overlap' in comparative ecological studies. Am. Nat. **100**, 419–424.

Hunter, M. D., and Price, P. W. (1992). Playing chutes and ladders: Heterogeneity and the relative roles of bottom-up and top-down forces in natural communities. Ecology **73**, 724–732.

Joern, A. (1979a). Feeding patterns in grasshoppers (Orthoptera: Acrididae): Factors influencing diet specialization. Oecologia **38**, 325–347.

Joern, A. (1979b). Resource utilization and community structure in assemblages of arid grassland grasshoppers (Orthoptera: Acrididae). Trans. Am. Entomol. Soc. **105**, 253–300.

Joern, A. (1986). Experimental study of avian predation on coexisting grasshopper populations (Orthoptera: Acrididae) in sandhills grassland. Oikos **46**, 243–249.

Joern, A. (1988). Foraging behavior and switching by the grasshopper sparrow Ammodramus savannarum searching for multiple prey in a heterogeneous environment. Am. Midl. Nat. **119**, 225–234.

Joern, A. (1989a). Insect herbivory in the transition to California annual grasslands: Did grasshoppers deliver the coup de grass? In "Grassland Structure and Function: California Annual Grasslands" (L. Huenneke and H. A. Mooney, eds.), pp. 117–134. Kluwer Academic Publ., Dordrecht, The Netherlands.

Joern, A. (1989b). Host plant quality: Demographic responses of range grasshoppers to stressed host plants. USDA-APHIS-PPQ Grasshopper IPM Proj. Annu. Rep. **FY-89**, 51–55.

Joern, A. (1990). Host plant quality: Demographic responses of range grasshoppers to stressed host plants. USDA-APHIS-PPQ Grasshopper IPM Proj. Annu. Rep. **FY-90**, 45–50.

Joern, A. (1992). Variable impact of avian predation on grasshopper assemblies in sandhills grassland. Oikos **64**, 458–463.

Joern, A., and Klucas, G. (1993). Intra- and interspecific competition between two abundant grasshopper species (Orthoptera: Acrididae) from a sandhills grassland. Environ. Entomol. **22**, 352–361.

Joern, A., and Lawlor, L. R. (1980). Food and microhabitat utilization by grasshoppers from arid grasslands: Comparisons with neutral models. *Ecology* **61**, 591–599.

Knowlton, N. (1992). Thresholds and multiple stable states in coral reef community dynamics. *Am. Zool.* **32**, 674–682.

Lack, D. (1954). "The Natural Regulation of Animal Numbers." Oxford Univ. Press (Clarendon), Oxford.

Leonardsson, K. (1994). Multiple density dependence in two sub-populations of the amphipod *Monoporeia affinis:* A potential for alternative equilibria. *Oecologia* **97**, 26–34.

Lockwood, J. A., and Lockwood, D. R. (1991). Rangeland grasshopper (Orthoptera: Acrididae) population dynamics: Insights from catastrophe theory. *Environ. Entomol.* **20**, 970–980.

Lomnicki, A. (1988). "Population Ecology of Individuals." Princeton Univ. Press, Princeton, NJ.

Messier, F. (1994). Ungulate population models with predation: A case study with the North American moose. *Ecology* **75**, 478–488.

Oksanen, L. (1990). Predation, herbivory, and plant strategies along gradients of primary productivity. *In* "Perspectives on Plant Competition" (J. P. Grace and D. Tilman, eds.), pp. 445–474. Academic Press, San Diego, CA.

Oksanen, L. (1991). Trophic levels and trophic dynamics: A consensus emerging? *TREE* **6**, 58–60.

Oksanen, L., Fretwell, S. D., Arruda, J., and Niemela, P. (1981). Exploitation ecosystems in gradients of primary productivity. *Am. Nat.* **118**, 240–261.

Otte, D. (1984). "The North American Grasshoppers," Vol. 2. Harvard Univ. Press, Cambridge, MA.

Otte, D., and Joern, A. (1977). On feeding patterns in desert grasshoppers and the evolution of specialized diets. *Proc. Natl. Acad. Sci. U.S.A.* **128**, 89–126.

Peterman, R., Clark, W. C., and Holling, C. S. (1979). The dynamics of resilience: Shifting stability domains in fish and insect systems. *In* "Population Dynamics" (R. M. Anderson, B. D. Turner, and L. R. Taylor, eds.), pp. 321–342. Blackwell, Oxford.

Price, P. W. (1984). Alternative paradigms in community ecology. *In* "A New Ecology: Novel Approaches to Interactive Systems" (P. W. Price, C. N. Slobodchikoff, and W. S. Gaud, eds.), pp. 354–383. Wiley, New York.

Price, P. W. (1992). Plant resources as the mechanistic basis for insect herbivore population dynamics. *In* "Effects of Resource Distribution on Plant–Animal Interactions" (M. D. Hunter, T. Ohgushi and P. W. Price, eds.), pp. 139–173. Academic Press, San Diego, CA.

Ricker, W. E. (1954). Stock and recruitment. *J. Fish. Res. Board Can.* **11**, 559–623.

Ricker, W. E. (1958). Production, reproduction, and yield. *Verh.—Int. Ver. Theor. Angew. Limnol.* **13**, 84–100.

Roughgarden, J. (1979). "Theory of Population Genetics and Evolutionary Ecology: An Introduction." Macmillan, New York.

Royama, T. (1993). "Analytical Population Dynamics." Chapman & Hall, New York.

Schmitz, O. J. (1992). Exploitation in model food chains with mechanistic consumer-resource dynamics. *Theor. Popul. Biol.* **41**, 161–183.

Schmitz, O. J. (1993). Trophic exploitation in grassland food chains: Simple models and a field experiment. *Oecologia* **93**, 327–335.

Schoener, T. W. (1973). Population growth regulated by intraspecific competition for energy or time: Some simple representations. *Theor. Popul. Biol.* **4**, 56–84.

Sinclair, A. R. E. (1989). Population regulation in animals. *In* "Ecological Concepts" (J. M. Cherett, ed.), pp. 197–241. Blackwell, Oxford.

Sinclair, A. R. E., Olsen, P. D., and Redhead, T. D. (1990). Can predators regulate small mammal populations? Evidence from house mouse outbreaks in Australia. *Oikos* **59**, 382–392.

Slobodkin, L. B., Smith, F. E., and Hairston, N. G. (1967). Regulation in terrestrial ecosystems, and the implied balance of nature. *Am. Nat.* **101**, 109–124.

Smith, R. W. (1965). A field population of *Melanoplus sanguinipes* (FAB.) (Orthoptera: Acrididae) and its parasites. *Can. J. Zool.* **43**, 179–201.

Southwood, T. R. E., and Comins, H. N. (1976). A synoptic population model. *J. Anim. Ecol.* **45**, 949–965.

Stiling, P. (1988). Density-dependent processes and key factors in insect populations. *J. Anim. Ecol.* **57**, 581–593.

Strong, D. R. (1984). Density-vague ecology and liberal population regulation in insects. *In* "A New Ecology: Novel Approaches to Interactive Systems" (P. W. Price, C. N. Slobodchikoff, and W. S. Gaud, eds.), pp. 313–327. Wiley, New York.

Strong, D. R. (1986a). Density-vague population change. *TREE* **1**, 39–42.

Strong, D. R. (1986b). Density vagueness: Abiding the variance in the demography of real populations. *In* "Community Ecology" (J. M. Diamond and T. J. Case, eds.), pp. 257–268. Harper & Row, New York.

Teeri, J. A., and Stowe, L. G. (1976). Climatic patterns and the distribution of C_4 grasses in North America. *Oecologia* **23**, 1–12.

Thompson, D. C., Torell, L. A., and Huddleston, E. W. (1989). Economic injury level— Warm season grasses. *USDA-APHIS-PPQ, Grasshopper IPM Proj. Annu. Rep.* **FY-89**, 117–123.

Varley, C. G., Gradwell, G. R., and Hassell, M. P. (1973). "Insect Population Ecology: An Analytical Approach." Univ. of California Press, Berkeley.

Wang, T., and Walgenbach, D. D. (1989). Economic thresholds of grasshoppers on cool season grasses. *USDA-APHIS-PPQ, Grasshopper IPM Proj. Annu. Rep.* **FY-89**, 124–131.

Wang, T., and Walgenbach, D. D. (1990). Rates of forage consumption and nymphal development, and fecundity and longevity of the migratory grasshopper (*Melanoplus sanguinipes*) when reared on three food sources. *USDA-APHIS-PPQ, Grasshopper IPM Proj. Annu. Rep.* **FY-90**, 152–158.

White, T. C. R. (1978). The importance of a relative shortage of food in animal ecology. *Oecologia* **33**, 71–86.

White, T. C. R. (1984). The availability of invertebrate herbivores in relation to the availability of nitrogen in stressed food plants. *Oecologia* **63**, 71–86.

White, T. C. R. (1993). "The Inadequate Environment: Nitrogen and the Abundance of Animals." Springer-Verlag, Berlin.

PART V

CONCLUSION

Chapter 19

Novelty and Synthesis in the Development of Population Dynamics

Peter W. Price and Mark D. Hunter

I. Introduction

Although population dynamics is a centerpiece in ecology, there is less emphasis in the field than should be expected. In a recent review, research papers in population ecology as a broad area outnumbered those on communities and ecosystems 5 to 1 in some major journals during the years 1987–1991. However, the subcategory of population dynamics/regulation was represented by only 5% of all papers [51.5% of all papers were in population ecology, 5.5% in population regulation, 9.4% in community ecology, and 9.6% in ecosystems ecology (Stiling, 1994)]. Areas favored by researchers in population ecology were competition (6.8% of papers), predation (6.3%), plant–herbivore interactions (8.4%), habitat selection (6.8%), and life-history strategies (9.0%).

If population dynamics is at the core of the ecological sciences, why is it so poorly represented in the current literature? The field necessitates an integration of most of the areas favored by ecologists mentioned here. In addition, such integration is essential for an adequate understanding of community ecology (Strong *et al.*, 1984; Colwell, 1984). There is a rich theoretical background on which to build in population dynamics, whereas in other areas, such as plant–herbivore interactions, theory appears to have been of minor concern. The pressing needs to understand the dynamics of pest species in agriculture and forestry, vectors of disease, the pathogens themselves, and the biology of common and rare species should all fuel an energetic discipline in population dynamics. Perhaps the cause of the underrepresentation of papers in population dynamics lies in the maturing of the science into a multifaceted discipline. Synthesis of ecological, behavioral, and evolutionary aspects of population dynamics is developing rapidly, with two consequences for the literature. First, relevant literature is likely to appear outside the main ecological journals. Second, integration and synthesis are perhaps more readily and usefully published in volumes such as this book.

It may be that synthesis in population dynamics has been slow to emerge because population change is more complicated than it first appears. After all, population change is determined ultimately by only four factors: birth, death, immigration, and emigration. This apparent simplicity is deceptive. It is easy to underestimate the complexity of biotic and abiotic interactions in the natural world that can influence these four population parameters. Indeed, we will argue in this chapter that the development of related fields such as plant–animal interactions, chemical ecology, and life-history evolution has proven to be a prerequisite for a realistic synthesis in population dynamics. These related fields provide the mechanistic basis, and therefore the predictive power, underlying the birth, death, and movement of organisms.

Nonetheless, synthesis in the field of population dynamics has deep historical roots. Of course, there has been a long tradition of empirical population study, such as by Howard (1897), which had obvious impact on the development of early theory by Lotka (1924). The development of life tables for field populations and their analysis gave a tremendous boost to the field (e.g., Morris and Miller, 1954; Varley and Gradwell, 1960). Major reputations were developed during this time of the 1950s and 1960s (cf. Southwood, 1968; Watson, 1970; Tamarin, 1978). But as the area of population dynamics prospered, the fledgling fields of evolutionary ecology (fostered by Robert MacArthur), coevolution, chemical ecology, life-history evolution, and plant–herbivore interactions were gaining ground, as noted in Chapter 1 (e.g., Sondheimer and Simeone, 1970). They flourished in the 1970s (e.g., Pianka, 1974; Gilbert and Raven, 1975; Rosenthal and Janzen, 1979; Collins, 1986). Our view is that these highly tractable fields eclipsed the core of population dynamics, which was bogging down: "the ecologist's phlogiston theory" (Krebs, 1979, p. 351) was proving to be intractable (McIntosh, 1985). "Because MacArthur's approach often began with the assumption that populations were at a steady state, the study of population dynamics was pushed into the background" (Kareiva, 1989, p. 71).

From these newer fields of ecology, the area of population dynamics has gained a new importance, and a new power. Its importance lies in the potential that population dynamics has to provide a central conceptual basis for the fusion of these newer fields, which seem to be growing apart rather than together. Fusion and synthesis is also inevitable when population dynamics encompasses behavior and phylogenetic relationships. By the same token, population dynamics is gaining enormous explanatory power as the newer fields reveal key mechanisms driving population change. "The ecologist's phlogiston theory" is being replaced by breaths of fresh air (and oxygen), as the newly synthetic science develops.

The new synthesis in population dynamics may be as important to ecology as a similar synthesis was to evolutionary theory (cf. Huxley, 1942; Mayr and Provine, 1980). Although the synthesis in population dynamics is incomplete, developments run parallel to the synthesis in evolution. Many disciplines in biology are becoming integrated under one umbrella. Scientists from many countries bring their own special talents and contributions. As the union proceeds, new debates are generated that accelerate the pace of science and discovery, and old debates are resolved.

In the rest of this chapter, we explore what we consider to be modern approaches to population dynamics. First, we consider the various elements that make up a synthetic approach to the study of population change. Section II is the backbone of the chapter and in it we offer a list of components that we regard as important to the study of population dynamics. Some of these elements are related fields of research, such as microbial ecology, whereas others are conceptual approaches, such as international cooperation in field research along important ecological gradients. Second, we describe how population dynamics has changed since its emergence as a field. Finally, we describe three broad scales of approaches to questions of population dynamics, and some of the difficulties in integrating the various elements of population biology together.

II. Elements of a New Synthesis

What are the elements of this new synthesis in population dynamics, the basis for building the science? In other words, what fields are providing us with a mechanistic understanding of the birth, death, and movement of organisms? We provide a necessarily brief overview of the disciplines from which synthesis in population dynamics should emerge.

A. Plant–Herbivore Interactions

Although it has deep roots in the agricultural and forestry literature, the study of plant–herbivore interactions has blossomed since the 1970s into a major research discipline. Strong motivation to understand the details of mechanisms, a forceful experimental initiative, and a coupling with evolutionary biology created a field with tremendous research energy. We learned that the world is not green for herbivores, and in fact provides marginal resources, at best, in many cases (e.g., White, 1993). The bottom of the food web in terrestrial systems is difficult to feast upon, and especially establishment of young is precarious, and perhaps central to the interaction between plant dynamics and herbivore life history and survival.

Realization of a strong bottom-up component in herbivore population dynamics dawned slowly. The life table approach to the study of dynamics rarely included the role of plant variation as a factor and female responses to this variation, except in research on bark beetles (Price *et al.*, 1990). Admittedly, there was long-standing interest in plant stress, but the area has been plagued by uncertainty about mechanisms and the lack of definitive research (Mattson and Haack, 1987a,b; Larsson, 1989; Waring and Cobb, 1992). It is now clear, however, that spatial and temporal variation in the quality of plants results in variation in birth, death, and movement rates at higher trophic levels (Denno and McClure, 1983; Hunter *et al.*, 1992).

Bottom-up effects in terrestrial trophic systems are now receiving a more balanced treatment in the study of population dynamics, such that the relative roles of all factors can be evaluated more effectively (cf. Hunter and Price, 1992).

B. Chemical Ecology

The mechanistic understanding of plant–herbivore interactions could not have advanced so rapidly without a phytochemical perspective. The fields became inextricably coupled, and both have prospered (e.g., Rosenthal and Berenbaum, 1991, 1992). The roles of constitutive and inducible plant defenses were explored. With increasing analytical power came the study of airborne chemicals such as plant volatiles, insect pheromones, and kairomones emitted by plants and herbivores. Sequestration of toxins by herbivores, the chemical ecology of oviposition stimulants, and host finding by parasites all became part of the mainstream of herbivore ecology (e.g., Sondheimer and Simeone, 1970; Nordlund *et al.*, 1981; Boethel and Eikenbary, 1986), and have been embraced by those working in marine systems also (e.g., Duffy and Hay, 1994). Of necessity, interactions among trophic levels, from plants to natural enemies, for example, needed more attention and emphasis, resulting in considerable research energy on three-trophic-level interactions (e.g., Price *et al.*, 1980; Barbosa and Letourneau, 1988).

The importance of the chemical perspective lies, first, in capturing one of the major modalities in which plants, herbivores, and carnivores interact, and second, in providing proximal mechanistic explanations of how natural food webs are integrated.

C. Microbial Ecology

Coupled with the emergence of chemical ecology and three-trophic-level interactions was an interest in the effects of plants on the pathogens of herbivores (e.g., Barbosa *et al.*, 1991). For example, did toxin sequestration by herbivores influence their resistance to pathogens? Do well-defended host plants confer pathogen resistance on herbivores? Are dynamical properties of herbivores over a landscape defined by complex interactions based on plants that promote pathogen resistance or susceptibility (e.g., Keating *et al.*, 1990; Schultz *et al.*, 1992; Foster *et al.*, 1992)? Even simple notions of density-dependent disease transmission have been shown to be surprisingly complex (Hunter and Schultz, 1993). In addition to their role as pathogens, microbes also mediate interactions among plants and herbivores. They are frequently involved as mutualistic symbionts of herbivores, providing enzymes and/or micronutrients related to refractive or deficient diets provided by plants (e.g., Jones, 1984; Martin, 1987). Endophytic fungi confer resistance in plants to herbivores (e.g., Clay, 1990, 1991). Fungal interactions with insects are widespread and diverse (e.g., Batra, 1979; Pirozynski and Hawksworth, 1988; Wilding *et al.*, 1989).

D. Life-History Evolution

As a part of evolutionary biology, the adaptive nature of life histories emerged as a major theme, probably founded in the clutch size debate initiated by Lack (e.g., 1947). Studies on insect life histories, and their adaptive variation, strengthened in the 1970s (e.g., Dingle, 1972, 1978) and matured in the 1980s (e.g., Denno and Dingle, 1981; Dingle and Hegmann, 1982; Dingle, 1984; Taylor and Karban, 1986; Roff, 1992).

Tracing the influence of evolutionary approaches to life history, and then to consequences for population dynamics, brings us into a nebulous area in the history of ecology. Nevertheless, during the 1980s the connection was made between macroevolutionary divergence of life histories and macroecological patterns in population dynamics (e.g., Rhoades, 1985; Nothnagle and Schultz, 1987; Wallner, 1987; Barbosa *et al.*, 1989). These initiatives have received increasing attention in the present decade (e.g., Price *et al.*, 1990; Hunter, 1991, 1995; Price, 1994).

The realization that phylogenetic constraints act on life-history evolution and population dynamics may seem obvious, but a strong phylogenetic approach to ecology has emerged belatedly (e.g., Harvey and Pagel, 1991; Mitter and Farrell, 1991). The new scale of macroevolution applied to population dynamics has fostered a novel and strongly comparative approach, and an incisive theme in macroecology.

E. Behavioral Ecology

Behavior was rarely discussed in studies of population dynamics in the 1960s except in terms of emigration or dispersal, and then it was usually evaluated indirectly as loss of individuals or reproductive shortfall. Life table analysis minimized the chances of recording relevant behavioral traits, for example, during oviposition (Price *et al.*, 1990). A notable exception to this general trend

was Wellington's (1957, 1960, 1962, 1964, 1977) emphasis on population quality variation and the behavioral differences in a population at high and low
densities. Remarkably, this fascinating research did not light a fire of research
energy, although Leonard (1970a,b) reinforced the validity of a variable behavioral element in population dynamics, and Chitty (1960) provided highly provoc-
ative hypotheses invoking behavioral mechanisms.

To date, predator-mediated changes in prey behavior that influence the fecundity of the prey have yet to be incorporated into population dynamics theory
in any meaningful way. Yet it is clear from work in aquatic systems that predators
can act to reduce the access of their prey to important resources (Werner et al.,
1983; Gilliam and Fraser, 1987). A similar approach in terrestrial population
dynamics is certainly needed (Schultz, 1992).

Researchers interested in the evolution of host-plant utilization and specificity in insect herbivores began to study variation in female choice of plant species
or plant individuals. Did mother know best in relation to food quality for her
progeny? Was there a linkage between female preference for oviposition sites
and larval performance measured by growth rate, survival, or ultimate size?
These questions were a natural development in evolutionary ecology and plant–
herbivore interactions in the early 1970s (cf. Singer, 1986; Thompson, 1988;
Courtney and Kibota, 1990; Leather, 1994). Coupling this preference–performance
approach with life-history traits (e.g., Leather, 1994), and then population dynamics, has brought female behavior into the limelight in helping to reveal major
differences in behavior that are important in the dynamic properties of populations and species (Preszler and Price, 1988; Price et al., 1990).

The importance of dispersal in influencing population dynamics has long
been recognized. In addition to merely redistributing individuals in a population,
density-dependent dispersal can regulate a local population if dispersers suffer
high mortality (Denno and Peterson, Chapter 6, this volume). It has now become
apparent that moderate movement between populations can also add to metapopulation stability (Gilpin and Hanski, 1991). Dispersal has been notoriously intractable to study. Only recently have we begun to make headway in measuring and
manipulating dispersal, often by manipulating the habitat, in such a way as to test
and refine theory (Bergelson and Kareiva, 1987; Kareiva, 1984, 1987; Walde,
1991, and Chapter 9, this volume; Roland and Taylor, Chapter 10, this volume).

F. Broad Comparative Ecology

"Study major, broad, repeatable patterns" admonished Tilman (1989, p. 90).
"Because the purpose of ecology is to understand the causes of patterns in nature,
we should start by studying the largest, most general, and most repeatable
patterns." Until recently, this was not the style in population dynamics. Quite the
reverse. Single species were studied often over long periods of time in a largely

idiosyncratic manner, with idiosyncratic results. Plot techniques frequently precluded the discovery of spatial pattern, on altitudinal gradients, for example, even for a single species (Price *et al.*, 1990). Understanding the dynamics of one species was an end unto itself, and a sufficient justification. Studies on the spruce budworm, the winter moth, the pine looper, the grey larch tortrix, psyllids on *Eucalyptus,* and many others were the fare of the day (e.g., Clark *et al.*, 1967; Southwood, 1968). The search for general patterns was not a preoccupation.

Broad patterns in nature involving population dynamics began to emerge in the 1970s. A remarkable attempt to find pattern was Southwood's (1975) broad analysis of life table data and key-factor analysis, which ultimately resulted in the synoptic population dynamics model by Southwood and Comins (1976). In this decade, T. C. R. White (1974, 1976, 1978) was also finding general pattern relating to plant stress and nutrient supply for herbivores. Later, Berryman began to classify kinds of outbreak dynamics, detecting patterns among a wide variety of herbivores (Berryman and Stark, 1985; Berryman, 1986, 1987). At a similar time came the comparative approaches to life-history traits of eruptive and noneruptive species (e.g., Rhoades 1985; Nothnagle and Schultz, 1987; Wallner, 1987; Barbosa *et al.*, 1989). More formal phylogenetically based analyses, advocated by Harvey and Pagel (1991), and illustrated for insect–plant relationships by Mitter and Farrell (1991), are in the making for insect–herbivore population dynamics (cf. Hunter, 1991, 1995).

G. Experimental Biology

Studies in population dynamics were generally descriptive and not experimental. A strong experimental approach was pioneered by Frank Morris. After many years of studying spruce budworm using plot techniques, life tables, and key-factor analysis (e.g., Morris, 1963), he began to emphasize process over description and correlation (Morris, 1969). Moving from spruce budworm to fall webworm, Morris searched for broad landscape-scale pattern in population densities in time and space (Morris and Bennett, 1967; Morris, 1964, 1971b), and the mechanisms that drove these patterns (e.g., Morris, 1967, 1971a, 1972; Morris and Fulton 1970a,b, summarized in Price, 1984). He studied maternal effects on subsequent generations (Morris, 1967) and the heritability of adaptations to warm and cool weather regimes (Morris and Fulton, 1970b). As with Wellington's pioneering research, it is strange to relate that Morris's research on *Hyphantria cunea* did not immediately stimulate others to adopt a mechanistic approach to population dynamics.

The increasing rigor of experimental approaches demanded by the developing fields of chemical ecology, plant–animal interactions, and behavioral ecology brought a renewed vigor to population ecology. Kareiva (1989, p. 71) noted a decline in experimental research during the "MacArthur Era," but in the more mechanistic areas of ecology, experimental research increased in frequency and importance. "In the 1970s, experimental field ecology came of age" (Colwell, 1984, p. 393). Again, rather belatedly, the experimental approach was espoused by a few interested in population dynamics: the California red scale and *Aphytis* experiments by Murdoch and associates (Murdoch, 1994; Murdoch and Walde, 1989; Murdoch *et al.*, 1985, 1992), the population introduction experiments by Myers (e.g., 1990), and the perturbation experiments by Cappuccino (1992) and Harrison (1991, 1994; Harrison *et al.*, 1995). These are very important developments that provide a clarity of results not found in correlational techniques.

H. Integration of Empirical and Theoretical Ecology

If we ask the question "Which comes first in research, empirical observations or theoretical constructs?" two positions seem to prevail. Theoretical ecologists will argue that theory is primary. "Let theory guide your observations" wrote Darwin, a phrase used in the first sentence of the book "Perspectives in Ecological Theory" (Roughgarden *et al.*, 1989, p. 3). "[A]t the time of their conception, the first models of species interactions provided insights unanticipated by field ecologists (e.g., predator–prey cycles)" (Kareiva, 1989, p. 69). But more empirically motivated ecologists would argue that Darwin was first and foremost an empiricist, and from field observation he was forced to explore a theoretical explanation for patterns of evolution in nature (e.g., Price, 1991). Also evident is the importance of early field research on the development of Lotka's theory. In 1897 Howard (p. 48) noted what Lotka later quoted in his book (1924, pp. 90–91): "With all very injurious lepidopterous larvae we constantly see a great fluctuation in numbers, their parasites rapidly increasing immediately after the increase of the host species, overtaking it numerically and reducing it to the bottom of another ascending period of development."

The building of theory is the essential motivation of scientists, but how it is developed is a matter of choice and opportunity, no doubt. The only requirement is that the theory must eventually explain broad and real patterns in nature. Therefore, theory without a factual foundation is trivial, when the goal of ecology is to understand major natural patterns. We would argue that a lack of communication and cooperation among so-called theoretical and empirical population ecologists has, and may continue, to hinder the development of synthesis in population dynamics. In other branches of ecology, such as food web analysis, numerical theory has become increasingly removed from the empirical foundation that it was supposed to represent (Paine, 1988) and has resulted in a reevaluation of the discipline as a whole (Pimm and Kitching, 1988; Polis, 1991; Polis and Winemiller, 1995). Theory built on top of theory is an unfortunate feature of much of the literature on population dynamics also. We might characterize the current situation as a theoretical hippo balanced on the head of an empirical pin. For example, disagreements on the degree to which apparent correlations among population densities at different trophic levels should be represented in mathematical population models (e.g., Matson and Berryman, 1992; Oksanen *et al.*, 1992; Gleeson, 1994; Sarnelle, 1994) would be more simply reconciled were there sufficient empirical studies available to assess the validity of competing approaches.

Though it is easy to class ecologists as theorists or empiricists, it should not actually be an interesting or useful dichotomy. All ecologists should be working toward general theory, with some using field studies and others using more hypothetical approaches. Preferably, any one research group will employ both avenues for the understanding of nature. Lately there has been a growing body of researchers and groups equipped to unite field research with ecological theory (e.g., Murdoch, 1994; Murdoch and Walde, 1989; Murdoch *et al.*, 1985, 1992; Kareiva, 1987; Kareiva and Andersen, 1988; Tilman, 1988, 1994; Turchin, 1990, 1993; Turchin and Taylor, 1992; Turchin *et al.*, 1991). Only in 1989 Kareiva lamented (p. 83) "the sad truth is that ecological theory exists largely in a world of its own, unnoticed by mainstream ecology" (see also Kareiva, 1994). He envisaged persistence of the dichotomy between empiricist and theoretician, but with increased communication. What we actually see is a blending of the two endeavors into individual research programs. This is where empiricism and theory belong: one science, one goal, one general approach, even though the eventual result will be pluralistic theory.

I. Genetics, Heritability, and the Evolution of Populations

A major flaw in the development of population dynamics as a discipline in ecology is the virtual absence of an evolutionary perspective. Although influential researchers such as Chitty (1960, 1967) and Carson (1968) emphasized many years ago a genetically based evolutionary scenario in the cycling of populations, the role of evolution in dynamics remains hardly explored. Even with Ford's (1957, 1964; Ford and Ford, 1930) pioneering studies on the marsh fritillary butterfly, demonstrating dramatic shifts in selection pressure and accelerated evolution during rapid population change, a major research program on population evolution never materialized. As mentioned earlier, Frank Morris's research on the fall webworm, with a strong component of population adaptation and heritability, remains almost in isolation. Had Morris trained graduate students perhaps a tradition would have been established.

The result is that a strong evolutionary approach to population dynamics has not entered into the mainstream of ecology. The need is obvious. Phylogenetic perspectives in the study of population dynamics will aid in bringing evolutionary approaches into the main arena of research. Evolutionary aspects of behavior, especially in reproducing females, may help in the understanding of macroevolutionary divergence of life histories and population dynamics. But still, we cannot assume that, as population densities change through time, they remain genetically constant. Neither can we assume that the expression of an organism's genome is constant across environments. Gene-by-environment interactions are likely to be responsible for much of the variation in the phenotype of organisms that we observe in nature. These include, of course, the phenotypic traits that are linked directly to population change, such as dispersal and fecundity (Rossiter, 1994).

J. International Integration

Many countries share common ecological problems and issues that foster international cooperation. For example, long-term studies of forest and woodland pests were well represented in Australia, Canada, England, Finland, Germany, The Netherlands, Switzerland, and the United States. While searching for pattern, a major opportunity lies in broadening the scope of interaction, not only in cool-temperate climates, but into Mediterranean and tropical climates, and their vegetation types. Broad altitudinal and latitudinal gradients could be explored in a comparative way, and perhaps related species might be compared on gradients from desert to montane forest. Broadly distributed taxa such as the herbivorous lady beetles (Coccinellidae: *Epilachna* spp.) provide an excellent opportunity for comparative studies in temperate and tropical regions (e.g., Iwao, 1970; Ohgushi, 1986; Ohgushi and Sawada, 1985; Abbas and Nakamura, 1985; Nakamura *et al.*, 1988; Inoue *et al.*, 1993). A broader international community of interaction, with broader comparable possibilities, will be essential for a major synthesis.

III. Novel Approaches and the Beginnings of Synthesis

A new synthesis of disciplines in population dynamics can be achieved by integration of many fields discussed here. We may well differ among ourselves on the complete list of subjects required in a revitalization of population dynamics and a break from the past. However, we probably can agree on many of the elements, and that those elements are largely represented in the literature, and many are to be found in this book.

A. Historical Perspective

This is not the first time between a pair of covers that we see so many features of a new synthesis in population dynamics. But it is important to recognize precursors in past volumes on the subject. Going back three decades, to the 1960s, two books, as examples, captured much of the enthusiasm and optimism of what was perhaps the heyday of population dynamics. Clark *et al.* (1967) represented effectively the *modus operandi* and the major debates of the times: life table studies and key-factor analysis, and the hypotheses of Nicholson and Bailey, Andrewartha and Birch, Milne, Chitty, and Pimentel. Process, or mechanistic, studies received short shrift because their shortage could only be lamented, although the weaknesses in plot techniques and life table analysis were not appreciated (cf. Price *et al.*, 1990). Southwood's (1968) symposium volume also featured major studies using life table analysis, such as by Baltensweiler, Klomp, and Varley and Gradwell. However, mechanistic approaches were represented well for aphid populations (Way), the link between competition and dispersal (Dempster), and behavioral aspects of the predator–prey interaction (Holling). Waloff noted the dynamic nature of the plant with its continually changing resource quality, and Holling rejoiced in "the new techniques of experimental analysis" and computer modeling. "Attention is thereby shifted from the classical population dynamics mechanisms that emphasize numbers destroyed, to processes that determine the quality of organisms, the duration of generations and historical and spatial sensitivity of the systems" (Holling, 1968, p. 57). Was a new age in population dynamics emerging? Were the elements for a synthesis already appreciated almost 30 years ago?

Certainly by the early 1970s the elements for synthesis were fully evident (e.g., den Boer and Gradwell, 1970; Watson, 1970): the genetics and behavior of voles (Krebs), reduced fecundity (Coulson), environmental and population heterogeneity (Birch, den Boer, Wolda), plant food quality for herbivores (Dixon), plant dynamics and its role in herbivore dynamics (van der Meijden, Dempster), and much experimental research. There were some retrospectives to the 1960s and earlier (e.g., Varley *et al.*, 1973; Tamarin, 1978), but it appears that Varley *et al.* (1973) was the last book on population dynamics to emphasize life table construction and key-factor analysis, and Hassell (1970, 1978), the junior author, was well on the way to mechanistic analysis of population change. By the end of the decade (Anderson *et al.*, 1979), the character of studies had continued to change: more modeling, more spatial dynamics, more evolutionary ecology, and more community dynamics. The classical long-term study was losing ground, although other kinds of long-term investigations had become important (Taylor and Taylor, Dixon, and Connell).

A strange hiatus ensued in the first half of the 1980s. Where had all the dynamics gone? Obviously, there was a synchronous, but rather patchy and fractionated motivation toward dynamical perspectives from various fields that were broadening in scope: behavior, life-history evolution, plant–herbivore interactions, and the rest, captured well in Barbosa and Schultz (1987). But at the turn of the decade, "Ecology for Tomorrow" (Kawanabe *et al.*, 1990) did not highlight the role of population dynamics, still reflecting perhaps some hysteresis in the rate of change in the field and recognition of that change.

B. What's New?

In a synthesis chapter, Colwell's (1984) title "What's new? Community ecology discovers biology" may well be applicable to population dynamics. The field is discovering and integrating the rest of biology. But, as Colwell questioned, is novelty involved? Is the old population dynamics mortally wounded? "Death of the 'Old Ecology'?: Epitaph for a straw man," he wrote (Colwell, 1984, p. 392). Given the foregoing historical perspective, perhaps our book title erects a straw man?

Most of all, what is new in this book is the emphasis on integrating elements toward a new synthesis. Most of the elements have been explored, but their blending has not received highly motivated attention. Though earlier authors have emphasized the need for more research in important areas, they did not emphasize their necessary integration. For example, Wilson (1968), at the end of the book "Insect Abundance," stressed many areas still in need of much more research today: "The population dynamics of innocuous and rare species" (p. 144); species at the edge of their range; cyclic and noncyclic species; genetic characteristics of populations; behavioral aspects of dynamics; and the importance of food supply. Although Wilson emphasized the need for laboratory experiments he did not emphasize experiments in the field. No mention was made of phylogenetic bases for divergence of dynamical types, and no synthesis of different approaches was advocated. Clark *et al.* (1967) anticipated the need for experimentation in the field and laboratory, studies on disease in natural populations, and the role of plant physiology and genetics. They recommended study of uncommon species in relatively undisturbed landscapes, and comparative studies of related species, and high- and low-density species in the same environments. But all of these items were very much on a kind of "wish list" for population dynamics. Some of these areas are now well developed in our current volume.

At the conclusion of the Oosterbeek Symposium (den Boer and Gradwell, 1970), Watt (1970) wrote on synthesis in the dynamics of populations. However, as was noted by those present, emphasis was on community development and structure, rather than on population dynamics, and instead of a tightly knit set of basic theorems, like a "bunch of bananas," it formed a long sequence "like a string of sausages" (den Boer and Gradwell, 1970, p. 578). In retrospect, the approach was not used apparently in the development of this field. Perhaps, Anderson *et al.* (1979) had learned an important lesson and highlighted developing areas rather than claiming any attempts at synthesis.

Admittedly, this book may appear as a string of sausages—a series of chapters that in themselves do not fulfill the goal of the book. But as elements in the synthesis they are fundamental. First, there is a strong theme of comparative studies: eruptive and noneruptive species, similar species in different localities, related species at typically different densities, and related species throughout their geographic range. Second, theoretical approaches are linked tightly to empirical data, and empirical studies test theory. Third, supply-side ecology, or the bottom-up supply of resources of appropriate quantity and quality, emerges repeatedly as a key to understanding dynamics. Fourth, field experiments provide unequivocal results, essential in weeding out the relative importance of bottom-up and top-down influences on dynamics. Fifth, a strong evolutionary perspective, through the comparison of life histories, provides a macroevolutionary backdrop to the understanding of ecological phenomena. The ecological play is set in a macroevolutionary theater—to twist the title of a famous book (Hutchinson, 1965). Sixth, the spatial structure of populations plays a key role in dynamics. On a finer scale, the dispersion of individuals within a population may influence their variability in performance and thus their propensity to fluctuate. On a larger scale, habitat structure influences the extinction/recolonization dynamics of metapopulations. Seventh, a return to a Wellingtonian emphasis on population quality and the importance of maternal effects in population cycles: the results of phenotypic plasticity. Eighth, behavior emerges as another theme involving dispersal, metapopulation dynamics, preference–performance linkage, aggregation of progeny, and responses to resource supply. Ninth, disease is treated in theoretical and empirical terms, and recognition of the subtle influence of pathogens as sublethal agents is reinforced. Tenth, beyond the scales involved in spatial structure mentioned earlier, we recognize that very different scales in evolution and ecology are necessary for a synthesis in population dynamics. Macroevolutionary patterns, including those resulting in major differences in life-history types, will have macroecological effects, such as the emergence of eruptive or latent population dynamics, common and rare species. Microevolution will play its role in a more ecological theater, as Hutchinson (1965) emphasized.

Are these the ten commandments on population dynamics? Obey and prosper? Certainly not. Other elements of importance require careful evaluation in terms of population dynamics. Just to name a few: the role of induced plant defenses in herbivore population dynamics (e.g., Tallamy and Raupp, 1991); nutritional ecology and the variable nature of resources (e.g., Slansky and Rodriguez, 1987); the genetics and evolution of populations as they change in density; and a strongly comparative approach to dynamics on latitudinal and climatic gradients from the Arctic to the tropics, from the deserts to rain forests. However, many significant components required for synthesis on population dynamics are represented in this book, and our sensitivity has been raised about the broad range of investigations necessary for a broadening theoretical development.

IV. The New Synthesis

A. Scale

Recognition of the several scales involved in a synthesis on population dynamics, and the evolutionary–ecological pathways involved, might be highlighted as an avenue for clarification. These scales may be defined as macrodynamics, mesodynamics, and microdynamics, and all are recognized in this book.

Macrodynamics involves those dynamical features of populations influenced strongly by macroevolutionary events. For example, the macroevolution of lineages with divergent life-history characteristics between families, genera, and species sets phylogenetic constraints on the performance of populations in ecological time (Price *et al.,* 1990; Price, 1994). The evolution of major differences in life histories, such as sexual reproduction in spring versus summer, changes dramatically the quality of resources for herbivores. Laying large clutches or single eggs has direct impacts on foraging among progeny. The evolution of life cycles with strong preference–performance linkage in which "mother knows best" is likely to result in population dynamics driven from below through resource supply. Lack of this kind of linkage in a phylogeny may well result in more generalized exploitation of resources, heavier exploitation of resources, and more eruptive kinds of population dynamics. The evolutionary background of species, and groups of related species, has a profound influence on their performance in ecological time.

Mesodynamics then becomes an intermediate scale, although involving large-scale ecological settings: the landscape, the gradient, and the metapopulation dynamics. After the search for macrodynamical patterns, the meso-scale offers many opportunities in the search for pattern in population dynamics inadequately developed to date. What are the dynamical properties of populations over the whole altitudinal or latitudinal range of a species? What are the dynamics of related species in temperate and tropical latitudes? How do the dynamics differ in the center and at the perimeter of an epidemic? Strongly comparative approaches to population dynamics are easily developed at this scale in the search for broad patterns in nature.

Microdynamics covers the local details and mechanisms of population cycles through time and local space, bottom-up and top-down influences on populations, maternal effects and qualitative change in populations, and the genetic components of dynamics, or microevolutionary aspects of population change. Microdynamics is the grist of the ecological mill. Detailed local studies provide new insights and understanding from which broader comparative studies may be developed. This is where the true synthesis begins, because by really understanding the mechanistic dynamics at one locality, the key ingredients of a broader perspective can be realized. Without this local and mechanistic view, the search for pattern will depend on correlation. And correlational approaches to population dynamics resulted in the ecologist's phlogiston theory.

B. Getting Empirical Facts into Theory

If ecologists aspire to generate a theory on population dynamics akin to a theory of evolution, or a theory of relativity, then the facts must prevail. Facts must guide the generation of theory. The facts must be derived from the real world of ecological interactions, and not from the $1 + 1 = 2$ factual convenience from the armchair, with all the connotations that "armchair" arouses. Of course, armchair theory has strong heuristic value, but ultimately such theory and reality must connect. Perhaps such theory will generate a search for empirical validation. On the other hand, there is no doubt that nature creates much more intricacy and subtlety than the human mind has ever imagined purely from the armchair. "Let theory guide your observations" as Darwin said, but before empirically based theory was available, he noted that "I worked on true Baconian principles, and without any theory collected facts on a wholesale scale" (1958, p. 42). Historians may debate the sequence in the development of Darwin's method, but the key question is from where theory is derived.

Perhaps the greatest challenge to the development of a new synthesis will be to incorporate the considerable natural heterogeneity inherent in biological systems. Twenty-five years of mechanistic studies have taught us that spatial and temporal variation in biotic and abiotic factors can dominate the population dynamics of animal and plant species: spatial variation in primary productivity can influence the range of dynamics possible for herbivore species (Oksanen *et al.*, 1981; Belovsky and Joern, this volume); phenotypic variation among individual plants can cause the interactions among insect herbivores on those plants to vary from competitive to commensal within the one geographic location (Hunter, 1992); the effect that the "microbial loop" has on population change at other trophic levels varies greatly between small northern dimictic (water mixes twice) and southeastern monomictic (water mixes once) lakes in the United States (Porter, 1995). Any synthetic theory of population dynamics must embrace rather than ignore these types of variability.

We began this chapter by emphasizing that the development of fields such as plant–animal interactions, chemical ecology, and life-history evolution was a prerequisite for a modern synthesis of population dynamics. This is because these related fields provide us with a mechanistic understanding of why rates of reproduction, mortality, immigration, and emigration vary in time and space. These parameters of population change are often represented as constants in mathematical population models, yet they clearly vary significantly, and in biologically meaningful ways (Rossiter *et al.*, 1988; Hunter and Schultz, 1993). The complexity of mathematical models may limit the amount of this variation that we can incorporate easily into our numerical framework, but mathematical ecology and experimental ecology must work simultaneously (e.g., Foster *et al.*, 1992) to determine what should be included and what can be simplified or ignored.

C. "A New Reductionism?"

In Schoener's (1986, p. 81) vision "for a mechanistic-ecological utopia" he notes that " 'best' is defined as biologically most appropriate rather than mathematically most convenient." He noted (p. 103) that "a pluralistic theory will have replaced an attempted universal one." Pluralism would involve some specificity according to biological traits such as life history, and environmental conditions. "In both observational and experimental approaches, a greater emphasis will be placed on discovering the mechanism of an interaction or process, not just its existence and strength" (Schoener, 1986, p. 103). "A new reductionism" was part of Schoener's title.

Sad to say, nothing is new in biology, all has been said before—Beckner (1974), Ayala and Dobzhansky (1974), and the rest. But, just as Eiseley (1958) noted, all the tenets of a Darwinian philosophy of evolution were preempted by the sages of an earlier time. However, Darwin clinched the ideas of many decades. We can always dream to have contributed toward a clinching pluralistic theory of population dynamics to be developed in the next decade or two.

Acknowledgments

Research resulting in the positions taken in this chapter was supported by grants from the National Science Foundation (DEB-7816152, DEB-8021754, BSR-83144594, BSR-8705302, BSR-8715090, BSR-8918083, BSR-9020317, and DEB-9318188), the United States Department of Agriculture (91-34103-6398), Northern Arizona University, a Fulbright Senior Scholar Award, 1993–1994, a NSERC International Fellowship, and personal funds. We are grateful for the research support and for a review by Naomi Cappuccino.

References

Abbas, I., and Nakamura, K. (1985). Adult population parameters and life tables of an epilachnine beetle (Coleoptera: Coccinellidae) feeding on bitter cucumber in Sumatra. *Res. Popul. Ecol.* **27,** 313–324.

Anderson, R. M., Turner, B. D., and Taylor, L. R., eds. (1979). "Population Dynamics." Blackwell, Oxford.

Ayala, F. J., and Dobzhansky, T., eds. (1974). "Studies in the Philosophy of Biology: Reduction and Related Problems." Univ. of California Press, Berkeley.

Barbosa, P., and Letourneau, D. K., eds. (1988). "Novel Aspects of Insect–Plant Interactions." Wiley, New York.

Barbosa, P., and Schultz, J. C., eds. (1987). "Insect Outbreaks." Academic Press, San Diego, CA.

Barbosa, P., Krischik, V., and Lance, D. (1989). Life-history traits of forest-inhabiting flightless Lepidoptera. *Am. Midl. Nat.* **122,** 262–274.

Barbosa, P., Krischik, V. A., and Jones, C. D., eds. (1991). "Microbial Mediation of Plant–Herbivore Interactions." Wiley, New York.

Batra, L. R., ed. (1979). "Insect–Fungus Symbiosis." Allanheld, Osmun, Montclair, NJ.

Beckner, M. (1974). Reduction, hierarchies and organicism. *In* "Studies in the Philosophy of Biology: Reduction and Related Problems" (F. J. Ayala and T. Dobzhansky, eds.), pp. 163–177. Univ. of California Press, Berkeley.

Bergelson, J., and Kareiva, P. (1987). Barriers to movement and the response of herbivores to alternative cropping patterns. *Oecologia* **71,** 457–460.

Berryman, A. A. (1986). "Forest Insects: Principles and Practice of Population Management." Plenum, New York.

Berryman, A. A. (1987). The theory and classification of outbreaks. *In* "Insect Outbreaks" (P. Barbosa and J. C. Schultz, eds.), pp. 3–30. Academic Press, San Diego, CA.

Berryman, A. A., and Stark, R. W. (1985). Assessing the risk of forest insect outbreaks. *Z. Angew. Entomol.* **99,** 199–208.

Boethel, D. J., and Eikenbary, R. D., eds. (1986). "Interactions of Plant Resistance and Parasitoids and Predators of Insects." Ellis Horwood, Chichester.

Cappuccino, N. (1992). The nature of population stability in *Eurosta solidaginis,* a non-outbreaking herbivore of goldenrod. *Ecology* **73,** 1792–1801.

Carson, H. L. (1968). The population flush and its genetic consequences. *In* "Population Biology and Evolution" (R. C. Lewontin, ed.), pp. 123–137. Syracuse Univ. Press, Syracuse, NY.

Chitty, D. (1960). Population processes in the vole and their relevance to general theory. *Can. J. Zool.* **38,** 99–113.

Chitty, D. (1967). The natural selection of self-regulating behaviour in animal populations. *Proc. Ecol. Soc. Aust.* **2,** 51–78.

Clark, L. R., Geier, P. W., Hughes, R. D., and Morris, R. F. (1967). "The Ecology of Insect Populations in Theory and Practice." Methuen, London.

Clay, K. (1990). Fungal endophytes of grasses. *Annu. Rev. Ecol. Syst.* **21,** 275–297.

Clay, K. (1991). Fungal endophytes, grasses, and herbivores. *In* "Microbial Mediation of Plant–Herbivore Interactions" (P. Barbosa, V. A. Krischik, and C. G. Jones, eds.), pp. 199–226. Wiley, New York.

Collins, J. P. (1986). Evolutionary ecology and the use of natural selection in ecological theory. *J. Hist. Biol.* **19,** 257–288.

Colwell, R. K. (1984). What's new? Community ecology discovers biology. *In* "A New Ecology: Novel Approaches to Interactive Systems" (P. W. Price, C. N. Slobodchikoff, and W. S. Gaud, eds.), pp. 387–396. Wiley, New York.

Courtney, S. P., and Kibota, T. T. (1990). Mother doesn't know best: Selection of hosts by ovipositing insects. *In* "Insect–Plant Interactions" (E. A. Bernays, ed.), Vol. 2, pp. 161–188. CRC Press, Boca Raton, FL.

Darwin, F., ed. (1958). "The Autobiography of Charles Darwin and Selected Letters." Dover, New York.

den Boer, P. J., and Gradwell, G. R., eds. (1970). "Dynamics of Populations." Centre for Agricultural Publication and Documentation, Wageningen.

Denno, R. F., and Dingle, H. (1981). "Insect Life History Patterns: Habitat and Geographic Variation." Springer, New York.

Denno, R. F., and McClure, M. S. (1983). "Variable Plants and Herbivores in Natural and Managed Ecosystems." Academic Press, New York.

Dingle, H. (1972). Migration strategies of insects. *Science* **175**, 1327–1335.

Dingle, H., ed. (1978). "Evolution of Insect Migration and Diapause." Springer, New York.

Dingle, H. (1984). Behavior, genes and life histories: Complex adaptations in uncertain environments. *In* "A New Ecology: Novel Approaches to Interactive Systems" (P. W. Price, C. N. Slobodchikoff, and W. S. Gaud, eds.), pp. 169–194. Wiley, New York.

Dingle, H., and Hegmann, J. P., eds. (1982). "Evolution and Genetics of Life Histories." Springer, New York.

Duffy, J. E., and Hay, M. E. (1994). Herbivore resistance to seaweed chemical defense: The roles of mobility and predation risk. *Ecology* **75**, 1304–1319.

Eiseley, L. (1958). "Darwin's Century: Evolution and the Men Who Discovered It." Doubleday, Garden City, NY.

Ford, E. B. (1957). "Butterflies," 3rd ed. Collins, London.

Ford, E. B. (1964). "Ecological Genetics," 2nd ed. Methuen, London.

Ford, H. D., and Ford, E. B. (1930). Fluctuation in numbers and its influence on variation in *Melitaea aurinia,* Rott. (Lepidoptera). *Trans. R. Entomol. Soc. London* **78**, 345–351.

Foster, M. A., Schultz, J. C., and Hunter, M. D. (1992). Modelling gypsy moth–virus–leaf chemistry interactions: Implications of plant quality for pest and pathogen dynamics. *J. Anim. Ecol.* **61**, 509–520.

Gilbert, L. E., and Raven, P. H., eds. (1975). "Coevolution of Animals and Plants." Univ. of Texas Press, Austin.

Gilliam, J. F., and Fraser, D. F. (1987). Habitat selection under predation hazard: Test of a model with foraging minnows. *Ecology* **68**, 1856–1862.

Gilpin, M., and Hanski, I., eds. (1991). "Metapopulation Dynamics: Empirical and Theoretical Investigations." Academic Press, London.

Gleeson, S. K. (1994). Density dependence is better than ratio dependence. *Ecology* **75**, 1834–1835.

Harrison, S. (1991). Local extinction and metapopulation persistence: An empirical evaluation. *Biol. J. Linn. Soc.* **42**, 73–88.

Harrison, S. (1994). Resources and dispersal as factors limiting a population of the tussock moth (*Orgyia vetusta*), a flightless defoliator. *Oecologia* (in press).

Harrison, S., Thomas, C. D., and Lewinsohn, T. M. (1995). Testing a metapopulation model of coexistence in the insect community on ragwort (*Senecio jacobaea*). *Am. Natur.* **145**, 546–562.

Harvey, P. H., and Pagel, M. D. (1991). "The Comparative Method in Evolutionary Biology." Oxford Univ. Press, Oxford.

Hassell, M. P. (1970). Parasite behaviour as a factor contributing to the stability of insect host–parasite interactions. *In* "Dynamics of Populations" (P. J. den Boer and G. R. Gradwell, eds.), pp. 366–378. Centre for Agricultural Publication and Documentation, Wageningen.

Hassell, M. P. (1978). "The Dynamics of Arthropod Predator–Prey Systems." Princeton Univ. Press, Princeton, NJ.

Holling, C. S. (1968). The tactics of a predator. *In* "Insect Abundance" (T. R. E. Southwood, ed.), pp. 47–58. Blackwell, Oxford.

Howard, L. O. (1897). A study of insect parasitism: A consideration of the parasites of the white-marked tussock moth, with an account of their habits and interrelations, and with descriptions of new species. *U.S., Dep. Agric., Tech. Ser.* **5**, 1–57.

Hunter, A. F. (1991). Traits that distinguish outbreaking and nonoutbreaking Macrolepidoptera feeding on northern hardwood trees. *Oikos* **60**, 275–282.

Hunter, A. F. (1995). The ecology and evolution of reduced wings in forest Macrolepidoptera. *Evol. Biol.* (in press).

Hunter, M. D. (1992). Interactions within herbivore communities mediated by the host plant: The keystone herbivore concept. *In* "Effects of Resource Distribution on Animal–Plant Interactions" (M. D. Hunter, T. Ohgushi, and P. W. Price, eds.), pp. 287–385. Academic Press, San Diego, CA.

Hunter, M. D., and Price, P. W. (1992). Playing chutes and ladders: Heterogeneity and the relative roles of bottom-up and top-down forces in natural communities. *Ecology* **73**, 724–732.

Hunter, M. D., and Schultz, J. C. (1993). Wound-induced defenses breached? Phytochemical induction protects an herbivore from disease. *Oecologia* **94**, 195–203.

Hunter, M. D., Ohgushi, T., and Price, P. W., eds. (1992). "Effects of Resource Distribution on Animal–Plant Interactions." Academic Press, San Diego, CA.

Hutchinson, G. E. (1965). "The Ecological Theater and the Evolutionary Play." Yale Univ. Press, New Haven, CT.

Huxley, J. S. (1942). "Evolution: The Modern Synthesis." Allen & Unwin, London.

Inoue, T., Nakamura, K., Salmah, S., and Abbas, I. (1993). Population dynamics of animals in unpredictably-changing tropical environments. *J. Biosci.* **18**, 425–455.

Iwao, S. (1970). Dynamics of numbers of a phytophagous lady-beetle, *Epilachna vigintioctomaculata,* living in patchily distributed habitats. *In* "Dynamics of Populations" (P. J. den Boer and G. R. Gradwell, eds.), pp. 129–147. Centre for Agricultural Publishing and Documentation, Wageningen.

Jones, C. G. (1984). Microorganisms as mediators of plant resource exploitation by insect herbivores. *In* "A New Ecology: Novel Approaches to Interactive Systems" (P. W. Price, C. N. Slobodchikoff, and W. S. Gaud, eds.), pp. 53–99. Wiley, New York.

Kareiva, P. (1984). Predator–prey dynamics in spatially-structured populations: Manipulating dispersal in a coccinellid–aphid interaction. *Lect. Notes Biomath.* **54**, 368–389.

Kareiva, P. (1987). Habitat fragmentation and the stability of predator–prey interactions. *Nature (London)* **321**, 388–391.

Kareiva, P. (1989). Renewing the dialogue between theory and experiments in population ecology. *In* "Perspectives in Ecological Theory" (J. Roughgarden, R. M. May, and S. A. Levin, eds.), pp. 68–88. Princeton Univ. Press, Princeton, NJ.

Kareiva, P. (1994). Space: The final frontier for ecological theory. *Ecology* **75**, 1.

Kareiva, P., and Andersen, M. (1988). Spatial aspects of species interactions: The wedding of models and experiments. *In* "Community Ecology" (A. Hastings, ed.), pp. 38–54. Springer, New York.

Kawanabe, H., Ohgushi, T., and Higashi, M., eds. (1990). Ecology for tomorrow. *Physiol. Ecol. Jpn.* **27** (Spec. No. March), 1–205.

Keating, S. T., Hunter, M. D., and Schultz, J. C. (1990). Leaf phenolic inhibition of the gypsy moth nuclear polyhedrosis virus: The role of polyhedral inclusion body aggregation. *J. Chem. Ecol.* **16**, 1445–1457.

Krebs, C. J. (1979). Small mammal ecology. *Science* **203**, 350–351.

Lack, D. (1947). The significance of clutch size. *Ibis* **89**, 302–352.

Larsson, S. (1989). Stressful times for the plant stress–insect performance hypothesis. *Oikos* **56**, 277–283.

Leather, S. R. (1994). Life history traits of insect herbivores in relation to host quality. *In* "Insect–Plant Interactions" (E. A. Bernays, ed.), Vol. 5, pp. 175–207. CRC Press, Boca Raton, FL.

Leonard, D. E. (1970a). Intrinsic factors causing qualitative changes in populations of *Porthetria dispar* (Lepidoptera: Lymantriidae). *Can. Entomol.* **62**, 239–249.

Leonard, D. E. (1970b). Intrinsic factors causing qualitative changes in populations of the gypsy moth. *Proc. Entomol. Soc. Ont.* **100**, 195–199.

Lotka, A. J. (1924). "Elements of Physical Biology." Williams & Wilkins, Baltimore (reprinted as "Elements of Mathematical Biology." Dover, New York, 1956).

Martin, M. M. (1987). "Invertebrate–Microbial Interactions: Ingested Fungal Enzymes in Arthropod Biology." Cornell Univ. Press, Ithaca, NY.

Matson, P. A., and Berryman, A. A. (1992). Special feature: Ratio-dependent predatory–prey theory. *Ecology* **73**, 1539–1566.

Mattson, W. J., and Haack, R. A. (1987a). The role of drought stress in provoking outbreaks of phytophagous insects. *In* "Insect Outbreaks" (P. Barbosa and J. C. Schultz, eds.), pp. 365–407. Academic, San Diego, CA.

Mattson, W. J., and Haack, R. A. (1987b). The role of drought in outbreaks of plant-eating insects. *BioScience* **37**, 110–118.

Mayr, E., and Provine, W. B., eds. (1980). "The Evolutionary Synthesis: Perspectives on the Unification of Biology." Harvard Univ. Press, Cambridge, MA.

McIntosh, R. P. (1985). "The Background of Ecology: Concept and Theory." Cambridge Univ. Press, Cambridge, UK.

Mitter, C., and Farrell, B. (1991). Macroevolutionary aspects of insect–plant relationships. *In* "Insect–Plant Interactions" (E. A. Bernays, ed.), Vol. 3, pp. 35–78. CRC Press, Boca Raton, FL.

Morris, R. F., ed. (1963). The dynamics of epidemic spruce budworm populations. *Mem. Entomol. Soc. Can.* **31**, 1–332.

Morris, R. F. (1964). The value of historical data in population research, with particular reference to *Hyphantria cunea* Drury. *Can. Entomol.* **96**, 356–368.

Morris, R. F. (1967). Influence of parental food quality on the survival of *Hyphantria cunea*. *Can. Entomol.* **99**, 24–33.

Morris, R. F. (1969). Approaches to the study of population dynamics. *USDA For. Serv. Res. Pap.* **NE-125**, 9–28.

Morris, R. F. (1971a). Observed and simulated changes in genetic quality in natural populations of *Hyphantria cunea*. *Can. Entomol.* **103**, 893–906.

Morris, R. F. (1971b). The influence of land use and vegetation on the population density of *Hyphantria cunea*. *Can. Entomol.* **103**, 1525–1536.

Morris, R. F. (1972). Predation by wasps, birds, and mammals on *Hyphantria cunea*. *Can. Entomol.* **104**, 1581–1591.

Morris, R. F., and Bennett, C. W. (1967). Seasonal population trends and extensive census methods for *Hyphantria cunea*. *Can. Entomol.* **99**, 9–17.

Morris, R. F., and Fulton, W. C. (1970a). Models for the development and survival of *Hyphantria cunea* in relation to temperature and humidity. *Mem. Entomol. Soc. Can.* **70**, 1–60.

Morris, R. F., and Fulton, W. C. (1970b). Heritability of diapause intensity in *Hyphantria cunea* and correlated fitness responses. *Can. Entomol.* **102**, 927–938.

Morris, R. F., and Miller, C. A. (1954). The development of life tables for the spruce budworm. *Can. J. Zool.* **32**, 283–301.

Murdoch, W. W. (1994). Population regulation in theory and practice. *Ecology* **75**, 271–287.

Murdoch, W. W., and Walde, S. J. (1989). Analysis of insect population dynamics. *In* "Towards a More Exact Ecology" (P. J. Grubb and J. B. Whittaker, eds.), pp. 113–140. Blackwell, Oxford.

Murdoch, W. W., Chesson, J., and Chesson, P. L. (1985). Biological control in theory and practice. *Am. Nat.* **125**, 344–366.

Murdoch, W. W., Briggs, C. J., Nisbet, R. M., Gurney, W. S. C., and Stewart-Oaten, A. (1992). Aggregation and stability in metapopulation models. *Am. Nat.* **140**, 41–58.

Myers, J. H. (1990). Population cycles of western tent caterpillars: Experimental introductions and synchrony of populations. *Ecology* **71**, 986–995.

Nakamura, K., Abbas, I., and Hasyim, A. (1988). Population dynamics of the phytophagous lady beetle, *Epilachna vigintioctopunctata*, in an eggplant field in Sumatra. *Res. Popul. Ecol.* **30**, 25–41.

Nordlund, D. A., Jones, R. L., and Lewis, W. J., eds. (1981). "Semiochemicals: Their Role in Pest Control." Wiley, New York.

Nothnagle, P. J., and Schultz, J. C. (1987). What is a forest pest? *In* "Insect Outbreaks" (P. Barbosa and J. C. Schultz, eds.), pp. 59–80. Academic Press, San Diego.

Ohgushi, T. (1986). Population dynamics of an herbivorous lady beetle, *Henosepilachna niponica*, in a seasonal environment. *J. Anim. Ecol.* **55**, 861–879.

Ohgushi, T., and Sawada, H. (1985). Population equilibrium with respect to available food resource and its behavioural basis in an herbivorous lady beetle, *Henosepilachna niponica*. *J. Anim. Ecol.* **54**, 781–796.

Oksanen, L., Fretwell, S. D., Arruda, J., and Niemela, P. (1981). Exploitation ecosystems in gradients of primary productivity. *Am. Nat.* **118**, 240–261.

Oksanen, L., Moen, J., and Lundberg, P. A. (1992). The time-scale problem in exploiter–victim models: Does the solution lie in ratio-dependent exploitation? *Am. Nat.* **140**, 938–960.

Paine, R. T. (1988). Food webs: Road maps of interactions or grist for theoretical development? *Ecology* **69**, 1648–1654.

Pianka, E. R. (1974). "Evolutionary Ecology." Harper & Row, New York.

Pimm, S. L., and Kitching, R. L. (1988). Food web patterns: Trivial flaws or the basis of an active research program? *Ecology* **69**, 1669–1672.

Pirozynski, K. A., and Hawksworth, D. L. (1988). "Coevolution of Fungi with Plants and Animals." Academic Press, London.

Polis, G. A. (1991). Complex trophic interactions in deserts: An empirical critique of food-web theory. *Am. Nat.* **138,** 123–155.

Polis, G. A., and Winemiller, K. (1995). "Food Webs: Integration of Patterns and Dynamics." Chapman & Hall, New York (in press).

Porter, K. G. (1995). Integrating the microbial loop and the classic food chain into a realistic pelagic food web. *In* "Food Webs: Integration of Patterns and Dynamics" (G. A. Polis and K. Winemiller, eds.). Chapman & Hall, New York (in press).

Preszler, R. W., and Price, P. W. (1988). Host quality and sawfly populations: A new approach to life table analysis. *Ecology* **69,** 2012–2020.

Price, P. W. (1984). "Insect Ecology," 2nd ed. Wiley, New York.

Price, P. W. (1991). Darwinian methodology and the theory of insect herbivore population dynamics. *Ann. Entomol. Soc. Am.* **84,** 465–473.

Price, P. W. (1994). Phylogenetic constraints, adaptive syndromes, and emergent properties: From individuals to population dynamics. *Res. Popul. Ecol.* **36,** 3–14.

Price, P. W., Bouton, C. E., Gross, P., McPherson, B. A., Thompson, J. N., and Weis, A. E. (1980). Interactions among three trophic levels: Influence of plants on interactions between insect herbivores and natural enemies. *Annu. Rev. Ecol. Syst.* **11,** 41–65.

Price, P. W., Cobb, N., Craig, T. P., Fernandes, G. W., Itami, J. K., Mopper, S., and Preszler, R. W. (1990). Insect herbivore population dynamics on trees and shrubs: New approaches relevant to latent and eruptive species and life table development. *In* "Insect–Plant Interactions" (E. A. Bernays, ed.), Vol. 2, pp. 1–38. CRC Press, Boca Raton, FL.

Rhoades, D. F. (1985). Offensive–defensive interactions between herbivores and plants: Their relevance in herbivore population dynamics and ecological theory. *Am. Nat.* **125,** 205–238.

Roff, D. A. (1992). "The Evolution of Life Histories: Theory and Analysis." Chapman & Hall, New York.

Rosenthal, G. A., and Berenbaum, M. R., eds. (1991). "Herbivores: Their Interactions with Secondary Metabolites," 2nd ed., Vol. 1. Academic Press, San Diego, CA.

Rosenthal, G. A., and Berenbaum, M. R., eds. (1992). "Herbivores: Their Interactions with Secondary Metabolites," 2nd ed., Vol. 2. Academic Press, San Diego, CA.

Rosenthal, G. A., and Janzen, D. H., eds. (1979). "Herbivores: Their Interaction with Plant Metabolites." Academic Press, New York.

Rossiter, M. C. (1994). Maternal effects hypothesis of herbivore outbreak. *BioScience* **44,** 752–763.

Rossiter, M. C., Schultz, J. C., and Baldwin, I. T. (1988). Relationships among defoliation, red oak phenolics and gypsy moth growth and reproduction. *Ecology* **69,** 267–277.

Roughgarden, J., May, R. M., and Levin, S. A., eds. (1989). "Perspectives in Ecological Theory." Princeton Univ. Press, Princeton, NJ.

Sarnelle, O. (1994). Inferring process from pattern: Trophic level abundances and imbedded interactions. *Ecology* **75,** 1835–1841.

Schoener, T. W. (1986). Mechanistic approaches to community ecology: A new reductionism? *Am. Zool.* **26,** 81–106.

Schultz, J. C. (1992). Factoring natural enemies into plant tissue availability to herbivores. *In* "Effects of Resource Distribution on Animal–Plant Interactions" (M. D. Hunter, T. Ohgushi, and P. W. Price, eds.), pp. 175–197. Academic Press, San Diego, CA.

Schultz, J. C., Hunter, M. D., and Appel, H. M. (1992). Antimicrobial activity of polyphenols mediates plant–herbivore interactions. *In* "Plant Polyphenols: Biogenesis, Chemical Properties, and Significance" (R. W. Hemingway and P. E. Laks, eds.), pp. 621–637. Plenum, New York.

Singer, M. C. (1986). The definition and measurement of oviposition preference in plant-feeding insects. *In* "Insect–Plant Relations" (J. Miller and T. A. Miller, eds.), pp. 66–94. Springer, New York.

Slansky, F., and Rodriguez, J. G., eds. (1987). "Nutritional Ecology of Insects, Mites, Spiders and Related Invertebrates." Wiley, New York.

Sondheimer, E., and Simeone, J. B., eds. (1970). "Chemical Ecology." Academic Press, London.

Southwood, T. R. E., ed. (1968). "Insect Abundance." Blackwell, Oxford.

Southwood, T. R. E. (1975). The dynamics of insect populations. *In* "Insects, Science, and Society" (D. Pimentel, ed.), pp. 151–199. Academic Press, New York.

Southwood, T. R. E., and Comins, H. N. (1976). A synoptic population model. *J. Anim. Ecol.* **45**, 949–965.

Stiling, P. (1994). What do ecologists do? *Bull. Ecol. Soc. Am.* **75**, 116–121.

Strong, D. R., Simberloff, D., Abele, L. G., and Thistle, A. B., eds. (1984). "Community Ecology: Conceptual Issues and the Evidence." Princeton Univ. Press, Princeton, NJ.

Tallamy, D. W., and Raupp, M. J., eds. (1991). "Phytochemical Induction by Herbivores." Wiley, New York.

Tamarin, R. H., ed. (1978). "Population Regulation." Dowden, Hutchinson & Ross, Stroudsburg, PA.

Taylor, F., and Karban, R. (1986). "The Evolution of Insect Life Cycles." Springer, New York.

Thompson, J. N. (1988). Evolutionary ecology of the relationship between oviposition preference and performance of offspring in phytophagous insects. *Entomol. Exp. Appl.* **47**, 3–14.

Tilman, D. (1988). "Plant Strategies and the Dynamics and Structure of Plant Communities." Princeton Univ. Press, Princeton, NJ.

Tilman, D. (1989). Discussion: Population dynamics and species interactions. *In* "Perspectives in Ecological Theory" (J. Roughgarden, R. M. May, and S. A. Levin, eds.), pp. 89–100. Princeton Univ. Press, Princeton, NJ.

Tilman, D. (1994). Competition and biodiversity in spatially structured habitats. *Ecology* **75**, 2–16.

Turchin, P. (1990). Rarity of density dependence or population regulation with lags? *Nature (London)* **344**, 660–663.

Turchin, P. (1993). Chaos and stability in rodent population dynamics: Evidence from nonlinear time-series analysis. *Oikos* **68**, 167–172.

Turchin, P., and Taylor, A. (1992). Complex dynamics in ecological time series. *Ecology* **73**, 289–305.

Turchin, P., Lorio, P. L., Taylor, A. D., and Billings, R. F. (1991). Why do populations of southern pine beetles (Coleoptera: Scolytidae) fluctuate? *Environ. Entomol.* **20,** 401–409.

Varley, G. C., and Gradwell, G. R. (1960). Key factors in population studies. *J. Anim. Ecol.* **29,** 399–401.

Varley, G. C., Gradwell, G. R., and Hassell, M. P. (1973). "Insect Population Ecology: An Analytical Approach." Blackwell, Oxford.

Walde, S. J. (1991). Patch dynamics of a phytophagous mite population: Effect of number of subpopulations. *Ecology* **75,** 1591–1598.

Wallner, W. E. (1987). Factors affecting insect population dynamics: Differences between outbreak and non-outbreak species. *Annu. Rev. Entomol.* **32,** 317–340.

Waring, G. L., and Cobb, N. S. (1992). The impact of plant stress on herbivore population dynamics. *In* "Insect–Plant Interactions" (E. A. Bernays, ed.), Vol. 4, pp. 167–226. CRC Press, Boca Raton, FL.

Watson, A., ed. (1970). "Animal Populations in Relation to Their Food Resources." Blackwell, Oxford.

Watt, K. E. F. (1970). Dynamics of populations: A synthesis. *In* "Dynamics of Populations" (P. J. den Boer and G. R. Gradwell, eds.), pp. 568–580. Centre for Agricultural Publishing and Documentation, Wageningen.

Wellington, W. G. (1957). Individual differences as a factor in population dynamics: The development of a problem. *Can. J. Zool.* **35,** 293–323.

Wellington, W. G. (1960). Qualitative changes in natural populations during changes in abundance. *Can. J. Zool.* **38,** 289–314.

Wellington, W. G. (1962). Population quality and the maintenance of nuclear polyhedrosis between outbreaks of *Malacosoma pluviale* (Dyar). *J. Insect Pathol.* **4,** 285–305.

Wellington, W. G. (1964). Qualitative changes in unstable environments. *Can. Entomol.* **96,** 436–451.

Wellington, W. G. (1977). Returning the insect to insect ecology: Some consequences for pest management. *Environ. Entomol.* **6,** 1–8.

Werner, E. E., Mittelbach, G. G., Hall, D. J., and Gilliam, J. F. (1983). Experimental tests of optimal habitat use in fish: The role of relative habitat profitability. *Ecology* **64,** 1525–1539.

White, T. C. R. (1974). A hypothesis to explain outbreaks of looper caterpillars, with special reference to populations of *Selidosema suavis* in a plantation of *Pinus radiata* in New Zealand. *Oecologia* **16,** 279–301.

White, T. C. R. (1976). Weather, food and plagues of locusts. *Oecologia* **22,** 119–134.

White, T. C. R. (1978). The importance of relative food shortage in animal ecology. *Oecologia* **33,** 71–86.

White, T. C. R. (1993). "The Inadequate Environment: Nitrogen and the Abundance of Animals." Springer-Verlag, Berlin.

Wilding, N., Collins, N. M., Hammond, P. M., and Webber, J. F., eds. (1989). "Insect–Fungus Interactions." Academic Press, London.

Wilson, F. (1968). Insect abundance: Prospect. *In* "Insect Abundance" (T. R. E. Southwood, ed.), pp. 143–158. Blackwell, Oxford.

Index

Abiotic mortality factors, leaf miner, 88
Acleris variana (black-headed budworm), 270
Acrobasis betulella (pyralid moth), 73–74
Acronicta, 55
Adaptive syndrome, 333
Adelges tsugae, 119–120
Adelgid, Adelges tsugae, 119–120
African migratory locust (Locusta migratoria), 270
Ageneotettix deorum, 362–363
Aggregation, see also Spatial variability
  egg survivorship and, 69–70
  feeding in, 65, 76–78
  herbivorous insects, 65–66
  larvae survival and, 70–72
  levels of, 65
  mortality and, 67, 87
  outbreak species and, 65–67, 69
  parasitism and, 70–72, 76
  population dynamics and, 78–79
  predation and, 70–72, 76
  survivorship and fecundity, 76
  temporal variability and, 7
Aggressive behavior, dispersal and, 118
Aglais urticae, 160
Agromyza, 83
Agromyzidae, 83
Aland islands, Glanville fritillary, 154–159
Alder, red (Alnus rubra), 242, 244–245
Aleiodes malacosomatus (braconid wasp), 205

Alnus rubra (red alder), 242, 244–245
Alsophila pometaria, 57
Ambystoma (salamander), 135
Anechura harmandi (earwig), 310–311, 312
Anthocharis cardamines, 160
Aonidiella aurantii (California redscale), 10
Aphantophus hyperantus, 160
Aphid, 30, 59
  cabbage (Brevicoryne brassicae), 122
  Cinara pinea, 120
  density-dependent dispersal, 114, 117, 118, 119, 120–124
  density experiments, 134, 136
  goldenrod
    Uroleucon caligatum, 67–68
    Uroleucon nigrotuberculatum, 67–68, 76, 78–79
  maternal effects, 269
  Megoura viciae, 119
Aphytis melinus, 10
Apple tree, Panonychus ulmi on, 174–183
Arboretum, Orland E. White, Cameraria hamadryadella eruptions, 96
Arctiidae, 50
Armyworm, fall (Spodoptera frugiperda), 57
Arroyo willow (Salix lasiolepis), 322–334
Aspen
  quaking (trembling) (Populus tremuloides), 92–95, 198, 200
Attraction, tests for, 20

Attractor, 25–26, 28
  chaotic, 25
  periodic, 25
  quasi-periodic, 25
Autumnal moth (*Epirrata autumnal*), 270–271
Avian predators, grasshoppers, 364, 370

Bacillus thuringiensis, 233, 234
Balsam poplar (*Populus balsamifera*), 198
Bark beetle, 11
  outbreaks, 339
  plant stress and, 341–342
  population dynamics, 339–353
Barros Colorado Island, Panama, rainfall, 23
Beetle
  bark, 11, 339–353
  carabid (*Pterostichus oblongipunctatus*), 137
  clerid, 11, 340–341
  goldenrod, 68–73, 76, 78
  lady, 11
    Epilachna niponica, 304–315
    Henosepilachna niponica, 135
  southern pine (*Dendroctonus frontalis*), 11, 32, 339–353
Behavioral ecology, 4, 393–394
Bet-hedging, 67; *see also* Risk-spreading
Betula papyrifera (paper birch), 73–76, 198
Biological control
  forest Lepidoptera, 237–240
  gypsy moth (*Lymantria dispar*), 238–239
  winter moth (*Operophtera brumata*), 239–240
Birch, paper (*Betula papyrifera*), 73–76, 198
Birch-feeding caterpillar, outbreak species, 73–76
Birth rate, density and, 229–230
Black-headed budworm (*Acleris variana*), 270
Blandy Farm, *Cameraria hamadryadella* eruptions, 96
Bombyx mori, 268
Bottom-up regulation, 7, 133, 294–295, 303, 312, 391–392
Boundedness
  regulation and, 6
  stochastic, Chesson, and population regulation definition, 26
  tests for, 20
Braconid wasp (*Aleiodes malacosomatus*), 205
Brenthis ino, 160
Brevicoryne brassicae (cabbage aphid), 122

British Columbia, field experiments, 237, 239, 240, 242
Budbreak, population density and, 94–95
Bud galler
  Euura lappo, 331
  Euura mucronata, 331–332
Budmoth, larch (*Zeiraphera diniana*), 34, 233, 270
Budworm
  black-headed (*Acleris variana*), 270
  spruce (*Choristoneura fumiferana*), 6, 195, 196, 206–207
Butterfly
  checkerspot (*Euphydryas editha*), 161, 163
  Finnish species, 154–161
  local population extinction, 162–163
  metapopulation dynamics, 10, 149–167
  migration rate, 164–166
  pollen-feeding heliconiine, 78
  stochastic extinction, 152, 166–167
  tiger swallowtail (*Papilio glaucus*), 6

Cabbage aphid (*Brevicoryne brassicae*), 122
Cage, for density-manipulation experiments, 137
California redscale (*Aonidiella aurantii*), 10
Callophrys rubi, 160
Calluna vulgaris, 164
Cameraria (leaf-mining moth), 136, 137
Cameraria hamadryadella, 95–99, 103, 104
  population crashes, 99
  population eruptions, 96–99
Cannibalism, 7, 139–140
Cape Cod, Massachusetts, gypsy moth introduction, 235
Carabid beetle (*Pterostichus oblongipunctatus*), 137
Carrying capacity, 151
Casebearer, larch (*Coleophora laricella*), 195
Case studies, 11
Caterpillar
  forest tent (*Malacosoma disstria*), 73–76, 196–207
  tent (*Malacosoma pluviale*), 230, 242–245
  western tent (*Malacosoma californicum pluviale*), 218, 235, 237, 242–243, 245
Catocala, 55
Cavelerius saccharivorus, 121–122
Ceiling, 25

Cellular automata model, 59
Chalcidoid wasp (*Eutetrastichus chlamytis*), 70–72
Chaos
  attractor, 25
  defined, 33
  identification of, 232
  quasi-, 34
Checkerspot butterfly (*Euphydryas editha*), 161, 163
Chemical ecology, 392
  galling sawfly (*Euura lasiolepis*), 324–325
Choristoneura fumiferana (spruce budworm), 6, 195, 196, 206–207
Chrysomelidae, 83
Cinara pinea, 120
Cirsium kagamontanum (thistle), 306
Clerid beetle (*Thanasimus dubius*), 11, 340–341, 347–351
Closed, equilibrium model, 140
Closed, nonequilibrium model, 140
Clossinia
  euphrosyne, 160
  freija, 160
  frigga, 160
Closterocerus tricinctus, 98
Clumping, 68
  outbreaking and, 7
  survival and, 69–72
Clustering of offspring, outbreak species, 54, 57, 58–59
Coccinellids, herbivorous, 11
Coleophora laricella (larch casebearer), 195
Coleoptera, 83
Colonization
  equilibrium and nonequilibrium models, 140
  incidence function model, 152–153
Community ecology, 4
Comparative ecology, 394–395
Comparative method, 6–8
Competition
  exploitative, 87
  interference, 7, 87, 139–140
Complex dynamics, 209
Compsilura cincinnata, 239
Compsilura concinnata (tachinid), 236
Contingent states test, 45–47
Convergence experiments, 10, 131
Cooking Lake, 198
Corn earworm (*Heliothis zea*), 57

Cotesia melanoscela (wasp), 235–236, 239
Cowbird (*Molothrus ater*), 370
Cropping experiments, 234–235
Crypsis, outbreak species, 48
Cyclical defoliation, 212–219
cyclic population dynamics, 232
Cyzenis albicans, 239–240

Damaged vegetation, *see also* Host-plant quality; Leaf abscission
  herbivore sensitivity to, 72–73, 74, 76
  herbivore survivorship and, 76–78
  leaf-miner avoidance of, 88
Damping
  over-, 24
  perfect, 24
  under-, 24
Death rate, density and, 229–230
Defoliation
  biological control, 237–240
  cyclical, 212–219
  forest tent caterpillar, 196–197, 204–206
  infectious disease and, 212–213
  population dynamics effects, 215–217
  spray programs for controlling, 232–234
Delayed density dependence, 218–219
Dendroctonus frontalis (southern pine beetle), 11, 32, 339–353
Density
  birth and death rates and, 229–230
  budbreak and, 94–95
  egg, *see* Egg density
  leaf-miner, 100–102
  low, population stability at, 306–308
  mortality and, 100–101
  parasitism and, 236–237
Density dependence, 12
  aggregation and, 66
  debate over, 19–22
  defined, 27, 229
  delayed, 218–219
  direct, 27, 32, 132, 133
  equilibrium and, 24
  evidence of, 131, 135
  hatchling production and, 368
  imperfect, 19–20
  inverse, 27, 133, 134
  larval interference, 94
Lygaeus equestris, 286–292

Density dependence (*continued*)
mortality sources and, 88–91
plant quality and quantity and, 378, 382
population model, 28
regulation and, 3–4, 19–22, 27, 131–132,
229, 380
reproductive processes, 308
return tendency, 27
role of, 131
spatial, 79, 236
statistical methods for detecting, 20–21
tests, 21, 27–30
time-series, 30–31, 36, 230–232
Density-dependent dispersal
consequences, 119–124
evidence for, 114–119, 124
experiments, 140
interhabitat, 117
interplant, 117
intraplant, 117
mechanisms mediating, 117–119
population dynamics and, 113–125
Density-dependent mortality, 7
leaf miner, 88–91, 95, 99, 100, 103–104
Density independence, 12, 380
aggregation and, 66
dispersal, 124
null model for, 29
stabilization of, 20
Density-manipulation experiments, 131–141,
177–184, 242–245
cages for, 137
equilibrium systems, 140–141
guidelines, 135–140
length of, 136
past studies, 132–135
return to equilibrium following, 176–177
spatial scale, 136–137
statistical analysis, 137–138
treatment level, 137
Density vagueness, 5, 20
Desert locust (*Schistocerca gregaria*), 270
Differential equation model
vs. complex simulation models, 220
host–parasitoid interactions, 209, 210–
212
Diprionid sawfly, 268
Diptera, 83
Direct density dependence, 27, 32, 132, 133

Disease, 8–9
aggregation and, 66
cyclic population dynamics and, 212–215
outbreak species, 59
viral, 220–224, 234, 243–245
Dispersal, 8, 394
density-dependent, 113–125, 140
density-independent, 124
manipulating, 10
models, 58–59
outbreak species, 57, 58–59
spider mite, 185
Dispersion, egg, outbreak species, 53–54, 58
Distribution, habitat and, 292–294
Distribution test, 47
Douglas fir tussock moth (*Orgyia pseudotsugata*), 10, 57, 234, 241, 271
Drought stress, bark beetle and, 342

Earwig (*Anechura harmandi*), 310–311, 312
Egg density, 308–309
offspring fitness and, 313–314
Egg dispersion, outbreak species, 53–54, 58
Egg hatch, *see also* Larvae
experiments, 242–245
maternal effects, 256–257
survivorship, aggregation and, 69–70
Egg mortality, leaf miner, 84–88, 94
Egg quality, 245
Egg resorption, 309–312, 314–315
Emergent properties, 333
Empirical ecology, theoretical ecology and,
395–396
Enargia infumata, 73–76
Environmental quality, maternal effects and,
252, 269–271
Eotetranychus sexmaculatus, 186
Ephydridae, 83
Epilachna niponica (lady beetle), 304–315
Epinoitia, 86
tedella, 86
Epirrata autumnal (autumnal moth), 270–271
Equilibrium
closed model, 140
as cloud of points, 25–26, 35–36
density variation and, 380–381
noise and, 26, 34
open model, 140

regulation and, 25–26, 229
Ricker curve, 371
Erebia embla, 160
Euphranta connexa (tephritid fly), 10, 281, 294
Euphydryas editha (checkerspot butterfly), 161, 163
Eurosta solidaginis (goldenrod galler), 134
Eutetrastichus chlamytis (chalcidoid wasp), 70–72
Euura
    amerinae, 331, 332–333
    atra (galling sawfly), 331
    carnivores, 332
    chemical ecology, 332–333
    lappo (bud galler), 331
    lasiolepis (galling sawfly), 321–335
    mucronata (bud galler), 331–332
    "*mucronata*" (galling sawfly), 331
    petiole galler, 331
    phylogenetic constraints hypothesis, 333–334
    plant vigor hypothesis, 334
    population dynamics, 333–334
    resource regulation, 332
    s-nodus?, 331
Evolution
    life-history, 393
    population dynamics and, 393, 397–398
Exema canadensis (goldenrod beetle), 68–73, 76, 78
Experimental biology, 395–396
Exploitative competition, 87
Exponentially stable equilibrium point, 34
Extinction
    equilibrium and nonequilibrium models, 140
    incidence function model, 152–153
    local butterfly populations, 162–163
    population scale and, 26
    predicting, 296
    randomly walking populations and, 4, 5
    stochastic, 152, 163, 166–167

Facilitation, galling sawfly, 325–326
Fall armyworm (*Spodoptera frugiperda*), 57
Fecundity
    maternal effects, 217–218, 256–257
    maximum, outbreak species, 52, 56
    solitary feeders, 76

Feeding
    in aggregations, outbreak species and, 65, 76–78
    phenology, outbreak species, 50
    site preemption, dispersal and, 118
Female behavior, galling sawfly, 325
Female mobility
    outbreak species, 52–53
    oviposition sites, 310
Finite population models, 26–27
Finland, butterfly metapopulation dynamics, 154–161
Floater, territorial animals and, 135
Floor, 25
Fly
    sarcophagid (*Sarcophaga* (=*Arachnidomyia*) *aldrichi*), 196–207
    tachinid
        Leschenaultia exul, 196
        Patelloa pachypyga, 196–207
    tephritid (*Euphranta connexa*), 10, 281, 294
Flying ability, outbreak species females, 52–53
Food-limited regulation, 361, 364–366
Forest fragmentation
    forest tent caterpillar and, 204–206
    parasitoid behavior and, 196–197, 200, 203–204
    population dynamics and, 195–207
    predator–prey systems in, 196
    spruce budworm and, 206–207
Forest Lepidoptera
    biological control of, 237–240
    experimental manipulation, 232–245
    sunspot fluctuation and, 241–242
Forest tent caterpillar (*Malacosoma disstria*), 73–76, 196–207
    forest fragmentation and, 204–206
    individual movement, 200
    large-scale dynamics, 204–207
    life tables, 199
    medium-scale dynamics, 203–204
    outbreak spread, 199–200
    parasitism, 196–207
    prolonged outbreaks, 204–206
    spatial distribution of, 204–206
    study methods, 197–202
    suppression of, 196
Fritillary, Glanville (*Melitaea cinxia*), 150, 154–159, 161, 163–167

Galler
  bud
    (*Euura lappo*), 331
    (*Euura mucronata*), 331–332
  goldenrod (*Eurosta solidaginis*), 134
  midrib (*Euura s-nodus?*), 331
  petiole (*Euura* sp.), 331
Galling sawfly
  Euura atra, 331
  Euura lasiolepis
    chemical ecology, 324–325
    facilitation and resource regulation, 325–326
    female behavior, 325
    oviposition behavior, 324
    population dynamics, 321–335
    predation, 326
    preference-performance linkage, 324, 329–331
    stability in natural populations, 326–328
Generalists, 7
Genetics, 397–398
Geometridae, 48, 54
Glanville fritillary (*Melitaea cinxia*), 150, 154–159, 161, 163–167
  metapopulation patterns and processes, 156–159
  metapopulation persistence, 154–156
GMLSM (Gypsy Moth Life System Model), 221–223
Goldenrod
  Solidago spp., 67–68
  Solidago rugosa, 69–70
Goldenrod aphid
  Uroleucon caligatum, 67–68
  Uroleucon nigrotuberculatum, 67–68, 72, 78–79
Goldenrod beetle
  Exema canadensis, 68–73, 76, 78
  Microrhopala vittata, 68–69
  Ophraella conferta, 68–70
  Trirhabda borealis, 68–72, 78
  Trirhabda virgata, 68–72, 78
Goldenrod galler (*Eurosta solidaginis*), 134
Gomphocerine grasshoppers, 366
Gonepteryx rhamni, 160
Gracillariidae, 83
Grasshopper
  Ageneotettix deorum, 362–363
  avian predators, 364, 370
  food-limited regulation, 364–366
  food resources, 361
  gomphocerine, 366
  Melanoplus femurrubrum, 362
  Melanoplus sanguinipes, 362, 366–379
  predation, 361, 362–365
  predator-limited regulation, 364–366
  rangeland, 11, 361–382
  regulating factors, 361–382
  study of, 360–361
Gregarious species
  aggregation in, 66
  egg dispersion, 53, 58
  larval behavior, 267–268
  maternal effects, 267–268
  offspring survival and, 69
  outbreak species, 53, 54, 56, 58
  patch survivorship of, 78
Grouse, red, 60
Growth-differentiation balance, 342–343
Gypsy moth (*Lymantria dispar*), 59, 195, 217, 218, 270
  biological control of, 238–239
  density-dependent experiments, 134
  inducing outbreaks, 235–237
  maternal effects, 252
  NPV and, 220–224
Gypsy Moth Life System Model (GMLSM), 221–223

Habitat, insect distribution and, 292–294
Habitat patch
  butterfly migration rate between, 164–165
  Glanville fritillary (*Melitaea cinxia*), 154–159
  incidence function model, 151–153
  metapopulation-level regulation and, 150, 151
  models, 151
  Panonychus ulmi density, 179–182
  stochastic extinction and, 166–167
Hatchling production, density dependence, 368
Haustellate seed-feeders, 114
Heliconiine butterfly, pollen-feeding, 78
Henosepilachna niponica (lady beetle), 135
Heodes birgaureae, 160
Herbivorous insects, 11–12, 41; *see also* Outbreak species
  aggregation in, 65–67, 69–72, 76–79, 87

bottom-up regulation, 7, 133, 294–295, 303–304, 312, 391–392
coccinellids, 11
host-plant defenses, 4, 66, 86–87, 88, 91, 93, 98, 215–217, 339–340, 391–392
host-plant quality and, 87, 88, 91, 102, 103, 104, 215–217, 378, 382
natural enemy dynamics, 8
opportunistic, 66, 74
oviposition behavior, 88, 92, 96, 104, 304–305, 309–315, 323, 325, 333–334, 394
parasitism, 8, 9, 66, 70–72, 76, 79, 86, 93, 98, 99, 103, 196–207, 210–212, 236–237, 239, 243–245
population growth, 251
population quality, 251
role of, 11–12
stealthy, 66, 69, 70–72, 74, 76–78
Heritability, 397–398
Hesperia comma, 163
Hesperiidae, 161
Heterocampa guttivita, 54
Heterocampini, 54
Heteroptera, density-dependent dispersal, 114, 116, 117
Homoptera, density-dependent dispersal, 114, 117
Horizontal factors, 7
leaf miner, 87–88, 91, 103
Host breadth, outbreak species, 51–52
Host–parasitoid model, 9, 210–212
Host–pathogen model, 212–215, 220–221
Host-plant defenses, 66
inducible, 215–217
leaf miner, 86–87, 88, 91, 93, 98
plant–herbivore interactions, 4, 391–392
southern pine beetle (*Dendroctonus frontalis*), 339–340
Host-plant-mediated dispersal, 117
Host-plant phenology, leaf-miner population dynamics and, 87, 92, 94–95, 104
Host-plant quality, *see also* Damaged vegetation
defoliation-induced, 215–217
density-dependent survival and, 378, 382
leaf abscission, 87, 88, 91, 102, 103, 104
leaf-miner eruption and, 98
Host-plant quantity, density-dependent survival and, 378, 382

Host suitability
models, 341–342
southern pine beetle (*Dendroctonus frontalis*), 339–340, 341–342
water availability, 342–343
Hymenoptera, 83, 93
Hyphantria cunea, 395

Incidence function model, 151–153, 157, 163
Infectious disease, cyclic population dynamics and, 212–215
Insecticide, 233–234
Interference competition, 7, 87, 139–140
Inverse density dependence, 27

k-factor analysis, 4
Kisatchie National Forest, 349

Lady beetle, 10, 11
Epilachna spp., 398
Epilachna niponica, 304–315
bottom-up regulation, 312
density-dependent processes, 308
egg resorption, 310–312, 314–315
female mobility, 310
low-density population stability, 306–308
oviposition behavior, 305–306, 309–315
reproductive season population stability, 308–309
reproductive success, 312–315
top-down regulation, 312
Henosepilachna niponica, 135
Lake Itasca State Park, *Phyllonorycter tremuloidiella* eruption, 92–95
Landscape structure, 206–207; *see also* Forest fragmentation
Larch budmoth (*Zeiraphera diniana*), 34, 233, 270
Larch casebearer (*Coleophora laricella*), 195
Larvae, *see also* Egg hatch; Offspring survival; Oviposition behavior; Survivorship
aggregation, 70–72
defenses, outbreak species, 50
gregarious behavior, 267–268
length, at maturity, outbreak species, 51
life-style, outbreak species, 54
mortality, leaf miners, 84–86, 93–95

Larvae (*continued*)
  oviposition behavior and, 304–305
  survival, 70–72, 304–305
Larval interference, 93, 94
Lateral regulation, 133, 135, 139–140
Latewood, 343
Lathrostizus euurae, 326
Leaf abscission, *see also* Damaged vegetation;
  Host-plant quality
  leaf-miner mortality and, 87, 88, 91, 102,
    103, 104
Leaf fall
  early, leaf-miner population crash and, 99
  late, leaf-miner eruption and, 96–98
Leafhopper, density-dependent dispersal, 116,
  124
Leaf miner, 7, 83–104
  abiotic mortality sources, 88, 91
  abundance patterns, 84
  annual population variability, 103–104
  Cameraria hamadryadella, 95–99, 103
  case studies, 92–102
  density-dependent mortality, 88–91, 95, 99,
    100–101, 103–104
  egg mortality, 84–88, 94
  eruption factors, 98–99, 102
  host-plant defenses, 86–87, 88, 91
  host-plant phenology, 87, 92, 94–95, 104
  incidence of, 84
  larval mortality, 84–86, 93–95
  latent, 100, 102
  leaf abscission and, 87, 88, 91, 102, 103,
    104
  mortality sources and variation, 84–95, 100
  natality variation, 91
  oviposition behavior/success, 88, 92, 96,
    104
  parasitoid community, 86
  Phyllonorycter tremuloidiella, 92–95, 103
  population crash factors, 99
  population density, 94–95
  pupae mortality, 84, 86, 94, 96
  pupal mass, 101–102
  regulation, 88–91, 102–104
  Stilbosis quadricustatella, 99–102
  univoltine, 99
Leaf mining, 83
Leaf-mining moth (*Cameraria* sp.), 136, 137
Leaf phenology, 242–243
Lepidobalanus, 95

Lepidoptera, 83
  forest
    biological control, 237–240
    experimental manipulation, 232–245
Leschenaultia exul (tachinid fly), 196
Leucoma salicis (satin moth), 239
Life-history evolution, 4, 251–271, 393
Life-history traits
  maternal effects (ME) score, 256–258
  outbreak species, 58
Limitation, tests, 20
Liriomyza, 83
Loblolly pine, water availability, 342–343
Local extinction, butterfly, 162–163
Local population
  persistence, 187–189
  regional dynamics and, 195–196
  spatial scale, 183
Local regulation, 140–141
Lochmaeus manteo, 54
Locust
  African migratory (*Locusta migratoria*), 270
  desert (*Schistocerca gregaria*), 270
  Locusta pardalina, 269
Locusta
  migratoria (African migratory locust), 270
  pardalina (locust), 269
Lyapunov exponent, 34
Lygaeid bug
  Cavelerius saccharivorus, 121–122
  density-dependent dispersal, 117
Lygaeus equestris
  abundance patterns, 283–286
  biotic interactions, 281–282
  breeding, 280–281
  density dependence, 286–292
  experimental methods, 282–283
  geographical variation, 296–297
  habitats, 280–282, 283
  long-term population change, 296
  mating, 281
  migratory rates, 293
  oviposition, 281
  population dynamics, 294–295
  population fluctuation, 286–292
Lymantria
  dispar (gypsy moth), 195, 217, 218, 220–
    224, 235–237, 238–239, 252, 270
  fumida (nun moth), 233–234
Lymantriidae, 50

Macrodynamics, 402
Macroevolution, 401, 402
Macrolepidoptera, 266
    egg dispersion, 53–54, 58
    feeding phenology, 48
    female flying ability, 52–53
    host breadth, 51–52
    larval defenses, 50
    larval length at maturity, 51
    larval life-style, 54
    maximum fecundity, 52, 56
    number of generations per year, 47–48
    offspring clustering, 54, 57, 58
    outbreak and nonoutbreak species, 41, 47, 56
    overwintering stage, 50
    taxonomic distribution of variation, 54–55
    trait definitions, 42–45
Maculinea arion, 160
Mainland population, butterfly, and meta-population level regulation, 151
Malacosoma
    californicum pluviale (western tent caterpillar), 218, 235, 237, 242–243, 245
    disstria (forest tent caterpillar), 73–76, 196–207
    pluviale (tent caterpillar), 230, 242–245
Markov chain, 152, 153
Mark–release–recapture studies, 164
Maternal effects, 8, 9, 217–218, 251–271
    defined, 252
    environmental quality and, 252, 269–271
    gregarious behavior and, 267–268
    hypothesis, 253–255
    mathematical model, 255
    outbreak species, 266–267
    plasticity, 269–269
    predicting, 256, 263–266, 268–270
Maternal effects score
    gregarious larval behavior and, 267
    life-history traits, 256–262
    predicting, 263–266, 268–269
Mathematical models, 209–224; *see also* Models
    complex, 220
    criticism, 209–210
    differential equation, 209, 210–212, 220
    host–parasitoid, 210–212
    host–pathogen, 212–215, 220–221
    Lotka-Volterra predator–prey, 210, 215, 219

    maternal effects, 217–218, 255
    May, 212
    Nicholson-Bailey, 210–212, 213
    parasitoid behavior, 201
    predator–prey, 210, 215, 219
    simple, 209–224
    simulation, 220
May model, 212
Mechanisms, 8–10
Megoura viciae, 119
Melanoplus
    femurrubrum, 362
    sanguinipes, 362, 366–379
        offspring survival, 366–368
        probability of individual being killed by predators, 368–370
        production of hatchlings per adult female, 368
Melitaea cinxia (Glanville fritillary), 150, 154–159, 161, 163–167
Mesodynamics, 402
Metapopulation
    defined, 5, 149, 173
    vs. patchy population, 293–294
Metapopulation dynamics, 5, 10, 131
    butterfly, 10, 149–167
    equilibrium, 140–141
    incidence function model, 151–153
    manipulating, 10
    models, 59, 140–141, 151–153, 174
    nonequilibrium, 140–141, 151
    patterns, 156–159
    persistence, 150–151, 154–156, 159–161, 184–189
    population stabilization, 10
    predator–prey systems, 173–189
    regulation, 10, 140–141, 150–151
    spatial scale, 183
    testing, 174
    value of approach, 167
Microbial ecology, 392–393
Microdynamics, 402–403
Microrhopala vittata (goldenrod beetle), 68–69
Migration rate
    butterfly, 164–166
    Lygaeus equestris, 293
    persistence and, 187–188, 189
Migratory locust, African (*Locusta migratoria*), 270

Ministik Hills, tent caterpillar–natural enemy
   study, 198
Mirid bug, density-dependent dispersal, 117,
   118
Mite
   Panonychus ulmi, 174–183, 189
   two-spotted (*Tetranychus urticae*), 185–188
   Typhlodromus pyri, 175–183, 189
Models, 9–10, 209–224; *see also* Mathematical models
   finite population, 26–27
   host suitability, 339–340, 341–342, 343
   metapopulation dynamics, 59, 140–141,
      151–153, 174
   outbreak species, 58–59
   patch, 151
   regulation, 370–376
   Ricker curve, 371–381
   water availability, 342–343
Monte Carol simulations, 133
Mortality
   density-dependent, *see* Density-dependent
      mortality
   egg, leaf miner, 84–88, 94
   horizontal factors, leaf miner, 87–88, 91
   leaf abscission and, 87, 88, 91, 102, 103,
      104
   population density and, 100–101
   pupae, leaf miner, 84, 94, 96
   vertical factors, leaf miner, 86–87, 88, 91,
      103
Mortality sources
   leaf miner, 84–86
      abiotic, 88, 91
      density dependence and, 88–91, 93–94
      horizontal, 84, 87–88, 91, 103
      regulatory role, 88–91
      vertical, 84, 86–87, 88, 91, 103
   population dynamics and, 93–95
Moth, 30, 59
   autumnal (*Epirrata autumnal*), 270–271
   Douglas fir tussock (*Orgyia pseudotsugata*),
      10, 57, 234, 241, 271
   gypsy (*Lymantria dispar*), 195, 217, 218,
      220–224, 235–237, 238–239, 252, 270
   leaf-mining (*Cameraria* sp.), 136, 137
   nun (*Lymantria fumida*), 233–234
   pyralid
      Acrobasis betulella, 73–74
      Ortholepis pasadamia, 73–74

satin (*Leucoma salicis*), 239
western tussock (*Orgyia vetusta*), 134
winter (*Operophtera brumata*), 195, 239–240
Mt. St. Helens, and tussock moth density, 241
MRR studies, *see* Mark–release–recapture
   studies
multivoltine species, 47–48

Natality, leaf miner, 91
Neodiprion swainei (sawfly), 268
Nerium oleander, 297
Nicholson-Bailey model, 210–212, 213
Nilaparvata lugens, 120, 121
Noctuidae, 48, 55
Noise, population dynamics and, 33–35
Nonequilibrium, 4–5, 131
   closed model, 140
   metapopulation, 151
   open model, 140
Nonlinear time-series model, 32–33
Nonoutbreak species, 41
   dynamics, 7, 8
   life-history traits, 58
   vs. outbreak species, 47, 56
   phylogenetic analysis, 41–47
Nonstationarity, 27
Notodontidae, 48, 54, 55
Nuclear polyhedral virus (NPV), 220–224
   in nun moth, 234
   phenology and, 243–245
   in tussock moth, 234
Nuclear polyhedrosis virus, *see* Nuclear polyhedral virus
Nun moth (*Lymantria fumida*), 233–234
Nymphalidae, 161

Odontota dorsalis, 84
Offspring clustering
   models, 58–59
   outbreak species, 54, 57, 58–59
Offspring fitness
   egg density and, 313–314
   oviposition site and, 313–314
   oviposition time and, 314–315
Offspring production, density dependence, 368
Offspring survival, 66–67; *see also* Larvae; Reproductive process; Survivorship
   gregarious species, 69

maternal quality and, 217
. *Melanoplus sanguinipes*, 366–368
oviposition behavior and, 304–305, 323
risk-spreading (bet-hedging) and, 67
Open, equilibrium model, 140
Open, nonequilibrium model, 140
Operophtera
bruceata, 57
brumata (winter moth), 195, 239–240
Ophraella conferta (goldenrod beetle), 68–70
Opportunistic herbivores, 66, 74
Orgyia
pseudotsugata (Douglas fir tussock moth), 10,
57, 234, 241, 271
vetusta (western tussock moth), 134
Orland E. White Arboretum, *Cameraria hamadryadella* eruptions, 96
Ortholepis pasadamia (pyralid moth), 73–74
Outbreak species, 7, 41–60; see also Temporal
variability
aggregation and, 65–67, 69
birch-feeding caterpillars, 73–76
clumping, 7
densities, 11
egg dispersion, 53–54, 58
factors affecting, 65–67
feeding, 50, 56, 65
female flying ability, 52–53
generations per year, 47–48
gregarious species, 58
host breadth, 51–52
incidence, 41
inducing outbreaks, 235–237
larval defenses, 50
larval length at maturity, 51
larval life-style, 54, 56
life-history traits, 58
limited dispersal, 57
maternal effects, 266–267
maximum fecundity, 52, 56
models, 58–59
vs. nonoutbreak species, 47, 56
offspring clustering, 54, 57, 58
overwintering, 50
phylogenetic analysis, 41–47, 56
as risk-concentrators, 67
solitary species, 58
southern pine beetle, 352–353
spatial behavior, 65, 67
spatial models, 58–59

taxonomic distribution of variation, 54–55
trait definitions, 42–45
Overdamping, 24
Overwintering, outbreak species, 50
Oviposition behavior, 304–305, 394; see also
Larvae; Reproductive process
female mobility, 310
galling sawfly, 323
leaf miner, 88, 92, 96, 104
offspring performance and, 323
phylogenetic constraints hypothesis, 333–
334
population stability and, 309–312
reproductive success and, 312–315
site selection, 313–314
Oviposition site
chemical interaction and, 325
offspring fitness and, 313–314
Oviposition time, offspring fitness and, 314–
315

Paired sister-group comparison, 47, 52, 53–54
Paleacrita vernata, 57
Panonychus ulmi, 174–183, 189
density, 175–176, 178–182
dispersal, 175
persistence, 176, 177
pest status, 175
population instability, 177
predators, 177
temporal variability, 176, 177
Paper birch (*Betula papyrifera*), 73–76, 198
Papilio
glaucus (tiger swallowtail butterfly), 6
machaon, 160
Parasetigena silvestris, 236
Parasitism, 8, 9
aggregation and, 66, 70–72, 76
forest fragmentation and, 196–197, 200,
203–204
forest tent caterpillar, 196–207
gypsy moth control, 239
leaf miner, 86, 93, 98, 99, 103
phenology and, 243–245
population density and, 236–237
simple mathematical model, 210–212
spatial patterns, 79
statistical model, 201
tent caterpillar, 244–245

Parental care, *see* Maternal effects
Patch, *see* Habitat patch
Patch models, 151
Patchy population, 293–294
Patelloa pachypyga (tachinid fly), 196–207
Pathogens, 8–9
    host–pathogen model, 212–215, 220–221
Pemphigine aphid, density-dependent dispersal,
    118
Perfect damping, 24
Periodic attractor, 25
Persistence, 26
    defining, 176, 177
    local population, 187–189
    metapopulation, 150–151, 184–189
    migration rates and, 184, 187–188, 189
    predator–prey systems, 184, 186–188
    spatial scale and, 186–187
Perturbation experiments, 10
Petiole galler (*Euura* sp.), 331
Phase polymorphism, locust, 270
Phenology, *see also* Weather
    experiments, 242–245
    feeding, 50
    host-plant, 87, 92, 94–95
    leaf miner and, 99
    nuclear polyhedral virus and, 243–245
    parasitism and, 243–245
    water stress, 343
Phyllonorycter, 83
    salicifoliella, 92
    tremuloidiella, 92–95, 103, 104
        initiation and termination of eruptions, 95
        mortality factors, 93–95
        population dynamics, 93–95
Phylogenetic analysis, 7, 41–47, 56
Phylogenetic constraints hypothesis, 333–334
Phytophagous insects, density-dependent dis-
    persal, 114–125
Phytoseiidae, 185
Phytoseiulus perseiulus, 185–188
Picea glauca (white spruce), 198
Pieris
    napi, 160
    rapae, 67
Pine
    loblolly, 342–343
    red (*Pinus resinosa*), 93
Pine beetle, southern (*Dendroctonus frontalis*),
    11, 32, 339–353

Pinus resinosa (red pine), 93
Plant, host, *see Host-plant* entries; Leaf abscis-
    sion; Leaf fall
Plantago lanceolata, 154
Plant growth, water availability and, 342–343,
    344–346
Planthopper
    density-dependent dispersal, 114, 120–124
    Nilaparvata lugens, 120, 121
    Prokelisia dolus, 120
    Prokelisia marginata, 120
Plant stress hypothesis, 341–342
Plant vigor hypothesis, 324, 334, 341
Plasticity, maternal effects, 269–269
Plebejus argus, 160, 163, 164
Pluralism, 404
Pollen-feeding
    heliconiine butterfly, 78
    spatial behavior and, 78
Polygonia calbum, 160
Poplar, balsam (*Populus balsamifera*), 198
Population crash, leaf miner, 99
Population cropping, 234–235
Population density, *see* Density
Population dynamics
    bark beetle, 339–353
    behavioral ecology and, 393–394
    bounded, 6
    broad comparative ecology and, 394–395
    chemical ecology and, 392
    comparative approach, 6–8
    cyclic, 212–215, 232
    density-dependent dispersal and, 113–125
    empirical and theoretical ecology and, 396–
        397
    evaluating geographic variation in, 240–241
    evolution and, 397–398
    experimental biology and, 395–396
    forest fragmentation and, 195–207
    forest insects, 195–207
    galling sawfly (*Euura lasiolepis*), 321–335
    genetics and, 397–398
    growth control, 3
    heritability and, 397–398
    historical perspective, 3–5, 398–400
    importance of, 389–390
    infectious disease and, 212–215
    insect quality, 245
    international integration, 398
    leaf-mining insects, 83–104

life-history evolution and, 251–271, 393
long-term, seed feeders, 279–297
maternal effects, 217–218, 251–271
mechanisms, 8–10
microbial ecology and, 392–393
modern approaches to, 391–404
mortality factors, 93–95
noise and, 33–35
periodic, 33
phylogenetic constraints hypothesis, 333–334
plant–herbivore interactions and, 391–392
plant vigor hypothesis, 334
pluralist synthesis, 12
predator–prey systems, 348–352
scale, 402–403
seed feeders, 294
spatial variability and, 78–79
stabilization, 10
study needs, 400
synthesis, 389–391, 400–404
theory, 403–404
time series evidence of, 5–8
weather and, 242–245
Population ecology, 4
Population models, *see also* Models
density-dependence tests, 28
finite, 26–27
Population peak, controlling
population cropping, 234–235
spray programs, 232–234
Population quality
maternal effects and, 253–255
variables, significance of, 251–253
Population regulation, *see* Regulation
Population scale, extinction and, 26
Population stability, 131
at low density, 306–308
galling sawfly, 326–328
oviposition behavior and, 309–312
reproductive season, 308–309
Populus
balsamifera (balsam poplar), 198
tremuloides (quaking aspen, trembling aspen), 92–95, 198, 200
Predator-limited regulation, 364–366
Predator–prey system, 8, 173–189
adult density without predation, 371–373
aggregation and, 66, 70–72, 76
in fragmented forests, 196

galling sawfly, 326
grasshopper, 361
leaf-miner, 98, 103
Lotka-Volterra model, 210, 215, 219
metapopulation dynamics, 173–189
migration rates and, 184
overexploitation by predator, 187
Panonychus ulmi and *Typhlodromus pyri*, 175–183, 189
persistence, 184–189
Phytoseiulus perseiulus and *Tetranychus urticae*, 185–188
population dynamics, 348–352
population instability, 177
predator density, 182
prey behavior, 394
prey density, 177, 182, 183
prey immigration levels, 181, 182, 183
probability of individual being killed, 368–370
seed feeders, 294
southern pine beetle, 340–341, 347–351
spatial patterns, 79
spider mite, 183–188
survival without predation, 266–268
Preference-performance pattern, 304–305, 394
galling sawfly, 323–324, 329–331
Prokelisia
dolus, 120
marginata, 120
Pseudophilotes baton, 160
Psyllid, density-dependent dispersal, 116
Pteromalus spp., 326
Pterostichus oblongipunctatus (carabid beetle), 137
Pupae mortality, leaf miner, 84, 94, 96
Pupal mass, leaf-miner density and, 101–102
Purus, 88
Pyralid moth
Acrobasis betulella, 73–74
Ortholepis pasadamia, 73–74
Pyrgus centaureae, 160

Quaking aspen (*Populus tremuloides*), 92–95, 198, 200
Quasi-chaos, 34
Quasi-period attractor, 25
Quasi-stationarity, 27

Quercus
  alba, 95–99
  geminata, 99–102
  macrocarpa, 96

Rainfall, as regulated system, 23
Random sequence, vs. random walk, 24
Random walk
  extinction and, 4, 5
  vs. random sequence, 24
  unbiased, 29
  as unregulated system, 23
Rangeland grasshoppers, 11, 361–382
Recruitment curve, 371
Red alder (*Alnus rubra*), 242, 244–245
Reddened spectra, 296
Red grouse, 60
Red pine (*Pinus resinosa*), 93
Redscale, California (*Aonidiella aurantii*), 10
Reductionism, 404
Regulation, 19–37
  in absence of predators, 266–268, 371–373
  bottom-up, 7, 133–134, 294–295, 303, 312, 391–392
  bounded populations, 6, 20
  conditions for, 27–28
  consensus on, 21–22
  defined, 10, 22–27, 229
  density dependence and, 3–4, 19–22, 27, 131–132, 229, 380
  density-manipulation experiments, 131–141, 232–245
  descriptive approach, critique, 292–232
  equilibrium and, 25–26, 140, 229
  evidence of, 5–6, 30–31, 246–247
  example, 23–24
  factors affecting, 3–4, 359–382
  floors and ceilings, 25
  food-limited, 364–366
  herbivorous insects and, 11–12
  lateral, 133, 135, 139–140
  leaf miners, 94–95, 102–104
  local, 140–141
  metapopulation, 10, 150–151
  models, 140–141, 370–376
  mortality sources as, 88–91
  nonequilibrium models and, 140–141
  predator-limited, 364–366

Ricker curve analysis, 371–376
  sources of, 135
  stationarity definition of, 22–25
  structure of, 31–35
  temporary lack of, 31
  testing for, 4, 10–11, 28
  time series, 5–6, 230–232
  top-down, 133, 134–135, 312
  weather, 23–24, 294–295
Reproductive process, *see also* Offspring survival; Oviposition behavior
  density dependence, 308
  population stabilization, 308–309
  success, oviposition behavior and, 312–315
Resource competition hypothesis, 304
Return tendency
  density dependence, 27–29
  time-series, 29, 36
Rheumaptera hastata, 54
Ricker curve, 371–381
  adult density without predation, 371–373
  grasshopper, 376–379
  operation of, 373–376
  strategy, 371
Ricker fisheries model, 219
Risk-concentrators, outbreak species as, 67
Risk-spreading, 20
  in goldenrod aphids, 68
  offspring performance and, 67
  in outbreak species, 69
Robinia pseudoacacia, 84
r-selected species, 6–7

Salamander (*Ambystoma*), 135
Salix
  alba, 331
  cinerea, 331
  lasiolepis (arroyo willow), 322–334
  pentandra, 331
Sap-feeding insects, density-dependent dispersal, 114–125
Sarcophaga (=*Arachnidomyia*) *aldrichi* (sarcophagid fly), 196–207
Sarcophagid fly (*Sarcophaga aldrichi*), 196–207
Satin moth (*Leucoma salicis*), 239
Saturniidae, 48
Satyridae, 162

Sawfly, 56, 58, 59, 66
  diprionid, 268
  galling
    Euura atra, 331
    Euura lasiolepis, 321–335
    Euura "mucronata", 331
    Neodiprion swainei, 268
  tenthredinid, 11
Scale crawler, density-dependent dispersal, 117
Scale insects, density-dependent dispersal, 117
Schistocerca gregaria (desert locust), 270
Scolitantides orion, 163
Seed bug, density-dependent dispersal, 114
Seed feeders, 11
  long-term population dynamics, 279–297
  population dynamics, 294
  as predators, 294
Semiothisa sexmaculata, 54
Simulation models, 220
Sister-group comparison, 47
  egg dispersion, 53–54
  host breadth, 52
Solidago, 196
  rugosa (goldenrod), 69–70
Solitary species
  egg dispersion, 53, 58
  feeders, survival and fecundity, 76
  larvae, survival, 70–72
  mortality of, 67
  outbreak species, 53, 54, 58
Southern pine, water availability, 342–343
Southern pine beetle (*Dendroctonus frontalis*),
    11, 32, 339–353
  host suitability, 339–340, 341–342
  outbreak cycles, 352–353
  predation, 340–341, 347–351
  reproductive success, 344–346
  tree growth and, 343
  water availability and, 343–347
Spatial models, outbreak species, 58–59
Spatial scale
  density dependence and, 79
  local populations, 183
  meaningfulness, 279, 280
  metapopulation, 183
  persistence and, 186–187
  temporal patterns and, 78–79
Spatial variability, 67; *see also* Aggregation
  density dependence, 236

leaf-miner mortality and, 87
  outbreak species, 65, 67
  population dynamics and, 78–79
  temporal variability and, 78–79
Specialists, variability of, 7
Species persistence, *see* Persistence
Spectra, reddened, 296
Sphingidae, 48
Spider mite
  dispersal behavior, 185
  metapopulation dynamics, 183–188
Spodoptera frugiperda (fall armyworm), 57
Spray programs, for controlling herbivorous
    insects, 232–234
Spring feeders, outbreak species, 47–50, 56
Spruce, white (*Picea glauca*), 198
Spruce budworm (*Choristoneura fumiferana*), 6,
    195, 196, 206–207
Stability
  chaos and, 33
  Lyapunov exponent, 34
  in mathematical models, 209
Stationarity, 22–25
  quasi-, 27
  regulation modeling and, 31
  temporal scale and, 27
Stationary probability distribution, 22, 23
Statistical methods, for detecting density de-
    pendence, 20–21
Stealthy herbivores, 66, 69, 74, 76–78
  Exema canadensis, 70–72
Stilbosis quadricustatella, 99–102, 104
  fitness components, 100–102
  mortality sources, 100
  regulation, 102
Stochastic boundedness, 5, 26
Stochastic extinction, butterfly, 152, 163, 166–
    167
Stochastic growth/decline model, 29
Stress
  drought, bark beetle and, 342
  water, 342–343
Subpopulations, metapopulation dynamics
    and, 5
Summer feeders, outbreak species, 50
Sunspot activity, 241–242
Survivorship, *see also* Larvae; Offspring sur-
    vival
  aggregation and, 76

Survivorship (*continued*)
on damaged plants, 76–78
fecundity and, 76
gregarious species, 78
solitary feeders, 76
Swallowtail butterfly, tiger (*Papilio glaucus*), 6
Sweden, cyclic and noncyclic dynamics, 241

Tachinid fly
Compsilura concinnata, 236
Leschenaultia exul, 196, 199
Patelloa pachypyga, 196–207
Taxonomic variation, outbreak species, 54–55
Temporal scale
meaningfulness, 279–280
stationarity and, 27
Temporal variability, 6–7, 67; *see also* Outbreak species
aggregation and, 7
assessing, 176, 177, 182
predictors of, 6–7
spatial variability and, 78–79
Tent caterpillar
forest (*Malacosoma disstria*), 73–76, 196–207
Malacosoma pluviale, 242–245
birth rate and density, 230
egg hatch, 242–245
nuclear polyhedral virus in, 243–244
parasitism and, 244–245
western (*Malacosoma californicum pluviale*), 218, 235, 237, 242–243, 245
Tenthredinidae, 83
Tenthredinid sawflies, 11
Tephritid fly, 10
Euphranta connexa, 10, 281, 294
Territorial behavior
dispersal and, 118
regulation and, 135
Tetranychus
kanzawai, 185
urticae (two-spotted mite), 185–188
Thanasimus dubius (clerid beetle), 341
life history, 350–351
southern pine beetle (*Dendroctonus frontalis*) and, 347–351
Theoretical ecology, empirical ecology and, 395–396
Thistle (*Cirsium kagamontanum*), 306
Thrip, density-dependent dispersal, 114, 117

Thymelicus lineola, 160
Thysanoptera, 114
Tiger swallowtail butterfly (*Papilio glaucus*), 6
Time-series analysis, 4
complex dynamics, 232
data length, 30–31
data return tendency, 29
data sets, 32
density data, statistical significance, 230–232
evidence of population dynamics in, 5–8
length of, 36
nonlinear model, 32–33
Top-down regulation, 133, 134–135, 312
Trait definitions, outbreak species, 42–45
Trembling aspen (*Populus tremuloides*), 92–95, 198, 200
Trichogramma sp., 100
Trirhabda
borealis (goldenrod beetle), 68–72, 78
virgata (goldenrod beetle), 68–72, 78
Tropidothorax leucopterus, 297
Tussock moth, 10
Douglas fir (*Orgyia pseudotsugata*), 10, 57, 234, 241, 271
western (*Orgyia vetusta*), 134
Two-spotted mite (*Tetranychus urticae*), 185–188
Typhlodromus occidentalis, 185–188
Typhlodromus pyri, 175–183, 189
density, 176
prey density, 177

Unbiased random walk, 29
Underdamping, 24
Univoltine species
leaf miner, 99
outbreaking, 47
Ricker curve, 371
Unregulated system, 22–23
Uroleucon
caligatum (goldenrod aphid), 67–68
nigrotuberculatum (goldenrod aphid), 67–68, 72, 76, 78–79

Vacciniina optilete, 160
Vancouver Island, winter moth introduction, 240

Variability, 6–7
  generalists, 7
  spatial, 65, 67, 78–79, 87, 236
  specialists, 7
  temporal, 6–7, 67, 78–79, 176, 177, 182
Vegetation, see Damaged vegetation; Host-plant entries; Plant entries
Veronica spicata, 154
Vertical mortality factors, leaf miner, 86–87, 88, 91, 103
Vincetoxicum hirundinaria, 297
  habitats, 280–282, 283
  hibernation sites, 283
  seed resources, 283, 294
Viral disease, see also Nuclear polyhedral virus (NPV)
  as regulation of forest Lepidoptera, 234

Wasp
  braconid (Aleiodes malacosomatus), 205
  chalcidoid (Eutetrastichus chlamytis), 70–72
  Cotesia melanoscela, 235–236, 239
  Trichogramma sp., 100
Water availability
  experiments, 343–347

model, 342–343
  southern pine beetle (Dendroctonus frontalis)
    outbreak cycle and, 343, 352–353
Water stress, 342–343
Weather, see also Phenology
  density experiments, 242–245
  experimental perturbations, 241–242
  food resources and, 294–295
  leaf-miner mortality and, 87, 88, 95
  as regulated system, 23–24
Western tent caterpillar (Malacosoma californicum pluviale), 218, 235, 237, 242–243, 245
Western tussock moth (Orgyia vetusta), 134
Whitefly, density-dependent dispersal, 116, 117
White spruce (Picea glauca), 198
Willow, arroyo (Salix lasiolepis), 322–334
Wing reduction, in outbreak species females, 52–53
Winter moth (Operophtera brumata), 195, 239–240

Zeiraphera diniana (larch budmoth), 34, 233, 270
Zetsellia mali, 178